Calculus Multivariable

Student Solutions Manual

2nd Edition

Brian E. Blank
Steven G. Krantz

Department of Mathematics
Washington University in St. Louis

Supplement Author
Donald G. Hartig
Mathematics Department
California Polytechnic State University
San Luis Obispo, California

WILEY
John Wiley & Sons, Inc.

Cover photo: ©Walter Bibikow/JAI/Corbis

ISBN-13   978-0470-64724-0

Printed in the United States of America

10 9 8 7 6 5 4 3 2 1

Printed and bound by Integrated Book Technology.

Contents

Chapter 9

Vectors

9.1 Vectors in the Plane

Problems for Practise

1–7. Calculate and plot the vectors.

1. $\overrightarrow{PQ} = \langle -1, -4 \rangle$

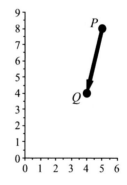

3. $\overrightarrow{PQ} = \langle -5, 1 \rangle$

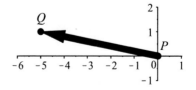

5. $\overrightarrow{PQ} = \langle 5, -11 \rangle$

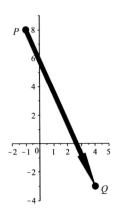

7. $\overrightarrow{PQ} = \langle 6, -6 \rangle$

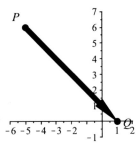

9. $\overrightarrow{PQ} = \langle 1, -2 \rangle$, $\|\overrightarrow{PQ}\| = \sqrt{1+4} = \sqrt{5}$
   $\overrightarrow{RS} = \langle 3, -6 \rangle$, $\|\overrightarrow{RS}\| = \sqrt{9+36} = \sqrt{45} = 3\sqrt{5}$
   The vectors are parallel because $\overrightarrow{RS} = 3\overrightarrow{PQ}$.

11. $\overrightarrow{PQ} = \langle -9, 3 \rangle$, $\|\overrightarrow{PQ}\| = \sqrt{81+9} = \sqrt{90} = 3\sqrt{10}$
    $\overrightarrow{RS} = \langle -2, 7 \rangle$, $\|\overrightarrow{RS}\| = \sqrt{4+49} = \sqrt{53}$
    The vectors are not parallel.

13, 15. Calculate the vectors and sketch.

$$2\mathbf{v} + \mathbf{w} = 2\langle 4, 2 \rangle + \langle 1, -3 \rangle$$
$$= \langle 9, 1 \rangle$$

$$\mathbf{v} - 2\mathbf{w} = \langle 4, 2 \rangle - 2\langle 1, -3 \rangle$$
$$= \langle 2, 8 \rangle$$

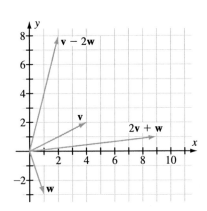

17. $P = (5, -7)$, $Q = (0, 5)$

(a) $\mathbf{v} = \overrightarrow{PQ} = \langle -5, 12 \rangle$

(b) $\|\mathbf{v}\| = \sqrt{25 + 144} = \sqrt{169} = 13$

(c) This is $-\mathbf{v} = \langle 5, -12 \rangle$.

(d) $\mathbf{dir}(\mathbf{v}) = \mathbf{v}/\|\mathbf{v}\| = \frac{1}{13}\langle -5, 12 \rangle$

(e) This is $-\mathbf{dir}(\mathbf{v}) = \frac{1}{13}\langle 5, -12 \rangle$.

19. $P = (-3, 7)$, $Q = (1, -9)$

(a) $\mathbf{v} = \overrightarrow{PQ} = \langle 4, -16 \rangle$

(b) $\|\mathbf{v}\| = \sqrt{16 + 256} = \sqrt{272} = 4\sqrt{17}$

(c) This is $-\mathbf{v} = \langle -4, 16 \rangle$.

(d) $\mathbf{dir}(\mathbf{v}) = \mathbf{v}/\|\mathbf{v}\| = \frac{1}{4\sqrt{17}}\langle -4, 16 \rangle = \langle -\frac{4}{\sqrt{17}}, \frac{16}{\sqrt{17}} \rangle$

(e) This is $-\mathbf{dir}(\mathbf{v}) = \langle \frac{4}{\sqrt{17}}, -\frac{16}{\sqrt{17}} \rangle$.

21. $\mathbf{v} = -7\mathbf{i} + 2\mathbf{j}$, $\mathbf{w} = -2\mathbf{i} + 9\mathbf{j}$

$$-4\mathbf{v} = -4(-7\mathbf{i} + 2\mathbf{j}) = 28\mathbf{i} - 8\mathbf{j}$$
$$3\mathbf{v} - 2\mathbf{w} = 3(-7\mathbf{i} + 2\mathbf{j}) - 2(-2\mathbf{i} + 9\mathbf{j}) = (-21 + 4)\mathbf{i} + (6 - 18)\mathbf{j} = -17\mathbf{i} - 12\mathbf{j}$$
$$4\mathbf{v} + 7\mathbf{j} = 4(-7\mathbf{i} + 2\mathbf{j}) + 7\mathbf{j} = -28\mathbf{i} + (8 + 7)\mathbf{j} = -28\mathbf{i} + 15\mathbf{j}$$

23. $\mathbf{v} = 6\mathbf{i} - 2\mathbf{j}$, $\mathbf{w} = 9\mathbf{i} - 3\mathbf{j}$

$$-4\mathbf{v} = -4(6\mathbf{i} - 2\mathbf{j}) = -24\mathbf{i} + 8\mathbf{j}$$
$$3\mathbf{v} - 2\mathbf{w} = 3(6\mathbf{i} - 2\mathbf{j}) - 2(9\mathbf{i} - 3\mathbf{j}) = (18 - 18)\mathbf{i} + (-6 + 6)\mathbf{j} = 0\mathbf{i} + 0\mathbf{j}$$
$$4\mathbf{v} + 7\mathbf{j} = 4(6\mathbf{i} - 2\mathbf{j}) + 7\mathbf{j} = 24\mathbf{i} + (-8 + 7)\mathbf{j} = 24\mathbf{i} - \mathbf{j}$$

25. $\mathbf{F} = \langle 3, 0 \rangle + \langle 0, 4 \rangle = \langle 3, 4 \rangle$, $\|\mathbf{F}\| = \sqrt{9 + 16} = \sqrt{25} = 5$.

27. $\mathbf{F} = \mathbf{i} - 2\mathbf{j} + 2\mathbf{i} + \mathbf{j} = 3\mathbf{i} - \mathbf{j}$, $\|\mathbf{F}\| = \sqrt{9 + 1} = \sqrt{10}$.

29. $\mathbf{u} = \langle \cos(\pi/6), \sin(\pi/6) \rangle = \langle \sqrt{3}/2, 1/2 \rangle$.

31. $\mathbf{u} = \langle \cos(\pi), \sin(\pi) \rangle = \langle -1, 0 \rangle$.

33. Since $\|\mathbf{v}\| = \sqrt{36 + \frac{25}{4}} = \frac{13}{2}$, $\mathbf{u} = \mathbf{dir}(\mathbf{v}) = \frac{2}{13}\mathbf{v} = \langle \frac{12}{13}, -\frac{5}{13} \rangle$. Therefore, with $\lambda = \frac{13}{2}$, $\mathbf{v} = \frac{13}{2}\langle \frac{12}{13}, -\frac{5}{13} \rangle$.

35. Since $\|\mathbf{v}\| = \sqrt{9 + 4} = \sqrt{13}$, $\mathbf{u} = \mathbf{dir}(\mathbf{v}) = \frac{1}{\sqrt{13}}\mathbf{v} = \frac{3}{\sqrt{13}}\mathbf{i} - \frac{2}{\sqrt{13}}\mathbf{i}$. Therefore, with $\lambda = \sqrt{13}$, $\mathbf{v} = \sqrt{13}\left(\frac{3}{\sqrt{13}}\mathbf{i} - \frac{2}{\sqrt{13}}\mathbf{i} \right)$.

37. $2(5\mathbf{i} + 3\mathbf{j}) - 3(\mathbf{i} - \mathbf{j}) = (10 - 3)\mathbf{i} + (6 + 3)\mathbf{j} = 7\mathbf{i} + 9\mathbf{j} = \langle 7, 9 \rangle$

39. $3\mathbf{j} - 2(-3\mathbf{i} + \mathbf{j}) - 2\mathbf{i} = (6 - 2)\mathbf{i} + (3 - 2)\mathbf{j} = 4\mathbf{i} + \mathbf{j} = \langle 4, 1 \rangle$

Further Theory and Practise

41. Since $\mathbf{v} = \langle a, 3 \rangle$ and $a^2 + 9 = 49$, $a = \sqrt{40} = 2\sqrt{10}$ (a is positive).

43. If $\langle a, b \rangle = \lambda \langle 5, 3b \rangle$, then (assuming $b \neq 0$), $\lambda = \frac{1}{3}$ and $a = \frac{5}{3}$. If $b = 0$, then a can be any number whatsoever.

45. Since $\|\mathbf{v}\| = 3\|\mathbf{w}\|$ and $\mathbf{v} = -\lambda \mathbf{w}$ for some $\lambda > 0$,

    $$3\|\mathbf{w}\| = \|\mathbf{v}\| = \|-\lambda \mathbf{w}\| = \lambda \|\mathbf{w}\|$$

    and $\lambda = 3$. Therefore, $\mathbf{v} = -3\mathbf{w} = -3\langle 2, c \rangle = \langle -6, -3c \rangle$, and $a = -6$.

47. Each of the four vectors $\pm\big(\langle 3, 5 \rangle \pm \langle -1, 7 \rangle\big)$ is a diagonal of the parallelogram so one of the possibilities for $\mathbf{v}$ is $\langle 2, 12 \rangle$. This is the only diagonal with positive components, so $a = 2$.

49. Each of the four vectors $\pm\big(\langle -7, 6 \rangle \pm \langle 3, 7 \rangle\big)$ is a diagonal of the parallelogram so one of the possibilities for $\mathbf{v}$ is $-\langle -10, -1 \rangle = \langle 10, 1 \rangle$. This is the only diagonal with positive components, so $a = 10$.

51. Observe that $\|\mathbf{w}\| = 2$ so $\mathbf{dir}(\mathbf{v}) = \mathbf{dir}(\mathbf{w}) = \frac{1}{2}\mathbf{w}$. Therefore, $\mathbf{v} = \|\mathbf{v}\|\mathbf{dir}(\mathbf{v}) = 3 \cdot \frac{1}{2}\mathbf{w} = \frac{3}{2}\langle 1, \sqrt{3} \rangle$ and $a = \frac{3}{2}$.

53. Observe that $\|\mathbf{w}\| = 5$ so $\mathbf{dir}(\mathbf{v}) = -\mathbf{dir}(\mathbf{w}) = -\frac{1}{5}\mathbf{w}$. Therefore, $\mathbf{v} = \|\mathbf{v}\|\mathbf{dir}(\mathbf{v}) = 10 \cdot \left(-\frac{1}{5}\mathbf{w}\right) = -2\langle 3, -4 \rangle$ and $a = -6$.

55. Mr. Woodman applies a force vector $\mathbf{v}$ (with magnitude 100). Let $\mathbf{w}$ denote the force vector that Mrs. Woodman applies. We first find $\|\mathbf{w}\|$. The wagon will move along the dotted line provided the horizontal component of $\mathbf{v}$ cancels the horizontal component of $\mathbf{w}$. Referring to Figure 14 in the text, $\|\mathbf{v}\| \sin(\pi/6) = \|\mathbf{w}\| \sin(\pi/4)$, so

    $$\|\mathbf{w}\| = \frac{\|\mathbf{v}\| \sin(\pi/6)}{\sin(\pi/4)} = \frac{100 \cdot \frac{1}{2}}{1/\sqrt{2}} = 50\sqrt{2}.$$

    The magnitude of the resultant force, $\|\mathbf{v} + \mathbf{w}\|$, is equal to the sum of the vertical components of $\mathbf{v}$ and $\mathbf{w}$,

    $$\|\mathbf{v} + \mathbf{w}\| = \|\mathbf{v}\| \cos(\pi/6) + \|\mathbf{w}\| \cos(\pi/4)$$
    $$= 100 \cdot \frac{\sqrt{3}}{2} + 50\sqrt{2} \cdot \frac{1}{\sqrt{2}}$$
    $$= 50(\sqrt{3} + 1).$$

57. The boat's velocity, with respect to the river, can be taken to be $\mathbf{v_r} = 50\mathbf{i}$. The current's pull can be represented by the velocity vector $\mathbf{v_c} = -20\mathbf{j}$. The resultant velocity of the boat (with respect to land) is

    $$\mathbf{v} = \mathbf{v_r} + \mathbf{v_c} = 50\mathbf{i} - 20\mathbf{j}.$$

    The boat takes 10 minutes to cross the river so when it docks on the other shore it will have been carried 200 meters downriver.

59. Let the resultant force be $\mathbf{F} = F_1\mathbf{i} + F_2\mathbf{j}$. The horizontal component, F_1, is the sum of the horizontal components of the three forces, and the vertical component, F_2, is the sum of the vertical components:

$$F_1 = -160\cos(30°) - 200 - 280\cos(45°)$$
$$= -160 \cdot \frac{\sqrt{3}}{2} - 200 - 280 \cdot \frac{1}{\sqrt{2}}$$
$$= -80\sqrt{3} - 200 - 140\sqrt{2} \approx -536.55$$

and

$$F_2 = 160\sin(30°) - 280\sin(45°)$$
$$= 160 \cdot \frac{1}{2} - 280 \cdot \frac{1}{\sqrt{2}}$$
$$= 80 - 140\sqrt{2} \approx -117.99$$

61. We minimize the function f defined as $f(\lambda) = \|\mathbf{v} + \lambda\mathbf{w}\|^2$. This is easier than minimizing $\lambda \to \|\mathbf{v} + \lambda\mathbf{w}\|$ and, because $x \to x^2$ is strictly increasing for x positive, it will give the same answer. Writing $\mathbf{v} = \langle v_1, v_2 \rangle$ and $\mathbf{w} = \langle w_1, w_2 \rangle$,

$$f(\lambda) = (v_1 + \lambda w_1)^2 + (v_2 + \lambda w_2)^2,$$

and we can see that f is a quadratic in λ. Its graph opens upward because the coefficient of λ^2 is $w_1^2 + w_2^2$, which is positive. f attains an absolute minimum for the value of λ satisfying

$$f'(\lambda) = 2w_1(v_1 + \lambda w_1) + 2w_2(v_2 + \lambda w_2) = 0.$$

That is, $\lambda = -(v_1 w_1 + v_2 w_2)/(w_1^2 + w_2^2)$.

63. Assume a and b are both not 0. Then the slope of the line from the origin $(0,0)$ to the point (a,b) is b/a and the slope of the line from $(0,0)$ to $(-b,a)$ is $a/(-b) = -a/b$, the negative reciprocal of b/a. The lines are perpendicular so the vector $\mathbf{w} = \langle -b, a \rangle$ is perpendicular to the vector $\mathbf{v} = \langle a, b \rangle$. If either a or b is 0, the vectors are parallel to the x and y axes making them perpendicular to each other.

All vectors that are perpendicular to $\mathbf{v} = \langle a, b \rangle$ have the form $\lambda\langle -b, a \rangle$.

65. The slope of the line ℓ is $-A/B$ and the slope of the vector $\mathbf{n}$ is B/A. These are negative reciprocals so the vector is perpendicular to the line.

Calculate $\overrightarrow{P_0Q}_x = \langle x - x_0, -\frac{Ax}{B} - \frac{D}{B} - y_0 \rangle$. This vector is perpendicular to the line ℓ when its slope is B/A. That is, $\frac{B}{A} = \frac{-\frac{Ax}{B} - \frac{D}{B} - y_0}{x - x_0}$. Consequently, $\left(\frac{B}{A} + \frac{A}{B}\right)x = \frac{B}{A}x_0 - y_0 - \frac{D}{B}$, and

$$x = \frac{AB\left(\frac{B}{A}x_0 - y_0 - \frac{D}{B}\right)}{A^2 + B^2} = \frac{B^2 x_0 - ABy_0 - AD}{A^2 + B^2}.$$

67. The vector $\overrightarrow{OP_t}$ can be expressed as the sum: $\overrightarrow{OP_t} = \overrightarrow{OP} + t\overrightarrow{PQ}$. Equivalently, $P_t = (p_1 + t(q_1 - p_1), p_2 + t(q_2 - p_2))$.

69. Observe that $\overrightarrow{A\alpha} = \frac{1}{2}(\overrightarrow{AB} + \overrightarrow{AC})$ and $\overrightarrow{B\beta} = \frac{1}{2}\overrightarrow{AC} - \overrightarrow{AB}$. Let λ and μ be the positive scalars such that $\overrightarrow{AM} = \lambda\overrightarrow{A\alpha}$ and $\overrightarrow{BM} = \mu\overrightarrow{B\beta}$. Since $\overrightarrow{AM} = \overrightarrow{AB} + \overrightarrow{BM}$, using the expressions above for $\overrightarrow{A\alpha}$ and $\overrightarrow{B\beta}$,

$$\lambda \cdot \tfrac{1}{2}(\overrightarrow{AB} + \overrightarrow{AC}) = \overrightarrow{AB} + \mu(\tfrac{1}{2}\overrightarrow{AC} - \overrightarrow{AB}),$$

implying that $\left(\tfrac{1}{2}\lambda - 1 + \mu\right)\overrightarrow{AB} = \tfrac{1}{2}(\mu - \lambda)\overrightarrow{AC}$. Since $\overrightarrow{AB}$ and $\overrightarrow{AC}$ are not parallel, both coefficients must be 0 yielding $\lambda = \mu$ and $\tfrac{1}{2}\lambda + \lambda = 1$. That is, $\lambda = \mu = \frac{2}{3}$.

Because the same relationship holds for $\overrightarrow{BM}$ and $\overrightarrow{CM}$, all three medians must intersect at the point M.

71. Since $\overrightarrow{P_tP_{t+3}} = \langle (t+3)^2 - t^2, (t+3)^3 - t^3 \rangle = \langle 6t + 9, 9t^2 + 27t + 27 \rangle$ this vector points in the same direction as $\langle 1, 7 \rangle$ when $9t^2 + 27t + 27 = 7 \cdot (6t + 9)$. This simplifies to $9t^2 - 15t - 36 = 0$ or $(3t - 4)(t - 3) = 0$. The answer is $t = 3$.

Calculator/Computer Exercises

73. Let $f(x) = e^x$ and $g(x) = 2 - x^2$. The graphs of f and g are displayed on the right. Using *Maple*'s *fsolve* procedure we find that they intersect at $(a, f(a))$ and $(b, f(b))$ where $a = -1.316$ and $b = 0.537$. The vector $\overrightarrow{PQ}$ is $\langle b, f(b) \rangle - \langle a, f(a) \rangle = \langle 1.853, 1.443 \rangle$.

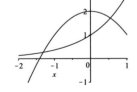

75. The graph of $f(t) = \sqrt{(t-1)^2 + (e^{-t} + t^2 - 4)}$ is displayed on the right. Using *Maple*'s *Minimize* procedure we find that it attains a minimum value of 0.9315 when $t = 1.8984$.

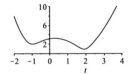

77. The vector $\mathbf{v}(t) = \langle 1, 3t^2 \rangle$ is tangent to the curve at the point $P_t = (t, t^3)$ and $\mathbf{T}(t) = \mathbf{dir}(\mathbf{v})(t) = \frac{\langle 1, 3t^2 \rangle}{\sqrt{1 + 9t^4}}$ is a unit vector tangent to the curve at P_t. Therefore, since $\overrightarrow{P_tQ_t} = \mathbf{T}(t)$, $Q_t = \left(t + \frac{1}{\sqrt{1+9t^4}}, t^3 + \frac{3t^2}{\sqrt{1+9t^4}}\right)$. A portion of the parametrized curve $t \mapsto Q_t$ is displayed on the right.

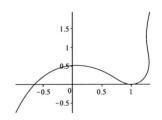

The t values for the x- and y-intercepts can be found using the plots of $\xi(t)$ and $\eta(t)$ shown on the right. The y-intercept corresponds to time t such that $\xi(t) = 0$, roughly $t = -0.5$. Using *Maple*'s *fsolve* procedure we find that $t = -0.6360$ and $y = \eta(t) = 0.5145$. The two x-intercepts correspond to times t such that $\eta(t) = 0$, roughly $t = -1$ and $t = 0$. Using *fsolve*, $t = -0.9813$ and $t = 0$. The x-intercepts are $\xi(-0.9813) = -0.6542$ and $\xi(0) = 1$.

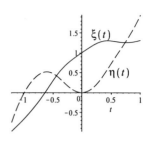

9.2   Vectors in Three-Dimensional Space

Problems for Practise

1, 3. The plots are shown below.

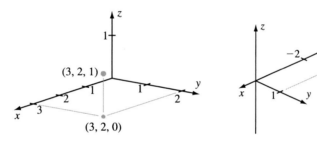

5. $\sqrt{(8-2)^2 + (3-0)^2 + (6-(-3))^2} = \sqrt{36+9+81} = \sqrt{126} = 3\sqrt{14}$

7. $\sqrt{(1-4)^2 + (1-4)^2 + (4-1)^2} = \sqrt{9+9+9} = \sqrt{27} = 3\sqrt{3}$

9. Complete the square in x by adding 1, complete the square in y by adding 4, and complete the square in z by adding 9 (to both sides of the equation):

$$(x+1)^2 + (y+2)^2 + (z+3)^2 = 8+1+4+9 = 22\,.$$

The center is $(-1, -2, -3)$, the radius is $\sqrt{22}$.

11. Divide both sides of the equation by 4:

$$x^2 + y^2 + z^2 + x + y + z = \tfrac{1}{2}\,.$$

Complete the square in x, y, z by adding 1/4 three times (to both sides of the equation):

$$(x+1/2)^2 + (y+1/2)^2 + (z+1/2)^2 = 1/2 + 3/4 = 5/4\,.$$

The center is $(-1/2, -1/2, -1/2)$, the radius is $\sqrt{5}/2$.

13. Write the equation in the form

$$x^2 + y^2 + z^2 - x = 0\,.$$

Complete the square in x by adding $1/4$ to both sides:

$$(x - 1/2)^2 + y^2 + z^2 = 1/4\,.$$

The center is $(1/2, 0, 0)$, the radius is $1/2$.

15. The $2z^2$ term prevents us from putting the equation into standard form for a sphere.

17. The $4xy$ term prevents us from putting the equation into standard form for a sphere.

19. Upon expanding $(x - y)^2$ the equation will have the term $-2xy$, making it impossible to put it into standard form for a sphere.

21. All points inside a sphere: $\{(x, y, z) : (x - 3)^2 + (y + 2)^2 + (z - 6)^2 < 16\}$.

23. All points on a sphere: $\{(x, y, z) : (x - \pi)^2 + (y - \pi)^2 + (z + \pi)^2 = \pi^2\}$.

25. All points outside a sphere: $\{(x, y, z) : (x - 2)^2 + (y - 1)^2 + z^2 > 4\}$.

27. $\overrightarrow{PQ} = \langle 1, -2, -6 \rangle$, $\|\overrightarrow{PQ}\| = \sqrt{1 + 4 + 36} = \sqrt{41}$
    $\overrightarrow{RS} = \langle 2, -4, -12 \rangle$, $\|\overrightarrow{RS}\| = \sqrt{4 + 16 + 144} = \sqrt{164} = 2\sqrt{41}$
    The vectors are parallel because $\overrightarrow{RS} = 2\overrightarrow{PQ}$.

29. $\overrightarrow{PQ} = \langle -1, 2, 2 \rangle$, $\|\overrightarrow{PQ}\| = \sqrt{1 + 4 + 4} = \sqrt{9} = 3$
    $\overrightarrow{RS} = \langle -5, 4, 2 \rangle$, $\|\overrightarrow{RS}\| = \sqrt{25 + 16 + 4} = \sqrt{45} = 3\sqrt{5}$
    The vectors are not parallel.

31-37. $\mathbf{v} = \langle 3, 2, 6 \rangle$, $\mathbf{w} = \langle 4, -1, 8 \rangle$

31. $-3\mathbf{w} = -3\langle 4, -1, 8 \rangle = \langle -12, 3, -24 \rangle$

33. $-4\mathbf{v} + 3\mathbf{w} = -4\langle 3, 2, 6 \rangle + 3\langle 4, -1, 8 \rangle = \langle 0, -11, 0 \rangle$

35. $2\mathbf{dir}(\mathbf{v}) = 2\frac{\langle 3, 2, 6 \rangle}{\sqrt{9 + 4 + 36}} = \frac{2}{7}\langle 3, 2, 6 \rangle = \langle \frac{6}{7}, \frac{4}{7}, \frac{12}{7} \rangle$

37. $\mathbf{v} + 3\mathbf{k} = \langle 3, 2, 6 \rangle + 3\langle 0, 0, 1 \rangle = \langle 3, 2, 9 \rangle$

39. $P = (1, 2, -1)$, $Q = (3, -1, 5)$

    (a) $\mathbf{v} = \overrightarrow{PQ} = \langle 2, -3, 6 \rangle$

    (b) This is $-\mathbf{v} = \langle -2, 3, -6 \rangle$.

    (c) $\|\mathbf{v}\| = \sqrt{4 + 9 + 36} = \sqrt{49} = 7$

    (d) $\mathbf{dir}(\mathbf{v}) = \frac{1}{7}\langle 2, -3, 6 \rangle = \langle \frac{2}{7}, -\frac{3}{7}, \frac{6}{7} \rangle$

    (e) $12\mathbf{dir}(\mathbf{v}) = 12\langle \frac{2}{7}, -\frac{3}{7}, \frac{6}{7} \rangle = \langle \frac{24}{7}, -\frac{36}{7}, \frac{72}{7} \rangle$

41. $P = (3, 1, -3)$, $Q = (4, 0, -5)$

    (a) $\mathbf{v} = \overrightarrow{PQ} = \langle 1, -1, -2 \rangle$

    (b) This is $-\mathbf{v} = \langle -1, 1, 2 \rangle$.

    (c) $\|\mathbf{v}\| = \sqrt{1 + 1 + 4} = \sqrt{6}$

    (d) $\mathbf{dir}(\mathbf{v}) = \frac{1}{\sqrt{6}} \langle 1, -1, -2 \rangle = \langle \frac{1}{\sqrt{6}}, -\frac{1}{\sqrt{6}}, -\frac{2}{\sqrt{6}} \rangle$

    (e) $12\,\mathbf{dir}(\mathbf{v}) = 12 \langle \frac{1}{\sqrt{6}}, -\frac{1}{\sqrt{6}}, -\frac{2}{\sqrt{6}} \rangle = \langle 2\sqrt{6}, -2\sqrt{6}, -4\sqrt{6} \rangle$

43-47. $\mathbf{v} = \langle 3, -4, 1 \rangle = 3\mathbf{i} - 4\mathbf{j} + \mathbf{k}$, $\mathbf{w} = \langle -5, 2, 0 \rangle = -5\mathbf{i} + 2\mathbf{j} + 0\mathbf{k}$

43. $-5\mathbf{v} = -5(3\mathbf{i} - 4\mathbf{j} + \mathbf{k}) = -15\mathbf{i} + 20\mathbf{j} - 5\mathbf{k}$

45. $\mathbf{w} + 2\mathbf{v} = -5\mathbf{i} + 2\mathbf{j} + 0\mathbf{k} + 2(3\mathbf{i} - 4\mathbf{j} + \mathbf{k}) = \mathbf{i} - 6\mathbf{j} + 2\mathbf{k}$

47. $\mathbf{v} - 3\mathbf{i} = 3\mathbf{i} - 4\mathbf{j} + \mathbf{k} - 3\mathbf{i} = -4\mathbf{j} + \mathbf{k}$

49. One diagonal is their sum $\langle 4, -6, 3 \rangle$ and the other diagonal is the first minus the second $\langle 2, 2, -1 \rangle$.

51. One diagonal is the second minus the first $\mathbf{i} + \mathbf{j} - 2\mathbf{k}$ and the other diagonal is the negative of their sum $-(-\mathbf{i} + 3\mathbf{j}) = \mathbf{i} - 3\mathbf{j}$.

Further Theory and Practise

53. The point in the xy-plane that is closest to the point $P_0 = (x_0, y_0, z_0)$ is the point $(x_0, y_0, 0)$. This is because the line from P_0 to $(x_0, y_0, 0)$ is perpendicular to the xy-plane. Similarly, the points $(0, y_0, z_0)$ and $(x_0, 0, z_0)$ are closest to P_0 in the yz-plane and the xz-plane respectively.

55. xz-plane $= \{(x, y, z) : y = 0\}$

57. Since the y-coordinate of the point $(1, 1, 1)$ is positive, this half-space is $\{(x, y, z) : y > 0\}$.

59. The distance from the point (x, y, z) to the xy-coordinate plane is $|z|$. Therefore, this set is $\{(x, y, z) : |z| > 5\}$.

61. The distance from a point (x, y, z) to the origin is $\sqrt{x^2 + y^2 + z^2}$ and the distance from (x, y, z) to $(2, 1, 2)$ is $\sqrt{(x - 2)^2 + (y - 1)^2 + (z - 2)^2}$. Therefore, this is the set of all points such that

$$\sqrt{x^2 + y^2 + z^2} > \sqrt{(x - 2)^2 + (y - 1)^2 + (z - 2)^2}.$$

Equivalently, $x^2 + y^2 + z^2 > (x - 2)^2 + (y - 1)^2 + (z - 2)^2$, which simplifies to $0 > 4 + 1 + 4 - 4x - 2y - 4z$, or $4x + 2y + 4z > 9$. Therefore, this is the set $\{(x, y, z) : 4x + 2y + 4z > 9\}$.

63. Let $x = \cos(\theta)\sin(\phi)$, $y = \sin(\theta)\sin(\phi)$, and $z = \cos(\phi)$. Observe that $x^2 + y^2 = \sin^2(\phi)$ and $x^2 + y^2 + z^2 = \sin^2(\phi) + \cos^2(\phi) = 1$. This is the set of all unit vectors in 3-space.

65. The car travels along the x-axis, at time t it is at the point $(40t, 0, 0)$. The student walks along the line $(0, y, 16)$, $-\infty < y < \infty$, at time t she is at $(0, 4t, 16)$. The distance between the student and the car at time t is $f(t) = \sqrt{(40t)^2 + (4t)^2 + 16^2} = \sqrt{1616t^2 + 256}$. Since $f'(t) = \frac{1616t}{\sqrt{1616t^2 + 256}}$, when $t = 2$ the distance is increasing at the rate of $f'(2) = \frac{404}{\sqrt{105}} \approx 39.4$ feet per second.

67. Let the center of the sphere be at the point (a, b, c) and its radius be r. All four of the given points must satisfy the equation

$$(x - a)^2 + (y - b)^2 + (z - c)^2 = r^2.$$

Substitute and simplify to obtain the following system of four equations in four unknowns.

$$a^2 + b^2 + c^2 - 10a - 4b - 6c + 38 = r^2$$
$$a^2 + b^2 + c^2 - 2a - 12b + 2c + 38 = r^2$$
$$a^2 + b^2 + c^2 - 6a + 4b - 10c + 38 = r^2$$
$$a^2 + b^2 + c^2 + 2a - 4b + 6c + 14 = r^2$$

Subtract the first equation from each of the other three to obtain a system of three equations for the constants a, b, c. This system simplifies to

$$\begin{array}{rrrrrcl} a & - & b & + & c & = & 0 \\ a & + & 2b & - & c & = & 0 \\ a & & & + & c & = & 2, \end{array}$$

which can be solved easily. For example, substitute $c = 2 - a$ into the first two equations, then solve for a and b to see that $a = -1, b = 2, c = 3$. Find r using any one of the four original equations, $r = 6$.

69. The sign of pq is positive when the charges are both positive or both negative. In this case, the force vector $\mathbf{F}$ (with initial point at Q) points directly *away* from the point P. Therefore, like charges repel. When the charges are opposite to one another, then pq is negative and the force vector points in the opposite direction: From Q towards P. Unlike charges attract.

71. When $x = 0$ the intersection with
    the yz-plane:

    $$y^2/9 + z^2/16 = 0\,,$$

    is an ellipse.
    When $y = 0$ the intersection with
    the xz-plane:

    $$x^2 + z^2/16 = 0\,,$$

    is also an ellipse.
    When $z = 0$ the intersection with
    the xy-plane:

    $$x^2 + y^2/9 = 0\,,$$

    is an ellipse also.

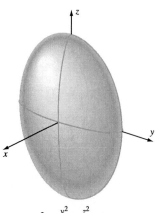

$$x^2 + \frac{y^2}{9} + \frac{z^2}{16} = 1$$

73. We must see if, given any vector $\langle a, b, c \rangle$, there are scalars x, y, z such that
    $x\langle 1, 1, 0 \rangle + y\langle 0, 1, 1 \rangle + z\langle 1, 0, 1 \rangle = \langle a, b, c \rangle$. Equivalently, the system of
    linear equations

    $$\text{1. } x + z = a \qquad \text{2. } x + y = b \qquad \text{3. } y + z = c$$

    has a solution for any scalars a, b, c. Subtract the third equation from the
    second to get $x - z = b - c$, then add this equation to the first to obtain
    $2x = a + b - c$ and $x = \frac{1}{2}(a + b - c)$. Knowing x, z can be found from
    equation 1, and y from equation 2. These vectors do span $\mathbb{R}^3$.

75. We must see if, given any vector $\langle a, b, c \rangle$, there are scalars x, y, z such that
    $x\langle 1, 1, 1 \rangle + y\langle 1, 1, 0 \rangle + z\langle -1, -1, 2 \rangle = \langle a, b, c \rangle$. Equivalently, the system of
    linear equations

    $$\text{1. } x + y - z = a \qquad \text{2. } x + y - z = b \qquad \text{3. } x + 2z = c$$

    has a solution for any scalars a, b, c. Clearly this is not the case. For
    example, if $a = 1$ and $b = 0$ there is no combination of x, y, z that makes
    equations 1 and 2 true. These vectors do not span $\mathbb{R}^3$.

Calculator/Computer Exercises

77. The lengths are equal when $4t^2 + t^2 + t^2 = t^4 + (1 - t)^2 + 1$. That
    is, $t^4 - 5t^2 - 2t + 2 = 0$. Using *Maple*'s *fsolve* procedure we find that
    $t = -1.8136, -1, 0.4707,$ and 2.3429.

79. The distance (squared) from P_t to the origin is $f(t) = (t - \cos(t))^2 +
    \sin^2(t) + t^2 \sin^2(t) = t^2 - 2t \cos(t) + t^2 \sin^2(t) + 1$. Using *Maple*'s *Minimize* procedure f attains the minimum value of 0.427978 when $t =
    0.476720$. Therefore, the minimum distance from the particle to the origin
    is $\sqrt{0.427978} = 0.654200$ units.

9.3   The Dot Product and Applications

Problems for Practise

1. $\langle 3, -2, 4 \rangle \cdot \langle 2, 1, 6 \rangle = 6 - 2 + 24 = 28$

3. $\langle 0, 4, 0 \rangle \cdot \langle 5, 1, 0 \rangle = 0 + 4 + 0 = 4$

5. $\langle 1, 4, 9 \rangle \cdot \langle 2, 3, -5 \rangle = 2 + 12 - 45 = -31$

7. $\cos(\theta) = \frac{\langle 1,1,0 \rangle \cdot \langle 0,1,1 \rangle}{\sqrt{1+1+0}\,\sqrt{0+1+1}} = \frac{1}{2} \implies \theta = \frac{\pi}{3}$

9. $\cos(\theta) = \frac{\langle 3,0,4 \rangle \cdot \langle 0,\sqrt{7},-5 \rangle}{\sqrt{9+0+16}\,\sqrt{0+7+25}} = -\frac{20}{\sqrt{800}} = -\frac{1}{\sqrt{2}} \implies \theta = \frac{3\pi}{4}$

11. $\cos(\theta) = \frac{\langle 2,-1,9 \rangle \cdot \langle -4,1,1 \rangle}{\sqrt{4+1+81}\,\sqrt{16+1+1}} = \frac{0}{\sqrt{86}\,\sqrt{18}} = 0 \implies \theta = \frac{\pi}{2}$

13. $\langle -3, 1, 5 \rangle \cdot \langle 4, -2, 3 \rangle = -12 - 2 + 15 = 1$. The vectors are not perpendicular because their dot product is not 0.

15. $\langle 2, -5, 8 \rangle \cdot \langle -2, 4, 3 \rangle = -4 - 20 + 24 = 0$. The vectors are perpendicular because their dot product is 0.

17. $\mathbf{P_v}(\mathbf{w}) = \frac{\langle 1,4,-4 \rangle \cdot \langle -3,7,2 \rangle}{\langle -3,7,2 \rangle \cdot \langle -3,7,2 \rangle} \langle -3, 7, 2 \rangle = \frac{17}{62} \langle -3, 7, 2 \rangle$, component: $\frac{17}{\sqrt{62}}$

   $\mathbf{P_w}(\mathbf{v}) = \frac{\langle -3,7,2 \rangle \cdot \langle 1,4,-4 \rangle}{\langle 1,4,-4 \rangle \cdot \langle 1,4,-4 \rangle} \langle 1, 4, -4 \rangle = \frac{17}{33} \langle 1, 4, -4 \rangle$, component: $\frac{17}{\sqrt{33}}$

19. $\mathbf{P_v}(\mathbf{w}) = \frac{\langle 4,8,12 \rangle \cdot \langle 2,4,6 \rangle}{\langle 2,4,6 \rangle \cdot \langle 2,4,6 \rangle} \langle 2, 4, 6 \rangle = 2\langle 2, 4, 6 \rangle = 4\langle 1, 2, 3 \rangle$, component: $4\sqrt{14}$

   $\mathbf{P_w}(\mathbf{v}) = \frac{\langle 2,4,6 \rangle \cdot \langle 4,8,12 \rangle}{\langle 4,8,12 \rangle \cdot \langle 4,8,12 \rangle} \langle 4, 8, 12 \rangle = \frac{1}{2}\langle 4, 8, 12 \rangle = 2\langle 1, 2, 3 \rangle$, component: $2\sqrt{14}$

21. $\mathbf{P_v}(\mathbf{w}) = \frac{\langle 0,1,0 \rangle \cdot \langle 1,0,1 \rangle}{\langle 1,0,1 \rangle \cdot \langle 1,0,1 \rangle} \langle 1, 0, 1 \rangle = 0\langle 1, 0, 1 \rangle$, component: 0

   $\mathbf{P_w}(\mathbf{v}) = \frac{\langle 1,0,1 \rangle \cdot \langle 0,1,0 \rangle}{\langle 0,1,0 \rangle \cdot \langle 0,1,0 \rangle} \langle 0, 1, 0 \rangle = 0\langle 0, 1, 0 \rangle$, component: 0

23. $\mathbf{v} = \langle 6, -2\sqrt{3}, 0 \rangle$, direction $= \frac{1}{\|\mathbf{v}\|}\mathbf{v} = \frac{1}{4\sqrt{3}}\langle 6, -2\sqrt{3}, 0 \rangle = \langle \sqrt{3}/2, -1/2, 0 \rangle$;

   $\cos(\alpha) = \sqrt{3}/2$, $\alpha = \pi/6$, $\cos(\beta) = -1/2$, $\beta = 2\pi/3$, $\cos(\gamma) = 0$, $\gamma = \pi/2$

25. $\mathbf{v} = \langle 0, 1, 0 \rangle$, direction $= \frac{1}{\|\mathbf{v}\|}\mathbf{v} = \langle 0, 1, 0 \rangle$;

   $\cos(\alpha) = 0$, $\alpha = \pi/2$, $\cos(\beta) = 1$, $\beta = 0$, $\cos(\gamma) = 0$, $\gamma = \pi/2$

27. $\mathbf{v} = \langle -6, 0, 6 \rangle$, direction $= \frac{1}{\|\mathbf{v}\|}\mathbf{v} = \frac{1}{6\sqrt{2}}\langle -6, 0, 6 \rangle = \langle -1/\sqrt{2}, 0, 1/\sqrt{2} \rangle$;

   $\cos(\alpha) = -1/\sqrt{2}$, $\alpha = 3\pi/4$, $\cos(\beta) = 0$, $\beta = \pi/2$, $\cos(\gamma) = 1/\sqrt{2}$, $\gamma = \pi/4$

29. $\langle 2, -4, 4 \rangle \cdot \langle -2, 1, -2 \rangle = -4 - 4 - 8 = -16$, $|\mathbf{v} \cdot \mathbf{w}| = 16$;

   $\|\langle 2, -4, 4 \rangle\| \cdot \|\langle -2, 1, -2 \rangle\| = \sqrt{36} \cdot \sqrt{9} = 6 \cdot 3$, $\|\mathbf{v}\|\,\|\mathbf{w}\| = 18$

31. $\langle 2, 0, 1\rangle \cdot \langle 4, \sqrt{3}, 1\rangle = 8 + 0 + 1 = 9$, $|\mathbf{v} \cdot \mathbf{w}| = 9$;

$\|\langle 2, 0, 1\rangle\| \cdot \|\langle 4, \sqrt{3}, 1\rangle\| = \sqrt{5} \cdot \sqrt{20} = 10$, $\|\mathbf{v}\| \|\mathbf{w}\| = 10$

33. $\mathbf{u}$ is a unit vector so

$$\begin{aligned}
\mathbf{P_u}(\mathbf{v}) &= (\mathbf{v} \cdot \mathbf{u})\mathbf{u} \\
&= \big(\langle 4, 1, -8\rangle \cdot \langle 2/3, 2/3, -1/3\rangle\big)\langle 2/3, 2/3, -1/3\rangle \\
&= \frac{18}{3}\langle 2/3, 2/3, -1/3\rangle \\
&= \langle 4, 4, -2\rangle
\end{aligned}$$

and $\mathbf{v} - \mathbf{P_u}(\mathbf{v}) = \langle 4, 1, -8\rangle - \langle 4, 4, -2\rangle = \langle 0, -3, -6\rangle$. This vector is perpendicular to $\mathbf{P_u}(\mathbf{v})$, (their dot product is 0).

35. $\mathbf{u}$ is a unit vector so

$$\begin{aligned}
\mathbf{P_u}(\mathbf{v}) &= (\mathbf{v} \cdot \mathbf{u})\mathbf{u} \\
&= \big(\langle \sqrt{12}, 0, \sqrt{48}\rangle \cdot \langle 1/\sqrt{3}, 1/\sqrt{3}, -1/\sqrt{3}\rangle\big)\langle 1/\sqrt{3}, 1/\sqrt{3}, -1/\sqrt{3}\rangle \\
&= \frac{2\sqrt{3} - 4\sqrt{3}}{\sqrt{3}}\langle 1/\sqrt{3}, 1/\sqrt{3}, -1/\sqrt{3}\rangle \\
&= -2\langle 1/\sqrt{3}, 1/\sqrt{3}, -1/\sqrt{3}\rangle \\
&= \langle -2/\sqrt{3}, -2/\sqrt{3}, 2/\sqrt{3}\rangle
\end{aligned}$$

and

$$\begin{aligned}
\mathbf{v} - \mathbf{P_u}(\mathbf{v}) &= \langle 2\sqrt{3}, 0, 4\sqrt{3}\rangle - \langle -2/\sqrt{3}, -2/\sqrt{3}, 2/\sqrt{3}\rangle \\
&= \langle 8/\sqrt{3}, 2/\sqrt{3}, 10/\sqrt{3}\rangle.
\end{aligned}$$

This vector is perpendicular to $\mathbf{P_u}(\mathbf{v})$, (their dot product is 0).

37. $\langle 3, 2, -1\rangle \cdot \langle s, 1, -4\rangle = 0$ when $3s + 2 + 4 = 0$. Therefore, $s = -2$.

39. $\langle 3, 1, 1\rangle \cdot \langle -7, s^2, 5\rangle = 0$ when $-21 + s^2 + 5 = 0$. Therefore $s^2 = 16$, so $s = \pm 4$.

Further Theory and Practice

41. The vector $\mathbf{u} = x\mathbf{i} + y\mathbf{j} + z\mathbf{k}$ is perpendicular to both $\mathbf{v}$ and $\mathbf{w}$ when $\mathbf{u} \cdot \mathbf{v} = 0$ and $\mathbf{u} \cdot \mathbf{w} = 0$. Therefore, x, y, and z must satisfy the following two equations.

$$\begin{array}{ccccccc}
x & + & 2y & + & z & = & 0 \\
x & - & y & & & = & 0
\end{array}$$

This implies that $x = y$ and $z = -3y$, so the vector $\mathbf{u}$ must be of the form $\langle t, t, -3t\rangle = t\langle 1, 1, -3\rangle$. Since $\mathbf{u}$ is a unit vector, $\mathbf{u} = \pm\frac{1}{\sqrt{11}}\langle 1, 1, -3\rangle$. All vectors perpendicular to both $\mathbf{v}$ and $\mathbf{w}$ are parallel to one another because they are scalar multiples of $\langle 1, 1, -3\rangle$.

43. $\mathbf{a} \cdot \mathbf{b} = 0 + 1/4 - 1/4 = 0$, $\mathbf{a} \cdot \mathbf{c} = \sqrt{3}/4 - \sqrt{3}/8 - \sqrt{3}/8 = 0$, and $\mathbf{b} \cdot \mathbf{c} = -\sqrt{3}/2 + \sqrt{3}/2 = 0$. Moreover, $\|a\| = \sqrt{3/4 + 1/8 + 1/8} = 1$, $\|b\| = \sqrt{1/2 + 1/2} = 1$, and $\|c\| = \sqrt{1/4 + 3/8 + 3/8} = 1$.

45. The component of the force $\mathbf{F}$ in the direction of motion is $\|\mathbf{F}\| \cos(30°) = 200 \cdot \frac{\sqrt{3}}{2} = 100\sqrt{3}$ lb. If the car moves 1000 ft, then the work done is $100000\sqrt{3}$ ft-lb.

47. Use the fact that $\|\mathbf{v}\|^2 = \mathbf{v} \cdot \mathbf{v}$.

$$\|\mathbf{v} + \mathbf{w}\|^2 + \|\mathbf{v} - \mathbf{w}\|^2 = (\mathbf{v} + \mathbf{w}) \cdot (\mathbf{v} + \mathbf{w}) + (\mathbf{v} - \mathbf{w}) \cdot (\mathbf{v} - \mathbf{w})$$
$$= \mathbf{v} \cdot \mathbf{v} + 2\,\mathbf{v} \cdot \mathbf{w} + \mathbf{w} \cdot \mathbf{w} + \mathbf{v} \cdot \mathbf{v} - 2\,\mathbf{v} \cdot \mathbf{w} + \mathbf{w} \cdot \mathbf{w}$$
$$= 2\,\mathbf{v} \cdot \mathbf{v} + 2\,\mathbf{w} \cdot \mathbf{w}$$
$$= 2\,\|\mathbf{v}\|^2 + 2\,\|\mathbf{w}\|^2$$

49. In each case we can simply calculate $\mathbf{P_w}(\mathbf{v}) = \frac{\mathbf{v} \cdot \mathbf{w}}{\mathbf{w} \cdot \mathbf{w}}\mathbf{w}$, so $\lambda = \frac{\mathbf{v} \cdot \mathbf{w}}{\mathbf{w} \cdot \mathbf{w}}$.

(a) $\lambda = \frac{\langle 1,-2,5\rangle \cdot \langle 1,2,3\rangle}{\langle 1,2,3\rangle \cdot \langle 1,2,3\rangle} = 6/7$

(b) $\lambda = \frac{\langle 1,-4,6\rangle \cdot \langle -1,0,2\rangle}{\langle -1,0,2\rangle \cdot \langle -1,0,2\rangle} = 11/5$

(c) $\lambda = \frac{\langle -3,4,8\rangle \cdot \langle 1,5,-3\rangle}{\langle 1,5,-3\rangle \cdot \langle 1,5,-3\rangle} = -1/5$

(d) $\lambda = \frac{\langle 1,1,3\rangle \cdot \langle 3,0,-4\rangle}{\langle 3,0,-4\rangle \cdot \langle 3,0,-4\rangle} = -9/25$

51. The angle between $\overrightarrow{OP}$ and $\mathbf{j}$ must be 30 degrees. Therefore, since $\overrightarrow{OP} \cdot \mathbf{j} = \|\overrightarrow{OP}\|\,\|\mathbf{j}\| \cos(30°)$, the equation is $\overrightarrow{OP} \cdot \mathbf{j} = (\sqrt{3}/2)\|\overrightarrow{OP}\|$.

53. Let the points be P, Q, and R respectively. Then the sides of the triangle are in the direction of $\mathbf{a} = \overrightarrow{PQ} = \langle 3, 2, -2\rangle$, $\mathbf{b} = \overrightarrow{QR} = \langle 2, 1, 4\rangle$, and $\mathbf{c} = \overrightarrow{PR} = \langle 5, 3, 2\rangle$. Since $\mathbf{a} \cdot \mathbf{b} = 0$, the angle at point Q is $90°$ so it is a right triangle.

55. Since $\mathbf{v} \cdot \mathbf{w} = 2\sqrt{ab}$ and $\|\mathbf{v}\| = \|\mathbf{w}\| = \sqrt{a+b}$, the Cauchy-Schwarz inequality—$|\mathbf{v} \cdot \mathbf{w}| \le \|\mathbf{v}\|\,\|\mathbf{w}\|$—has the form $2\sqrt{ab} \le a + b$. The rearrangement

$$\sqrt{ab} \le \frac{a+b}{2}$$

is called the Geometric-Arithmetic Mean inequality.

57. Observe that

$$\|\mathbf{v} + \mathbf{w}\|^2 = (\mathbf{v} + \mathbf{w}) \cdot (\mathbf{v} + \mathbf{w})$$
$$= \mathbf{v} \cdot \mathbf{v} + 2\,\mathbf{v} \cdot \mathbf{w} + \mathbf{w} \cdot \mathbf{w}$$
$$= \|\mathbf{v}\|^2 + 2\,\mathbf{v} \cdot \mathbf{w} + \|\mathbf{w}\|^2.$$

Apply the Cauchy-Schwarz inequality to the middle term to see that

$$\|\mathbf{v} + \mathbf{w}\|^2 \le \|\mathbf{v}\|^2 + 2\|\mathbf{v}\|\|\mathbf{w}\| + \|\mathbf{w}\|^2$$
$$= (\|\mathbf{v}\| + \|\mathbf{w}\|)^2.$$

Now take the square root on both sides to obtain $\|\mathbf{v} + \mathbf{w}\| \le \|\mathbf{v}\| + \|\mathbf{w}\|$.

59. The first identity can be verified easily by expanding and simplifying the left side, then expanding and simplifying the right side.

If $\mathbf{v} = \langle v_1, v_2, v_3 \rangle$ and $\mathbf{w} = \langle w_1, w_2, w_3 \rangle$, then the left side of the first identity is identically equal to the left side of the second identity.

The Cauchy-Schwarz Inequality follows because the right hand side of the second identity is non-negative, implying that $\|\mathbf{v}\|^2\|\mathbf{w}\|^2 - (\mathbf{v} \cdot \mathbf{w})^2 \ge 0$. Therefore, $\|\mathbf{v}\|^2\|\mathbf{w}\|^2 \ge (\mathbf{v} \cdot \mathbf{w})^2$ and (taking square roots) $\|\mathbf{v}\|\|\mathbf{w}\| \ge |\mathbf{v} \cdot \mathbf{w}|$.

61. Observe that

$$\langle a, b \rangle \cdot T(\langle a, b \rangle) = \langle a, b \rangle \cdot \langle a\cos(\theta) - b\sin(\theta), a\sin(\theta) + b\cos(\theta) \rangle$$
$$= (a^2 + b^2)\cos(\theta).$$

Moreover, because $\|T(\langle a, b \rangle)\| = \|\langle a, b \rangle\|$ (shown below), the angle between $\langle a, b \rangle$ and $T(\langle a, b \rangle)$ is

$$\arccos\left(\frac{\langle a, b \rangle \cdot T(\langle a, b \rangle)}{\|\langle a, b \rangle\| \|T(\langle a, b \rangle)\|} \right) = \arccos\left(\frac{(a^2 + b^2)\cos(\theta)}{\|\langle a, b \rangle\|^2} \right)$$
$$= \arccos(\cos(\theta))$$
$$= \theta.$$

To verify that $\|T(\langle a, b \rangle)\| = \|\langle a, b \rangle\|$ observe that

$$\|T(\langle a, b \rangle)\|^2 = (a\cos(\theta) - b\sin(\theta))^2 + (a\sin(\theta) + b\cos(\theta))^2$$
$$= a^2\cos^2(\theta) + b^2\sin^2(\theta) + a^2\sin^2(\theta) + b^2\cos^2(\theta)$$
$$= (a^2 + b^2)\cos^2(\theta) + (a^2 + b^2)\sin^2(\theta)$$
$$= a^2 + b^2.$$

Calculator/Computer Exercises

63. Define the three "vertex vectors" $\mathbf{a} = \langle 1, 1, 2 \rangle$, $\mathbf{b} = \langle 2, 0, 3 \rangle$ and $\mathbf{c} = \langle 1, 2, -5 \rangle$. Using *Maple*, the three vertex angles, in radians, are

$$\alpha = \arccos\left(\frac{(\mathbf{b} - \mathbf{a}) \cdot (\mathbf{c} - \mathbf{a})}{\|\mathbf{b} - \mathbf{a}\| \|\mathbf{c} - \mathbf{a}\|} \right) = 2.282595666$$

$$\beta = \arccos\left(\frac{(\mathbf{c} - \mathbf{b}) \cdot (\mathbf{a} - \mathbf{b})}{\|\mathbf{c} - \mathbf{b}\| \|\mathbf{a} - \mathbf{b}\|} \right) = 0.7004490075$$

$$\gamma = \arccos\left(\frac{(\mathbf{a} - \mathbf{c}) \cdot (\mathbf{b} - \mathbf{c})}{\|\mathbf{a} - \mathbf{c}\| \|\mathbf{b} - \mathbf{c}\|} \right) = 0.1585479832$$

Their sum is $\alpha + \beta + \gamma = 3.141592657$. Note that $\pi \approx 3.141592654$.

65. The graph of the function f is displayed on the right. Using *Maple*'s *Minimize* procedure it attains a minimum value of 3.898694341 at $t = 0.2038883547$.

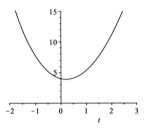

9.4   The Cross Product and Triple Product

Problems for Practise

1. $\mathbf{v} = \langle 3, 3, -5 \rangle$, $\mathbf{w} = \langle -1, 3, -2 \rangle$

$$\mathbf{v} \times \mathbf{w} = \det\left(\begin{bmatrix} \mathbf{i} & \mathbf{j} & \mathbf{k} \\ 3 & 3 & -5 \\ -1 & 3 & -2 \end{bmatrix}\right) = \langle 9, 11, 12 \rangle$$

$$\mathbf{v} \cdot (\mathbf{v} \times \mathbf{w}) = \langle 3, 3, -5 \rangle \cdot \langle 9, 11, 12 \rangle = 0$$

$$\mathbf{w} \cdot (\mathbf{v} \times \mathbf{w}) = \langle -1, 3, -2 \rangle \cdot \langle 9, 11, 12 \rangle = 0$$

3. $\mathbf{v} = \langle 4, -3, 6 \rangle$, $\mathbf{w} = \langle 2, 1, 2 \rangle$

$$\mathbf{v} \times \mathbf{w} = \det\left(\begin{bmatrix} \mathbf{i} & \mathbf{j} & \mathbf{k} \\ 4 & -3 & 6 \\ 2 & 1 & 2 \end{bmatrix}\right) = \langle -12, 4, 10 \rangle$$

$$\mathbf{v} \cdot (\mathbf{v} \times \mathbf{w}) = \langle 4, -3, 6 \rangle \cdot \langle -12, 4, 10 \rangle = 0$$

$$\mathbf{w} \cdot (\mathbf{v} \times \mathbf{w}) = \langle 2, 1, 2 \rangle \cdot \langle -12, 4, 10 \rangle = 0$$

5. $\mathbf{v} = \langle -4, -2, 3 \rangle$, $\mathbf{w} = \langle 7, 1, -5 \rangle$

$$\mathbf{v} \times \mathbf{w} = \det\left(\begin{bmatrix} \mathbf{i} & \mathbf{j} & \mathbf{k} \\ -4 & -2 & 3 \\ 7 & 1 & -5 \end{bmatrix}\right) = \langle 7, 1, 10 \rangle$$

$$\mathbf{v} \cdot (\mathbf{v} \times \mathbf{w}) = \langle -4, -2, 3 \rangle \cdot \langle 7, 1, 10 \rangle = 0$$

$$\mathbf{w} \cdot (\mathbf{v} \times \mathbf{w}) = \langle 7, 1, -5 \rangle \cdot \langle 7, 1, 10 \rangle = 0$$

7. $\mathbf{v} = \langle 1, 6, 8 \rangle$, $\mathbf{w} = \langle 1, 2, 8 \rangle$

$$\mathbf{v} \times \mathbf{w} = \det\left(\begin{bmatrix} \mathbf{i} & \mathbf{j} & \mathbf{k} \\ 1 & 6 & 8 \\ 1 & 2 & 8 \end{bmatrix}\right) = \langle 32, 0, -4 \rangle$$

$$\mathbf{v} \cdot (\mathbf{v} \times \mathbf{w}) = \langle 1, 6, 8 \rangle \cdot \langle 32, 0, -4 \rangle = 0$$

$$\mathbf{w} \cdot (\mathbf{v} \times \mathbf{w}) = \langle 1, 2, 8 \rangle \cdot \langle 32, 0, -4 \rangle = 0$$

9. $\mathbf{v} = \langle 2, 1, 3 \rangle$, $\mathbf{w} = \langle -3, -2, -5 \rangle$;

$$\mathbf{v} \times \mathbf{w} = \det\left(\begin{bmatrix} \mathbf{i} & \mathbf{j} & \mathbf{k} \\ 2 & 1 & 3 \\ -3 & -2 & -5 \end{bmatrix}\right) = \langle 1, 1, -1 \rangle;$$

$\|\mathbf{v} \times \mathbf{w}\| = \sqrt{3}$, $\mathbf{dir}(\mathbf{v} \times \mathbf{w}) = \langle \frac{1}{\sqrt{3}}, \frac{1}{\sqrt{3}}, -\frac{1}{\sqrt{3}} \rangle$

11. $\mathbf{v} = \langle 3, -1, 1 \rangle$, $\mathbf{w} = \langle 1, 1, 1 \rangle$

$$\mathbf{v} \times \mathbf{w} = \det\left(\begin{bmatrix} \mathbf{i} & \mathbf{j} & \mathbf{k} \\ 3 & -1 & 1 \\ 1 & 1 & 1 \end{bmatrix}\right) = \langle -2, -2, 4 \rangle$$

$\|\mathbf{v} \times \mathbf{w}\| = \sqrt{24} = 2\sqrt{6}$, $\mathbf{dir}(\mathbf{v} \times \mathbf{w}) = \langle -\frac{1}{\sqrt{6}}, -\frac{1}{\sqrt{6}}, \frac{2}{\sqrt{6}} \rangle$

13. $\mathbf{v} = \langle 1, 2, 2 \rangle$, $\mathbf{w} = \langle 2, -1, 1 \rangle$

$$\mathbf{v} \times \mathbf{w} = \det\left(\begin{bmatrix} \mathbf{i} & \mathbf{j} & \mathbf{k} \\ 1 & 2 & 2 \\ 2 & -1 & 1 \end{bmatrix}\right) = \langle 4, 3, -5 \rangle$$

$\|\mathbf{v} \times \mathbf{w}\| = \sqrt{50} = 5\sqrt{2}$, $\mathbf{dir}(\mathbf{v} \times \mathbf{w}) = \frac{\sqrt{2}}{10}\langle 4, 3, -5 \rangle$

15. $\mathbf{v} \times \mathbf{w} = \det\left(\begin{bmatrix} \mathbf{i} & \mathbf{j} & \mathbf{k} \\ 2 & 1 & 2 \\ 3 & 2 & 3 \end{bmatrix}\right) = \langle -1, 0, 1 \rangle$

$Area = \frac{1}{2} \cdot |\mathbf{v} \times \mathbf{w}| = \frac{1}{2}\sqrt{2}$

17. $\mathbf{v} \times \mathbf{w} = \det\left(\begin{bmatrix} \mathbf{i} & \mathbf{j} & \mathbf{k} \\ -3 & 4 & 1 \\ -2 & 0 & 1 \end{bmatrix}\right) = \langle 4, 1, 8 \rangle$

$Area = \frac{1}{2} \cdot |\mathbf{v} \times \mathbf{w}| = \frac{1}{2}\sqrt{81} = \frac{9}{2}$

19. $\mathbf{v} \times \mathbf{w} = \det\left(\begin{bmatrix} \mathbf{i} & \mathbf{j} & \mathbf{k} \\ 2 & 1 & 1 \\ 0 & -3 & 1 \end{bmatrix}\right) = \langle 4, -2, 6 \rangle$

$Area = |\mathbf{v} \times \mathbf{w}| = \sqrt{56} = 2\sqrt{14}$

21. $\mathbf{v} \times \mathbf{w} = \det\left(\begin{bmatrix} \mathbf{i} & \mathbf{j} & \mathbf{k} \\ 1 & 0 & 2 \\ 1 & -3 & 0 \end{bmatrix}\right) = \langle 6, 2, -3 \rangle$

$Area = |\mathbf{v} \times \mathbf{w}| = \sqrt{49} = 7$

23. $\mathbf{v} \times \mathbf{w} = \det\left(\begin{bmatrix} \mathbf{i} & \mathbf{j} & \mathbf{k} \\ 2 & -1 & -3 \\ 3 & -2 & -3 \end{bmatrix}\right) = \langle -3, -3, -1 \rangle$

$$\mathbf{u} \times (\mathbf{v} \times \mathbf{w}) = \det\left(\begin{bmatrix} \mathbf{i} & \mathbf{j} & \mathbf{k} \\ -4 & 1 & 0 \\ -3 & -3 & -1 \end{bmatrix}\right) = \langle -1, -4, 15 \rangle$$

$$\mathbf{u} \times \mathbf{v} = \det\left(\begin{bmatrix} \mathbf{i} & \mathbf{j} & \mathbf{k} \\ -4 & 1 & 0 \\ 2 & -1 & -3 \end{bmatrix}\right) = \langle -3, -12, 2 \rangle$$

$$(\mathbf{u} \times \mathbf{v}) \times \mathbf{w} = \det\left(\begin{bmatrix} \mathbf{i} & \mathbf{j} & \mathbf{k} \\ -3 & -12 & 2 \\ 3 & -2 & -3 \end{bmatrix}\right) = \langle 40, -3, 42 \rangle$$

25. $\mathbf{v} \times \mathbf{w} = \det\left(\begin{bmatrix} \mathbf{i} & \mathbf{j} & \mathbf{k} \\ 2 & 1 & 2 \\ -3 & 4 & -3 \end{bmatrix}\right) = \langle -11, 0, 11 \rangle$

$$\mathbf{u} \times (\mathbf{v} \times \mathbf{w}) = \det\left(\begin{bmatrix} \mathbf{i} & \mathbf{j} & \mathbf{k} \\ 1 & 1 & 2 \\ -11 & 0 & 11 \end{bmatrix}\right) = \langle 11, -33, 11 \rangle$$

$$\mathbf{u} \times \mathbf{v} = \det\left(\begin{bmatrix} \mathbf{i} & \mathbf{j} & \mathbf{k} \\ 1 & 1 & 2 \\ 2 & 1 & 2 \end{bmatrix}\right) = \langle 0, 2, -1 \rangle$$

$$(\mathbf{u} \times \mathbf{v}) \times \mathbf{w} = \det\left(\begin{bmatrix} \mathbf{i} & \mathbf{j} & \mathbf{k} \\ 0 & 2 & -1 \\ -3 & 4 & -3 \end{bmatrix}\right) = \langle -2, 3, 6 \rangle$$

27. $\mathbf{u} \times \mathbf{v} = \det\left(\begin{bmatrix} \mathbf{i} & \mathbf{j} & \mathbf{k} \\ 3 & 0 & 1 \\ 2 & -1 & -3 \end{bmatrix}\right) = \langle 1, 11, -3 \rangle$

$$\mathbf{v} \times \mathbf{w} = \det\left(\begin{bmatrix} \mathbf{i} & \mathbf{j} & \mathbf{k} \\ 2 & -1 & -3 \\ -1 & -3 & 2 \end{bmatrix}\right) = \langle -11, -1, -7 \rangle$$

$$(\mathbf{u} \times \mathbf{v}) \cdot \mathbf{w} = \langle 1, 11, -3 \rangle \cdot \langle -1, -3, 2 \rangle = -40$$

$$\mathbf{u} \cdot (\mathbf{v} \times \mathbf{w}) = \langle 3, 0, 1 \rangle \cdot \langle -11, -1, -7 \rangle = -40$$

29. $\mathbf{u} \times \mathbf{v} = \det\left(\begin{bmatrix} \mathbf{i} & \mathbf{j} & \mathbf{k} \\ 1 & 1 & -2 \\ 2 & 1 & -2 \end{bmatrix}\right) = \langle 0, -2, -1 \rangle$

$$\mathbf{v} \times \mathbf{w} = \det\left(\begin{bmatrix} \mathbf{i} & \mathbf{j} & \mathbf{k} \\ 2 & 1 & -2 \\ 3 & 2 & 3 \end{bmatrix}\right) = \langle 7, -12, 1 \rangle$$

$$(\mathbf{u} \times \mathbf{v}) \cdot \mathbf{w} = \langle 0, -2, -1 \rangle \cdot \langle 3, 2, 3 \rangle = -7$$

$$\mathbf{u} \cdot (\mathbf{v} \times \mathbf{w}) = \langle 1, 1, -2 \rangle \cdot \langle 7, -12, 1 \rangle = -7$$

31. Use a determinant to calculate the triple scalar product.

$$(\mathbf{u} \times \mathbf{v}) \cdot \mathbf{w} = \det \left(\begin{bmatrix} 1 & -2 & 4 \\ 2 & 0 & 1 \\ 3 & 1 & 1 \end{bmatrix} \right) = 1 \cdot (0-1) - (-2) \cdot (2-3) + 4 \cdot (2-0)$$
$$= 5$$

33. Use a determinant to calculate the triple scalar product.

$$(\mathbf{u} \times \mathbf{v}) \cdot \mathbf{w} = \det \left(\begin{bmatrix} 1 & 1 & 1 \\ 3 & 0 & 2 \\ -2 & -3 & 3 \end{bmatrix} \right) = 1 \cdot (0+6) - 1 \cdot (9+4) + 1 \cdot (-9-0)$$
$$= -16$$

35. Vectors are coplanar when the triple scalar product is zero.

$$(\mathbf{u} \times \mathbf{v}) \cdot \mathbf{w} = \det \left(\begin{bmatrix} 1 & -2 & 4 \\ 2 & 0 & 1 \\ 5 & -2 & 6 \end{bmatrix} \right) = 1 \cdot (0+2) - (-2) \cdot (12-5) + 4 \cdot (-4-0)$$
$$= 0$$

37. Vectors are coplanar when the triple scalar product is zero.

$$(\mathbf{u} \times \mathbf{v}) \cdot \mathbf{w} = \det \left(\begin{bmatrix} 1 & 1 & 1 \\ 3 & 0 & 2 \\ 8 & 2 & 6 \end{bmatrix} \right) = 1 \cdot (0-4) - 1 \cdot (18-16) + 1 \cdot (6-0)$$
$$= 0$$

39. The volume is the absolute value of the scalar triple scalar product. Since

$$(\mathbf{u} \times \mathbf{v}) \cdot \mathbf{w} = \det \left(\begin{bmatrix} 2 & 1 & 1 \\ 1 & 1 & -1 \\ 1 & -1 & -1 \end{bmatrix} \right) = 2 \cdot (-2) - 1 \cdot 0 + 1 \cdot (-2)$$
$$= -6,$$

the volume is 6.

41. The volume is the absolute value of the scalar triple scalar product. Since

$$(\mathbf{u} \times \mathbf{v}) \cdot \mathbf{w} = \det \left(\begin{bmatrix} 3 & 2 & 1 \\ 1 & 2 & 3 \\ 1 & 4 & -2 \end{bmatrix} \right) = 3 \cdot (-16) - 2 \cdot (-5) + 1 \cdot 2$$
$$= -36,$$

the volume is 36.

Further Theory and Practice

43. Since $\mathbf{u} \times (\mathbf{v} \times \mathbf{w}) = \langle 8, 18, 7 \rangle$, we want to find scalars s and t such that $\langle 8, 18, 7 \rangle = s\langle 2, 0, 1 \rangle + t\langle 5, -3, 2 \rangle$. Equate the second components on each side of this equation: $18 = -3t$, to see see that $t = -6$. Substitute this value for t into the equation and equate the third components on each side: $7 = s - 12$, to see that $s = 19$.

45. Since $\mathbf{u} \times (\mathbf{v} \times \mathbf{w}) = \langle 11, -2, 9 \rangle$, we want to find scalars s and t such that $\langle 11, -2, 9 \rangle = s\langle 3, 0, 2 \rangle + t\langle 4, 2, 1 \rangle$. Equate the second components on each side of this equation: $-2 = 2t$, to see see that $t = -1$. Substitute this value for t into the equation and equate the third components on each side: $9 = 2s - 1$, to see that $s = 5$.

47. Since $\mathbf{v} \times \mathbf{w} = \det\left(\begin{bmatrix} \mathbf{i} & \mathbf{j} & \mathbf{k} \\ 2 & 1 & 2 \\ 1 & -2 & -2 \end{bmatrix}\right) = \langle 2, 6, -5 \rangle$, $\|\mathbf{v} \times \mathbf{w}\|^2 = 4 + 36 + 25 = 65$, and $\|\mathbf{v}\|^2\|\mathbf{w}\|^2 - (\mathbf{v} \cdot \mathbf{w})^2 = 9 \cdot 9 - (-4)^2 = 65$.

49. Since $\mathbf{v} \times \mathbf{w} = \det\left(\begin{bmatrix} \mathbf{i} & \mathbf{j} & \mathbf{k} \\ 2 & 3 & 1 \\ 2 & 3 & -1 \end{bmatrix}\right) = \langle -6, 4, 0 \rangle$, $\|\mathbf{v} \times \mathbf{w}\|^2 = 36 + 16 = 52$, and $\|\mathbf{v}\|^2\|\mathbf{w}\|^2 - (\mathbf{v} \cdot \mathbf{w})^2 = 14 \cdot 14 - 12^2 = 52$.

51. (a) $\mathbf{v} \times \mathbf{w} = \det\left(\begin{bmatrix} \mathbf{i} & \mathbf{j} & \mathbf{k} \\ 2 & 1 & 1 \\ 1 & 2 & -1 \end{bmatrix}\right) = \langle -3, 3, 3 \rangle$

   (b) $\sin(\theta) = \frac{\|\mathbf{v} \times \mathbf{w}\|}{\|\mathbf{v}\| \|\mathbf{w}\|} = \frac{3\sqrt{3}}{\sqrt{6}\sqrt{6}} = \frac{\sqrt{3}}{2}$

   (c) $\mathbf{v} \cdot \mathbf{w} = 3$

   (d) $\cos(\theta) = \frac{\mathbf{v} \cdot \mathbf{w}}{\|\mathbf{v}\| \|\mathbf{w}\|} = \frac{3}{6} = \frac{1}{2}$

   (e) The values of $\sin(\theta)$ and $\cos(\theta)$ are consistent, both imply that $\theta = \frac{\pi}{3}$.

53. (a) $\mathbf{v} \times \mathbf{w} = \det\left(\begin{bmatrix} \mathbf{i} & \mathbf{j} & \mathbf{k} \\ 2 & -2 & 1 \\ 8 & 4 & 1 \end{bmatrix}\right) = \langle -6, 6, 24 \rangle$

   (b) $\sin(\theta) = \frac{\|\mathbf{v} \times \mathbf{w}\|}{\|\mathbf{v}\| \|\mathbf{w}\|} = \frac{6\sqrt{18}}{3 \cdot 9} = \frac{2\sqrt{2}}{3}$

   (c) $\mathbf{v} \cdot \mathbf{w} = 9$

   (d) $\cos(\theta) = \frac{\mathbf{v} \cdot \mathbf{w}}{\|\mathbf{v}\| \|\mathbf{w}\|} = \frac{9}{27} = \frac{1}{3}$

   (e) The values of $\sin(\theta)$ and $\cos(\theta)$ are consistent, $\sin^2(\theta) + \cos^2(\theta) = \frac{8}{9} + \frac{1}{9} = 1$.

55. Let $\mathbf{u} = \langle u_1, u_2, u_3 \rangle$, $\mathbf{v} = \langle v_1, v_2, v_3 \rangle$, and $\mathbf{w} = \langle w_1, w_2, w_3 \rangle$. Then,

displaying the first components only, we have

$$\mathbf{u} \times (\mathbf{v} + \mathbf{w}) = \det\left(\begin{bmatrix} \mathbf{i} & \mathbf{j} & \mathbf{k} \\ u_1 & u_2 & u_3 \\ v_1 + w_1 & v_2 + w_2 & v_3 + w_3 \end{bmatrix}\right)$$

$$= \langle u_2(v_3 + w_3) - u_3(v_2 + w_2), \dots \rangle$$

$$= \langle u_2 v_3 - u_3 v_2, \dots \rangle + \langle u_2 w_3 - u_3 w_2, \cdots \rangle$$

$$= \det\left(\begin{bmatrix} \mathbf{i} & \mathbf{j} & \mathbf{k} \\ u_1 & u_2 & u_3 \\ v_1 & v_2 & v_3 \end{bmatrix}\right) + \det\left(\begin{bmatrix} \mathbf{i} & \mathbf{j} & \mathbf{k} \\ u_1 & u_2 & u_3 \\ w_1 & w_2 & w_3 \end{bmatrix}\right)$$

$$= \mathbf{u} \times \mathbf{v} + \mathbf{u} \times \mathbf{w}.$$

57. Let $\mathbf{v} = \langle v_1, v_2, v_3 \rangle$ and $\mathbf{w} = \langle w_1, w_2, w_3 \rangle$. Then

$$(\lambda\mathbf{v}) \times (\mu\mathbf{w}) = \det\left(\begin{bmatrix} \mathbf{i} & \mathbf{j} & \mathbf{k} \\ \lambda v_1 & \lambda v_2 & \lambda v_3 \\ \mu w_1 & \mu w_2 & \mu w_3 \end{bmatrix}\right)$$

$$= \langle \lambda\mu\left(v_2 w_3 - v_3 w_2\right), \lambda\mu\left(v_3 w_1 - v_1 w_3\right), \lambda\mu\left(v_1 w_2 - v_2 w_1\right)\rangle$$

$$= \lambda\mu \langle v_2 w_3 - v_3 w_2, v_3 w_1 - v_1 w_3, v_1 w_2 - v_2 w_1 \rangle$$

$$= \lambda\mu \det\left(\begin{bmatrix} \mathbf{i} & \mathbf{j} & \mathbf{k} \\ v_1 & v_2 & v_3 \\ w_1 & w_2 & w_3 \end{bmatrix}\right)$$

$$= (\lambda\mu)\,\mathbf{v} \times \mathbf{w}.$$

59. Assume that $\mathbf{v} \times \mathbf{w} = \mathbf{v} + \mathbf{w}$. If $\mathbf{w} = \mathbf{0}$, then $\mathbf{v} = \mathbf{0}$ also. Assume that $\mathbf{w} \neq \mathbf{0}$. Then $\mathbf{v} \neq \mathbf{0}$ also and it must be the case that $\mathbf{v} \times \mathbf{w} = \mathbf{0}$ as well. Otherwise, $\mathbf{v} \times \mathbf{w}$ would be a nonzero vector that is perpendicular to itself. Consequently, $\mathbf{v} = -\mathbf{w}$.

61. The magnitude of the torque is $\|\overrightarrow{PQ} \times \mathbf{F}\| = \|\overrightarrow{PQ}\|\,\|\mathbf{F}\|\sin(90°) = \frac{8}{12} \cdot 60 = 40$ foot-pounds.

63. Let T be a point on the line that is different from S. The triangle $\triangle RST$ is a right triangle having the line seqment from R to T as its hypotenuse. Therefore, $|\overline{RT}| > |\overline{RS}|$, and $d(R, \ell) = \|\overrightarrow{RS}\|$.

The area of the triangle $\triangle PQR$ is $\frac{1}{2}\|\overrightarrow{PQ} \times \overrightarrow{PR}\|$. This area is also given by the formula $\frac{1}{2} \cdot \text{base} \cdot \text{height}$: $\frac{1}{2}\|\overrightarrow{PQ}\|\,d(R, \ell)$. Set these two area formulas equal to one another and solve for $d(R, \ell)$ to obtain

$$d(R, \ell) = \frac{\|\overrightarrow{PQ} \times \overrightarrow{PR}\|}{\|\overrightarrow{PQ}\|}.$$

65. Observe that $\mathbf{u}\cdot(\mathbf{v}\times(s\mathbf{u}+t\mathbf{v})) = s\mathbf{u}\cdot(\mathbf{v}\times\mathbf{u}) + t\mathbf{u}\cdot(\mathbf{v}\times\mathbf{v}) = 0$. Therefore, by Theorem 7, $\mathbf{u}$, $\mathbf{v}$, and $s\mathbf{u}+t\mathbf{v}$ are co-planar.

67. This can be handled by carefully expanding the left side of the identity to obtain the sum of 12 terms; 6 of them will be "+" terms and 6 will be "−" terms. A similar expansion of the right side:

$$(v_1p_1 + v_2p_2 + v_3p_3)(w_1q_1 + w_2q_2 + w_2q_3)$$
$$- (v_1q_1 + v_2q_2 + v_3q_3)(w_1p_1 + w_2p_2 + w_2p_3),$$

will produce 9 "+" terms and 9 "−" terms. However, 3 of the "+" terms cancel 3 of the "−" terms to leave the same 12 terms that are on the left side.

Calculator/Computer Exercises

69. The equation simplifies to

$$7x + 22y + 9z = 78 .$$

The solution set appears to be a plane. It contains the point P_0 and the vector $\mathbf{v}\times\mathbf{w} = \langle -7, -22, -9 \rangle$ is perpendicular to the plane. The picture displays the plane and the vector $-\frac{1}{10}\mathbf{v}\times\mathbf{w}$ with its tail at the point P_0.

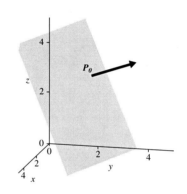

71. The graph of f is shown on the right. f appears to attain a minimum value near $t = 1$. Using *Maple*'s *Minimize* procedure we find that the minimum value is -7.161655 attained at $t = 0.770073$.

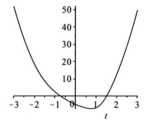

9.5   Lines and Planes in Space

Problems for Practice

1. All normals to the plane are parallel to the vector $\mathbf{v} = \langle 1, -3, 4 \rangle$. They are of the form $\lambda\mathbf{v}$ for some real number $\lambda \neq 0$.

When the coordinates of P, Q, and R are substituted into the left side of the equation, its value is 2.

$$\mathbf{n} = \det\left(\begin{bmatrix} \mathbf{i} & \mathbf{j} & \mathbf{k} \\ -1 & 1 & 1 \\ 3 & 1 & 0 \end{bmatrix}\right) = \langle -1, 3, -4 \rangle\,;\ \mathbf{n} = -\mathbf{v}\,.$$

3. All normals to the plane are parallel to the vector $\mathbf{v} = \langle 3, 0, -4 \rangle$. They are of the form $\lambda\,\mathbf{v}$ for some real number $\lambda \neq 0$.

When the coordinates of P, Q, and R are substituted into the left side of the equation, its value is 1.

$$\mathbf{n} = \det\left(\begin{bmatrix} \mathbf{i} & \mathbf{j} & \mathbf{k} \\ 4 & 1 & 3 \\ -8 & 2 & -6 \end{bmatrix}\right) = \langle -12, 0, 16 \rangle\,;\ \mathbf{n} = -4\mathbf{v}\,.$$

5. The vectors $\mathbf{v} = \langle 2, 2, 1 \rangle$ and $\mathbf{w} = \langle 1, -1, -3 \rangle$ are parallel to the plane. (These are the coefficients of s and t respectively.) Consequently, the vector

$$\mathbf{v} \times \mathbf{w} = \det\left(\begin{bmatrix} \mathbf{i} & \mathbf{j} & \mathbf{k} \\ 2 & 2 & 1 \\ 1 & -1 & -3 \end{bmatrix}\right) = \langle -5, 7, -4 \rangle$$

is normal to the plane, and all other normals are scalar multiples of this vector.

7. The vectors $\mathbf{v} = \langle 2, 0, 1 \rangle$ and $\mathbf{w} = \langle 0, -1, -2 \rangle$ are parallel to the plane. (These are the coefficients of s and t respectively.) Consequently, the vector

$$\mathbf{v} \times \mathbf{w} = \det\left(\begin{bmatrix} \mathbf{i} & \mathbf{j} & \mathbf{k} \\ 2 & 0 & 1 \\ 0 & -1 & -2 \end{bmatrix}\right) = \langle 1, 4, -2 \rangle$$

is normal to the plane, and all other normals are scalar multiples of this vector.

9. The equation has the form $-3x + 7y + 9z = D$. Since $P = (1, 4, 6)$ lies on the plane, $D = -3\cdot 1 + 7\cdot 4 + 9\cdot 6 = 79$. The equation is $-3x + 7y + 9z = 79$.

11. The equation has the form $2x + 2y + 5z = D$. Since $P = (-5, 9, 2)$ lies on the plane, $D = 2\cdot(-5) + 2\cdot 9 + 5\cdot 2 = 18$. The equation is $2x + 2y + 5z = 18$.

13. Name the points P, Q, and R, respectively. Then the vectors $\mathbf{v} = \overrightarrow{PQ} = \langle 3, -10, 2 \rangle$ and $\mathbf{w} = \overrightarrow{PR} = \langle 4, 0, 3 \rangle$ lie in the plane. Therefore, the vector

$$\mathbf{v} \times \mathbf{w} = \det\left(\begin{bmatrix} \mathbf{i} & \mathbf{j} & \mathbf{k} \\ 3 & -10 & 2 \\ 4 & 0 & 3 \end{bmatrix}\right) = \langle -30, -1, 40 \rangle$$

is normal to the plane and its equation is of the form $-30x - y + 40z = D$. Since $P = (0, 1, 3)$ lies on the plane, $D = (\mathbf{v} \times \mathbf{w}) \cdot \overrightarrow{OP} = 119$. The equation is $-30x - y + 40z = 119$.

15. Name the points P, Q, and R, respectively. Then the vectors $\mathbf{v} = \overrightarrow{PQ} = \langle -1, -3, -12 \rangle$ and $\mathbf{w} = \overrightarrow{PR} = \langle -7, 7, -4 \rangle$ lie in the plane. Therefore, the vector

$$\mathbf{v} \times \mathbf{w} = \det\left(\begin{bmatrix} \mathbf{i} & \mathbf{j} & \mathbf{k} \\ -1 & -3 & -12 \\ -7 & 7 & -4 \end{bmatrix} \right) = \langle 96, 80, -28 \rangle$$

is normal to the plane and so is $\langle 24, 20, -7 \rangle$ so its equation is of the form $24x + 20y - 7z = D$. Since $P = (2, -6, 8)$ lies on the plane, $D = 24 \cdot 2 - 20 \cdot 6 - 7 \cdot 8 = -128$. The equation is $24x + 20y - 7z = -128$.

17. The components of the vector $\mathbf{m}$ are the coefficients in the parametrization:
$$x = 2 - 3t, \ y = 1 + t, \ z = 9 + 7t.$$

19. The components of the vector $\mathbf{m}$ are the coefficients in the parametrization:
$$x = 2t, \ y = 1 - t, \ z = t.$$

21. The line is in the direction of the vector $\mathbf{m} = \langle -5, 8, 2 \rangle$. The components of this vector appear in the denominators of fractions in the symmetric equations:
$$\frac{x - 2}{-5} = \frac{y - 1}{8} = \frac{z - 5}{2}.$$

23. The line is in the direction of the vector $\mathbf{m} = \langle 1, 1, 9 \rangle$. The components of this vector appear in the denominators of fractions in the symmetric equations:
$$\frac{x - 0}{1} = \frac{y - 1}{1} = \frac{z - 0}{9}.$$

This simplifies to
$$x = y - 1 = \frac{z}{9}.$$

25. The components of the vector $\mathbf{m} = \langle 1, -3, 7 \rangle$ are the coefficients in the parametrization:
$$x = 7 + t, \ y = -4 - 3t, \ z = 6 + 7t.$$

27. The components of the vector $\mathbf{m} = \langle 1, -1, 1/3 \rangle$ are the coefficients in the parametrization:
$$x = -1 + t, \ y = -1 - t, \ z = -8 + \frac{1}{3}t.$$

29. The line is in the direction of the vector $\mathbf{m} = \langle 1, 1, -1 \rangle$. The components of this vector appear in the denominators of the fractions in the symmetric equations:
$$\frac{x-1}{1} = \frac{y-0}{1} = \frac{z-1}{-1}.$$
This simplifies to
$$x - 1 = y = 1 - z.$$

31. The line is in the direction of the vector $\mathbf{m} = \langle 2, -1, -1 \rangle$. The components of this vector appear in the denominators of the fractions in the symmetric equations:
$$\frac{x-1}{2} = \frac{y-2}{-1} = \frac{z-1}{-1}.$$
This simplifies to
$$\frac{x-1}{2} = 2 - y = 1 - z.$$

33. We can begin by setting $z = t$ and writing the two equations in the following form.
$$\begin{array}{ccccccc} x & - & 3y & = & 2 & - & 5t \\ 2x & + & 6y & = & 4 & - & 3t \end{array}$$

Multiply the first equation by 2 and add it to the second to solve for x: $4x = 8 - 13t$. Thus $x = 2 - 13t/4$. To solve for y multiply the first equation by 2 and subtract it from the second to obtain $12y = 7t$ so $y = 7t/12$. This yields the parametrization

$$x = 2 - \frac{13}{4}t$$
$$y = \frac{7}{12}t$$
$$z = t.$$

The formulas can be simplified by replacing t with $12t$ to obtain the parametrization $x = 2 - 39t$, $y = 7t$, $z = 12t$.

35. Let $z = t$ and write the two equations in the following form.
$$\begin{array}{ccccccc} 4x & + & 6y & = & 1 & + & t \\ 4x & & & = & & - & t \end{array}$$

Solve the second equation for x: $x = -t/4$, and subtract the second equation from the first to isolate y: $6y = 1 + 2t$ so $y = 1/6 + t/3$. The formulas can be simplified by replacing t with $12t$ to obtain the parametrization $x = -3t$, $y = 1/6 + 4t$, $z = 12t$.

37. Let $z = t$ and write the two equations in the following form.

$$\begin{aligned} x &- y &= 2 &- t \\ x &+ y &= 6 &- 3t \end{aligned}$$

Add the equations: $2x = 8 - 4t$, and $x = 4 - 2t$. Subtract the first equation from the second: $2y = 4 - 2t$, and $y = 2 - t$. The symmetric equations are $(x-4)/2 = y - 2 = -z$.

39. Let $z = t$ and write the two equations in the following form.

$$\begin{aligned} 2x &+ y &= 2 &+ t \\ 4x & &= 3 &+ t \end{aligned}$$

According to the second equation, $x = 3/4 + t/4$. Subtract the second equation from twice the first equation to see that $2y = 1 + t$, and $y = 1/2 + t/2$. Replace t with $4t$ to obtain the parametric equations $x = 3/4 + t$, $y = 1/2 + 2t$, $z = 4t$. This parametrization yields the symmetric equations

$$x - \frac{3}{4} = \frac{y - 1/2}{2} = \frac{z}{4}.$$

Add $3/4$ to each term to obtain the symmetric equations in the text.

41. Substitute values for $x, y,$ and z as determined by the parametric equations into the equation for the plane to obtain $-21t - 21 = 0$. The line meets the plane when $t = -1$. Now substitute $t = -1$ into the parametric equations to obtain the point where the line intersects the plane: $(4, -2, -2)$.

43. Substitute values for $x, y,$ and z as determined by the parametric equations into the equation for the plane to obtain $t + 6 = 4$. The line meets the plane when $t = -2$. Now substitute $t = -2$ into the parametric equations to obtain the point where the line intersects the plane: $(2, 2, 7)$.

45. Substitute $x = z/2$ and $y = z/2 - 1$ into the equation for the plane to obtain $4z + 3 = 19$. Therefore, the line intersects the plane when $z = 4$. Substitute $z = 4$ into the symmetric equations for the line to obtain $x = 2$ and $y = 1$. The line intersects the plane at the point $(2, 1, 4)$.

47. Substitute $y = 4 - x$ into the first of the symmetric equations to obtain $(x + 1)/2 = 5 - x$. This shows that $x = 3$. Consequently, $y = 1$ (using the first equation above) and $z = 0$ (using the second symmetric equation). The point of intersection is $(3, 1, 0)$.

49. The vector $\mathbf{v} = \langle 4, 0, -3 \rangle$ and the vector $\mathbf{w} = \langle 2, 1, 2 \rangle$ are normals to the planes so the angle θ between them satisfies the equation

$$\cos(\theta) = \frac{\mathbf{v} \cdot \mathbf{w}}{\|\mathbf{v}\| \, \|\mathbf{w}\|} = \frac{2}{15}.$$

51. The vector $\mathbf{v} = \langle -5, -3, -4 \rangle$ and the vector $\mathbf{w} = \langle 1, 1, 1 \rangle$ are normals to the planes so the angle θ between them satisfies the equation

$$\cos(\theta) = \frac{\mathbf{v} \cdot \mathbf{w}}{\|\mathbf{v}\| \, \|\mathbf{w}\|} = \frac{-12}{\sqrt{50}\sqrt{3}} = -\frac{2\sqrt{6}}{5}.$$

53. Let $Q = (1, 0, 1)$. The point $P = (3, 2, 1)$ is on the plane and the vector $\mathbf{v} = \langle 2, 1, 1 \rangle$ is normal to the plane. The distance from Q to the plane equals the length of the projection of the vector $\overrightarrow{PQ} = \langle -2, -2, 0 \rangle$ in the direction of $\mathbf{v}$:

$$\text{distance} = \left\| \frac{\overrightarrow{PQ} \cdot \mathbf{v}}{\mathbf{v} \cdot \mathbf{v}} \mathbf{v} \right\| = \frac{6}{6}\sqrt{6} = \sqrt{6}.$$

55. Let $Q = (1, 1, 1)$. The point $P = (1, 1, 10)$ is on the plane and the vector $\mathbf{v} = \langle 2, 2, 1 \rangle$ is normal to the plane. The distance from Q to the plane equals the length of the projection of the vector $\overrightarrow{PQ} = \langle 0, 0, 9 \rangle$ in the direction of $\mathbf{v}$:

$$\text{distance} = \left\| \frac{\overrightarrow{PQ} \cdot \mathbf{v}}{\mathbf{v} \cdot \mathbf{v}} \mathbf{v} \right\| = \frac{9}{9}\sqrt{9} = 3.$$

57. Since x must be -7, the value of the parameter must be $t = -2$. For this t value, the point on the line has y-coordinate 6 and z-coordinate 2. The point $(-7, 6, 2)$ is on the line.

59. Since x must be 1, the value of the parameter must be $t = 1$. However, for this t value, the point on the line has y-coordinate 5 and z-coordinate -2. The point $(1, 5, 1)$ is not on the line.

61. The lines intersect provided the following system of equations has a solution.

$$
\begin{aligned}
3t - 5 &= s + 1 \\
-4t - 5 &= s - 6 \\
t + 1 &= s + 5
\end{aligned}
$$

Subtract the second equation from the first to obtain $7t = 7$ so $t = 1$. The first two equations will then be satisfied if $s = -3$. Since this will also satisfy the third equation, these two lines intersect. They do so at the point $(-2, -9, 2)$.

63. The lines intersect provided the following system of equations has a solution.

$$
\begin{aligned}
t - 1 &= 3s + 7 \\
-2t + 14 &= -s + 4 \\
2t + 3 &= -2s + 11
\end{aligned}
$$

Add the second equation to the third to obtain $17 = -3s + 15$ so $s = -2/3$. Substitute this value into the first equation to see that $t = 6$. Since these values of s and t do not satisfy the second equation (or the third equation), the two lines do not intersect.

65. Use the distance formula in Theorem 7. The line is parallel to the vector $\mathbf{m} = \langle 1, 1, 1 \rangle$. The point $Q = (1, 2, 3)$ is on the line and $P = (0, 0, 0)$.

Therefore, $\mathbf{m} \times \overrightarrow{PQ} = \det \left(\begin{bmatrix} \mathbf{i} & \mathbf{j} & \mathbf{k} \\ 1 & 1 & 1 \\ 1 & 2 & 3 \end{bmatrix} \right) = \langle 1, -2, 1 \rangle$, and

$$\mathbf{n} = \mathbf{m} \times (\mathbf{m} \times \overrightarrow{PQ}) = \det \left(\begin{bmatrix} \mathbf{i} & \mathbf{j} & \mathbf{k} \\ 1 & 1 & 1 \\ 1 & -2 & 1 \end{bmatrix} \right) = \langle 3, 0, -3 \rangle,$$

so the distance is $\|\mathbf{P_n}(\overrightarrow{PQ})\| = \frac{|\overrightarrow{PQ} \cdot \mathbf{n}|}{\|n\|} = \frac{6}{3\sqrt{2}} = \sqrt{2}$.

67. Use Theorem 7. The line is parallel to the vector $\langle 1, -1, 1/2 \rangle$ so let $\mathbf{m} = \langle 2, -2, 1 \rangle$. The point $Q = (0, 1, 0)$ is on the line and $P = (1, 0, -1)$.

Therefore, $\mathbf{m} \times \overrightarrow{PQ} = \det \left(\begin{bmatrix} \mathbf{i} & \mathbf{j} & \mathbf{k} \\ 2 & -2 & 1 \\ -1 & 1 & 1 \end{bmatrix} \right) = \langle -3, -3, 0 \rangle$, and

$$\mathbf{n} = \mathbf{m} \times (\mathbf{m} \times \overrightarrow{PQ}) = \det \left(\begin{bmatrix} \mathbf{i} & \mathbf{j} & \mathbf{k} \\ 2 & -2 & 1 \\ -3 & -3 & 0 \end{bmatrix} \right) = \langle 3, -3, -12 \rangle,$$

so the distance is $\|\mathbf{P_n}(\overrightarrow{PQ})\| = \frac{|\overrightarrow{PQ} \cdot \mathbf{n}|}{\|n\|} = \frac{18}{9\sqrt{2}} = \sqrt{2}$.

69. $\mathbf{v} \times \mathbf{w} = \det \left(\begin{bmatrix} \mathbf{i} & \mathbf{j} & \mathbf{k} \\ 2 & 0 & 1 \\ 3 & 1 & 1 \end{bmatrix} \right) = \langle -1, 1, 2 \rangle$;

$\mathbf{u} \times (\mathbf{v} \times \mathbf{w}) = \det \left(\begin{bmatrix} \mathbf{i} & \mathbf{j} & \mathbf{k} \\ 1 & -2 & 4 \\ -1 & 1 & 2 \end{bmatrix} \right) = \langle -8, -6, -1 \rangle$;

$(\mathbf{u} \cdot \mathbf{w}) \mathbf{v} - (\mathbf{u} \cdot \mathbf{v}) \mathbf{w} = 5 \langle 2, 0, 1 \rangle - 6 \langle 3, 1, 1 \rangle = \langle -8, -6, -1 \rangle$.

71. $\mathbf{v} \times \mathbf{w} = \det \left(\begin{bmatrix} \mathbf{i} & \mathbf{j} & \mathbf{k} \\ 3 & 0 & 2 \\ -2 & -3 & 3 \end{bmatrix} \right) = \langle 6, -13, -9 \rangle$;

$\mathbf{u} \times (\mathbf{v} \times \mathbf{w}) = \det \left(\begin{bmatrix} \mathbf{i} & \mathbf{j} & \mathbf{k} \\ 1 & 1 & 1 \\ 6 & -13 & -9 \end{bmatrix} \right) = \langle 4, 15, -19 \rangle$;

$(\mathbf{u} \cdot \mathbf{w}) \mathbf{v} - (\mathbf{u} \cdot \mathbf{v}) \mathbf{w} = -2 \langle 3, 0, 2 \rangle - 5 \langle -2, -3, 3 \rangle = \langle 4, 15, -19 \rangle$.

Further Theory and Practice

73. The plane has the equation $3x+4y+z = D$ where $D = 3 \cdot 1 + 4 \cdot 2 + 1 \cdot 3 = 14$.

75. Since a normal is $\langle 1, 2, 3 \rangle$, the plane has the equation $x + 2y + 3z = D$ where $D = 1 \cdot 1 + 2 \cdot 2 + 3 \cdot 3 = 14$. The equation for the plane can also be obtained by using the left hand side of the given equation and replacing the 1 on the right hand side with 0.

77. The intercepts are $x = 3$, $y = 2$, and $z = 6$.
See the picture on the right.

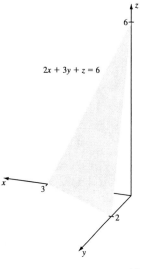

79. The intercepts are $x = 5$, $y = 3$, and $z = 1$.
See the picture on the right.

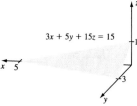

81. (a) The second normal is twice the first. The planes are parallel.

    (b) The second normal is thrice the first. The planes are parallel.

    (c) The second normal is not 4 times the first. The planes are not parallel.

    (d) The second normal is -2 times the first. The planes are parallel.

    (e) The second normal is twice the first. The planes are parallel.

83. The line ℓ lies in the plane $\mathcal{V}$ if it intersects the plane and its direction vector $\mathbf{m}$ is perpendicular to a normal for the plane.

    If ℓ does not lie in $\mathcal{V}$, then it intersects $\mathcal{V}$ when its direction vector $\mathbf{m}$ is not perpendicular to a normal for the plane.

    (a) $\mathbf{m} = \langle -3, -1, -1 \rangle$ and $\mathbf{n} = \langle 1, -5, 2 \rangle$ is a normal to the plane. These vectors are perpendicular. Since the point $Q = (4, -6, 2)$ is on the line but not on the plane, the line does not intersect the plane.

(b) $\mathbf{m} = \langle 3, 1, -1 \rangle$ and $\mathbf{n} = \langle 3, -7, 2 \rangle$ is a normal. These vectors are perpendicular so the line is parallel to the plane. Since the point $Q = (-6, 2, 5)$ is on the line and also on the plane, the line lies in the plane.

(c) $\mathbf{m} = \langle 4, 1, -3 \rangle$ and $\mathbf{n} = \langle 1, 1, -2 \rangle$ is a normal. These vectors are not perpendicular so the line intersects the plane in one point. Substitute the parametric equations into the equation of the plane to obtain the equation $11t - 12 = 10$, so $t = 2$. When t is assigned this value the parametrization of the line evaluates to the point $(8, 8, 3)$.

(d) $\mathbf{m} = \langle 2, 3, 2 \rangle$ and $\mathbf{n} = \langle -3, -2, 6 \rangle$ is a normal. These vectors are perpendicular so the line is parallel to the plane. Since the point $Q = (1, -4, 8)$ is on the line but not on the plane, the line does not intersect the plane.

85. Use Theorem 8. The first line passes through $P_1 = (1, 0, 0)$ and is parallel to the vector $\mathbf{m}_1 = \langle 1, 2, 1 \rangle$. The second line passes through $P_2 = (0, 0, 0)$ and is parallel to the vector $\mathbf{m}_2 = \langle 1, 1, 1 \rangle$. Since

$$\mathbf{n} = \mathbf{m}_1 \times \mathbf{m}_2 = \det \left(\begin{bmatrix} \mathbf{i} & \mathbf{j} & \mathbf{k} \\ 1 & 2 & 1 \\ 1 & 1 & 1 \end{bmatrix} \right) = \langle 1, 0, -1 \rangle,$$

the distance between the lines is $\| \mathbf{P_n}(\overrightarrow{P_1 P_2}) \| = \frac{|\overrightarrow{P_1 P_2} \cdot \mathbf{n}|}{\|\mathbf{n}\|} = \frac{1}{\sqrt{2}} = \frac{1}{2}\sqrt{2}$.

87. Use Theorem 8. The first line passes through $P_1 = (1, -1, 0)$ and is parallel to the vector $\mathbf{m}_1 = \langle 1, 1, -1 \rangle$. The second line passes through $P_2 = (0, 0, 0)$ and is parallel to the vector $\langle 1, 1, 1/2 \rangle$ and also to $\mathbf{m}_2 = \langle 2, 2, 1 \rangle$. Since

$$\mathbf{n} = \mathbf{m}_1 \times \mathbf{m}_2 = \det \left(\begin{bmatrix} \mathbf{i} & \mathbf{j} & \mathbf{k} \\ 1 & 1 & -1 \\ 2 & 2 & 1 \end{bmatrix} \right) = \langle 3, -3, 0 \rangle,$$

the distance between the lines is $\| \mathbf{P_n}(\overrightarrow{P_1 P_2}) \| = \frac{|\overrightarrow{P_1 P_2} \cdot \mathbf{n}|}{\|\mathbf{n}\|} = \frac{6}{3\sqrt{2}} = \sqrt{2}$.

89. The vector $\langle 1, -1, 1 \rangle$ is normal to the plane so the line ℓ, parametrized with $x = 8 + t$, $y = -2 - t$, $z = 4 + t$, passes through the point Q and is perpendicular to the plane. Substitute for x, y, z in the equation for the plane to obtain $3t + 14 = 4$ and the line intersects the plane when $t = -10/3$. This is the point $P = (14/3, 4/3, 2/3)$. The distance from P to Q is $\sqrt{\left(8 - \frac{14}{3}\right)^2 + \left(-2 - \frac{4}{3}\right)^2 + \left(4 - \frac{2}{3}\right)^2} = \frac{10\sqrt{3}}{3}$.

91. The vector $\langle 2, -5, 7 \rangle$ is normal to the plane so the line ℓ, parametrized with $x = 2 + 2t$, $y = -3 - 5t$, $z = 6 + 7t$, passes through the point Q and is perpendicular to the plane. Substitute for x, y, z in the equation

for the plane to obtain $78t + 61 = 4$ and the line intersects the plane when $t = -19/26$. This is the point $P = (7/13, 17/26, 23/26)$. The distance from P to Q is

$$\sqrt{\left(2 - \frac{7}{13}\right)^2 + \left(-3 - \frac{17}{26}\right)^2 + \left(6 - \frac{23}{26}\right)^2} = \frac{19\sqrt{78}}{26}.$$

93. The line can be parametrized as $x = 1+t$, $y = -8-4t$, and $z = 7+3t$. The square of the distance from $P = (5, 3, 3)$ to the point on the line defined as $Q_t = (t+1, -8-4t, 7+3t)$ is $f(t) = (t+1-5)^2 + (-8-4t-3)^2 + (7+3t-3)^2$ which simplifies to $f(t) = 26t^2 + 104t + 153$. This is a parabola opening upward with a minimum at t satisfying $f'(t) = 52t + 104 = 0$. That is, $t = -2$. The minimum distance is $\sqrt{f(-2)} = \sqrt{49} = 7$.

95. Observe that

$$\begin{aligned}
(\mathbf{u} \times \mathbf{v}) \times \mathbf{w} &= -\mathbf{w} \times (\mathbf{u} \times \mathbf{v}) \\
&= \mathbf{w} \times (\mathbf{v} \times \mathbf{u}) \\
&\overset{(9.5.10)}{=} (\mathbf{w} \cdot \mathbf{u})\mathbf{v} - (\mathbf{w} \cdot \mathbf{v})\mathbf{u} \\
&= -(\mathbf{v} \cdot \mathbf{w})\mathbf{u} + (\mathbf{u} \cdot \mathbf{w})\mathbf{v}.
\end{aligned}$$

Calculator/Computer Exercises

97. Using *Maple*, the first entry below loads the plots and VectorCalculus packages, defines the center C and the point P as vectors, computes the vector $\overrightarrow{CP}$, and verifies that the distance from P to the center of the sphere is 3.

```
> with(plots): with(VectorCalculus):
  C,P := <2,1,2>,<3,3,4>:
  P - C;
  Norm(P-C);
```

$$e_x + 2e_y + 2e_z$$

$$3$$

Now plot the sphere, and name the plot S for use later.

```
> implicitplot3d( Norm(<x,y,z>-C)=3, x=-2..6, y=-3..4, z=-2..6,
           scaling=constrained, style=patchcontour,
           axes=normal, orientation=[-60,70]);
  S := %:
```

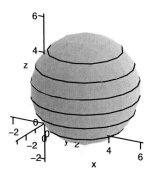

Finally, two perpendicular vectors in the tangent plane are defined to use in its parametrization. Then the sphere and the parametrized tangent plane at the point P are displayed. A slightly different viewing angle is used.

```
> u,v := <2,0,-1>,<0,1,-1>:  #Perpendicular in tangent plane.
  display( S, plot3d(P+s*u+t*v, s=-1..1, t=-1..1, grid=[3,3],
                   color=blue), orientation=[-20,65]);
```

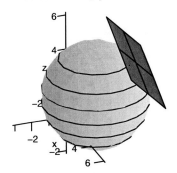

99. Using *Maple* define the set *Param* of parametric equations for the line ℓ, and let *eqn* be the equation of the surface. Then substitute the parametrization equations into the surface equation, and solve for t (approximate solutions).

```
> Param := {x = 1 - t, y = 1 - 2*t, z = 3*t}:
  eqn := z = x^4 + y^4:
  subs( Param, eqn);
  T := fsolve( %, t );
```

$$3t = (1 - t)^4 + (1 - 2t)^4$$

$$T := 0.1911729503, 1.186758051$$

Substitute the t values into the set of parametrization equations to get one point, then the other.

```
> OnePoint, subs(t=T[1],Param);
```

`TheOther, subs(t=T[2],Param);`

$OnePoint, \{x = 0.8088270497, y = 0.6176540994, z = 0.5735188509\}$

$TheOther, \{x = -.186758051, y = -1.373516102, z = 3.560274153\}$

Chapter 10

Vector-Valued Functions

10.1 Vector-Valued Functions—Limits, Derivatives, and Continuity

Problems for Practice

1. This is a curve in the xy plane. Observe that if we write $\mathbf{r}(t) = x\,\mathbf{i} + y\,\mathbf{j}$, then $x = t$ and $y = t^2$, so $y = x^2$.

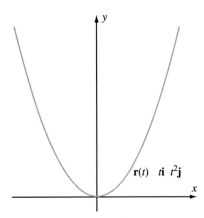

3. This is a curve in the plane $z = 2$. It is the parabola in Exercise 1 shifted up 2 units.

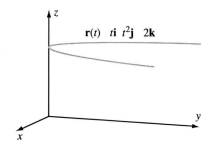

5. This is a curve in the plane $x = 1$. Since $y^2 + z^2 = 1$, the curve is a circle.

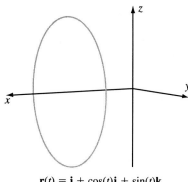

$$\mathbf{r}(t) = \mathbf{i} + \cos(t)\mathbf{j} + \sin(t)\mathbf{k}$$

7. The particle moves in straight line paths in the plane $z = 1$. The motion is directly above the line $y = x$ (in the xy plane). It starts at the point $(1,1,1)$, moves straight to $(0,0,1)$, then doubles back through $(1,1,1)$ to end up at $(4,4,1)$.

9. The trajectory describes an elongated sine wave in the plane $y = x$. It oscillates from $(0,0,0)$ up to $(\pi/2, \pi/2, 1)$, back down to $(\pi, \pi, 0)$, then down to $(3\pi/2, 3\pi/2, -1)$, and ends up at $(2\pi, 2\pi, 0)$.

11. Since $t \mapsto \cos(\pi t)$ and $t \mapsto 2^t$ are continuous for all t, so is the function $\mathbf{r}$. Therefore, $\lim_{t \to -1} \mathbf{r}(t) = \mathbf{r}(-1) = \cos(-\pi)\mathbf{i} + 2^{-1}\mathbf{j} = -\mathbf{i} + (1/2)\mathbf{j}$.

13. The three component functions are continuous at $c = 2$. Therefore, $\lim_{t \to 2} \mathbf{r}(t) = \mathbf{r}(2) = \langle 2^2, 2^2, e^{2\ln(1)} \rangle = \langle 4, 4, 1 \rangle$.

15. According to Theorem 5, $\mathbf{r}'(t) = \frac{1}{2}t^{-1/2}\mathbf{i} + 2t\,\mathbf{k}$.

17. According to Theorem 5, $\mathbf{r}'(t) = \sec^2(t)\,\mathbf{i} + \csc^2(t)\,\mathbf{j} + \sec(t)\tan(t)\,\mathbf{k}$.

19. According to Theorem 5, $\mathbf{r}'(t) = \mathbf{i} - (t+5)^{-2}\mathbf{j} - 2(t-7)^{-3}\,\mathbf{k}$.

21. According to Theorem 5,

$$\mathbf{r}'(t) = 2t\,\mathbf{i} - \frac{3}{2}t^{1/2}\mathbf{j} + \frac{5}{2}(t+1)^{3/2}\,\mathbf{k}.$$

Therefore,

$$\mathbf{r}''(t) = 2\,\mathbf{i} - \frac{3}{4}t^{-1/2}\,\mathbf{j} + \frac{15}{4}(t+1)^{1/2}\,\mathbf{k}\,.$$

23. According to Theorem 5,

$$\mathbf{r}'(t) = -e^{-t}\,\mathbf{i} - \frac{2t+4t^3}{t^2+t^4}\,\mathbf{j}$$
$$= -e^{-t}\,\mathbf{i} - \frac{2+4t^2}{t+t^3}\,\mathbf{j}\,.$$

Therefore,

$$\mathbf{r}''(t) = e^{-t}\,\mathbf{i} - \frac{4t^4+2t^2+2}{(t+t^3)^2}\,\mathbf{j}\,.$$

25. According to Theorem 5,

$$\mathbf{r}'(t) = -\frac{1}{3}t^{-4/3}\,\mathbf{i} + t(1+t^2)^{-3/2}\,\mathbf{j} + \frac{1}{4}t^{-3/4}\,\mathbf{k}\,.$$

Therefore,

$$\mathbf{r}''(t) = \frac{4}{9}t^{-7/3}\,\mathbf{i} - (2t^2-1)(1+t^2)^{-5/2}\,\mathbf{j} - \frac{3}{16}t^{-7/4}\,\mathbf{k}\,.$$

27. Integrating term-by-term, we have

$$\int (t^2\,\mathbf{i} - t^{-1/2}\,\mathbf{j} + \cos(2t)\,\mathbf{k})\,dt = \frac{1}{3}t^3\,\mathbf{i} - 2t^{1/2}\,\mathbf{j} + \frac{1}{2}\sin(2t)\,\mathbf{k} + \mathbf{C}\,.$$

29. Integrating term-by-term, using integration by parts on the log term ($u = \ln(t)$) and the half-angle identity on the cosine term, we have

$$\int (\ln(t)\,\mathbf{i} - \cos^2(t)\,\mathbf{j} + \tan(t)\,\mathbf{k})\,dt = (t\ln(t) - t)\,\mathbf{i} - \frac{1}{2}(t + \sin(t)\cos(t))\,\mathbf{j} - \ln(|\cos(t)|)\,\mathbf{k} + \mathbf{C}\,.$$

31. Integrate to see that the function $\mathbf{F}$ is of the form $\mathbf{F}(t) = t^2\,\mathbf{i} - t^3\,\mathbf{j} + t^4\,\mathbf{k} + \mathbf{C}$. Substitute $t = 2$ to obtain

$$5\mathbf{i} - 2\mathbf{j} + \mathbf{k} = 4\mathbf{i} - 8\mathbf{j} + 16\mathbf{k} + \mathbf{C}\,,$$

implying that

$$\mathbf{C} = \mathbf{i} + 6\mathbf{j} - 15\mathbf{k}\,.$$

Therefore,

$$\mathbf{F}(t) = (t^2 + 1)\,\mathbf{i} + (6 - t^3)\,\mathbf{j} + (t^4 - 15)\,\mathbf{k}\,.$$

33. Integrate to see that $\mathbf{F}(t) = \sin(t)\,\mathbf{i} + \frac{1}{2}\cos(2t)\,\mathbf{j} + \sec(t)\,\mathbf{k} + \mathbf{C}$. Substitute $t = 0$ to obtain

$$3\mathbf{i} + \frac{3}{2}\mathbf{j} - 2\mathbf{k} = \frac{1}{2}\mathbf{j} + \mathbf{k} + \mathbf{C},$$

implying that

$$\mathbf{C} = 3\mathbf{i} + \mathbf{j} - 3\mathbf{k}.$$

Therefore,

$$\mathbf{F}(t) = (\sin(t) + 3)\,\mathbf{i} + \left(\frac{1}{2}\cos(2t) + 1\right)\mathbf{j} + (\sec(t) - 3)\,\mathbf{k}.$$

35–51. $\mathbf{f}(t) = e^t\,\mathbf{i} - \cos(t)\,\mathbf{j} + \ln(1 + t^2)\,\mathbf{k}$, $\mathbf{g}(t) = t^3\,\mathbf{i} - t\,\mathbf{k}$, $\phi(t) = 1 + 5t$, $\lambda = 3$

$\mathbf{f}'(t) = e^t\,\mathbf{i} + \sin(t)\,\mathbf{j} + \frac{2t}{1+t^2}\,\mathbf{k}$, $\qquad \mathbf{g}'(t) = 3t^2\,\mathbf{i} - \mathbf{k}$, $\phi'(t) = 5$

35. $\mathbf{r}'(t) = \lambda\,\mathbf{f}'(\lambda t) = 3e^{3t}\,\mathbf{i} + 3\sin(3t)\,\mathbf{j} + \frac{18t}{1+9t^2}\,\mathbf{k}$

37. $\mathbf{r}'(t) = \phi'(t)\mathbf{g}'(\phi(t)) = 5(3(1 + 5t)^2\,\mathbf{i} - \mathbf{k}) = 15(1 + 5t)^2\,\mathbf{i} - 5\mathbf{k}$

39. $\mathbf{r}'(t) = \big((\phi'(t)\mathbf{k}) \times \mathbf{f}(t) + (\phi(t)\mathbf{k}) \times \mathbf{f}'(t)\big) \cdot (\lambda\mathbf{j})$

$= \left((5\mathbf{k}) \times (e^t\mathbf{i} - \cos(t)\mathbf{j} + \ln(1 + t^2)\mathbf{k}) + ((1 + 5t)\mathbf{k}) \times (e^t\mathbf{i} + \sin(t)\mathbf{j} + \frac{2t}{1+t^2}\mathbf{k})\right) \cdot (3\mathbf{j})$

$= \big(5e^t\mathbf{j} + 5\cos(t)\mathbf{i} + (1 + 5t)e^t\mathbf{j} - (1 + 5t)\sin(t)\mathbf{i}\big) \cdot (3\mathbf{j})$

$= 5e^t \cdot 3 + (1 + 5t)e^t \cdot 3 = 3e^t(6 + 5t)$

41. $\mathbf{r}'(t) = -3t^{-4}\mathbf{g}'(t^{-3}) = -3t^{-4}(3t^{-6}\,\mathbf{i} - \mathbf{k}) = -9t^{-10}\,\mathbf{i} + 3t^{-4}\mathbf{k}$

43. Observe that $\psi(t) = \mathbf{g}(t) \cdot \mathbf{g}(t)$. Therefore, using the product rule,

$$\psi'(t) = \mathbf{g}(t) \cdot \mathbf{g}'(t) + \mathbf{g}'(t) \cdot \mathbf{g}(t) = 2\,\mathbf{g}(t) \cdot \mathbf{g}'(t) = 2(3t^5 + t).$$

45. Expand the dot product to obtain an explicit formula for the function ψ: $\psi(t) = 1 + 5t^3 e^t - 5t\ln(1 + t^2)$. Consequently,

$$\psi'(t) = 5t^3 e^t + 15t^2 e^t - 5\ln(1 + t^2) - \frac{10t^2}{1 + t^2}.$$

47. $\mathbf{r}'(t) = \phi'(t)\,\mathbf{i} \times \mathbf{g}(t) + (\phi(t)\,\mathbf{i} + \lambda\,\mathbf{k}) \times \mathbf{g}'(t)$

$= 5\,\mathbf{i} \times (t^3\,\mathbf{i} - t\,\mathbf{k}) + ((1 + 5t)\,\mathbf{i} + 3\,\mathbf{k}) \times (3t^2\,\mathbf{i} - \mathbf{k})$

$= 5t\,\mathbf{j} + (1 + 5t)\,\mathbf{j} + 9t^2\mathbf{j} = (9t^2 + 10t + 1)\,\mathbf{j}$

Further Theory and Practice

49. Since $\mathbf{i} \cdot \mathbf{f}(t) = e^t$ and $\mathbf{i} \times \mathbf{g}(t) = t\,\mathbf{j}$, $\mathbf{r}(t) = te^t\,\mathbf{j}$. Therefore,

$$\mathbf{r}'(t) = (t + 1)e^t\,\mathbf{j}.$$

51. $\psi(t) = (\mathbf{f}(t) \times \mathbf{f}(t)) \cdot \mathbf{g}(t) = 0$. Therefore, $\psi'(t) = 0$.

53. The limit is $\mathbf{i}+\mathbf{j}+\mathbf{k}$ because all three component functions approach 1 as $t \to 0$. This is clearly true for the $\mathbf{k}$ component, $(1-t)^t$. L'Hôpital's rule can be applied to the other two:

$$\lim_{t\to0} |t|^t = \lim_{t\to0} e^{t\ln(|t|)} = \exp\left(\lim_{t\to0} \frac{\ln(|t|)}{t^{-1}}\right) = \exp\left(\lim_{t\to0} \frac{t^{-1}}{-t^{-2}}\right) = 1;$$

$$\lim_{t\to0} \cos(t)^{1/t} = \lim_{t\to0} e^{\ln(\cos(t))/t} = \exp\left(\lim_{t\to0} \frac{\ln(\cos(t))}{t}\right)$$
$$= \exp\left(\lim_{t\to0} \frac{-\tan(t)}{1}\right) = 1.$$

55. Apply L'Hôpital's rule to the first component function:

$$\lim_{t\to\pi} \frac{\sin(t)}{t-\pi} = \lim_{t\to\pi} \frac{\cos(t)}{1} = -1.$$

The other two component functions are continuous at $t = \pi$. Therefore,

$$\lim_{t\to\pi} \left\langle \frac{\sin(t)}{t-\pi}, \frac{\ln(t^2)}{\cos(t)}, |\sec(t)| \right\rangle = \langle -1, -\ln(\pi^2), 1 \rangle.$$

57. The function ϕ is continuous at all points except $t = 5$ and the function $t \mapsto -7t^2$ is continuous everywhere. Consequently, $\mathbf{r}$ is continuous everywhere except $t = 5$.

59. The function $\mathbf{r}$ is continuous everywhere. Observe that its $\mathbf{j}$- and $\mathbf{k}$-component functions are clearly continuous everywhere. Its $\mathbf{i}$-component function ϕ is also continuous everywhere because

$$\lim_{t\to0^-} \frac{\sin(t)}{t} = 1 = \lim_{t\to0^+} \cos(t).$$

61. A polynomial has degree $n-1$ if, and only if, its nth derivative is the lowest order derivative that is identically 0. Therefore, if $N-1$ is the highest degree of the three polynomials, then the Nth derivative of $\mathbf{r}$ is identically 0. This is the least N that will suffice.

63. Observe that

$$\|\mathbf{r}(t)\|^2 = \cos^4(t) + \cos^2(t)\sin^2(t) + \sin^2(t)$$
$$= \cos^2(t)\left(\cos^2(t) + \sin^2(t)\right) + \sin^2(t)$$
$$= \cos^2(t) + \sin^2(t)$$
$$= 1.$$

65. $\mathbf{r}(t) = |t - c|\,\mathbf{i}$

67. If $\lim_{t \to c} \mathbf{r}(t) = \mathbf{L}$, then the nth component of $\mathbf{L}$ is uniquely determined as $L_n = \lim_{t \to c} r_n(t)$.

69. According to Theorem 4 b, the function $\phi(t) = \mathbf{r}(t) \cdot \mathbf{r}(t)$ is continuous at c. It follows that $t \mapsto \|\mathbf{r}(t)\|$ is continuous at c also, because it is the composition of ϕ and the square root function.

    The converse is false. For example, if $f(t) = \begin{cases} -1 & , & t < 0 \\ 1 & , & t \geq 0 \end{cases}$, and $\mathbf{r}(t) = f(t)\,\mathbf{i}$, then $\|\mathbf{r}(t)\| = 1$ for all t and the function $t \mapsto \|\mathbf{r}(t)\|$ is continuous everywhere. However, the function $\mathbf{r}$ is not continuous at $t = 0$.

71. Apply the formal ϵ-δ definition of limit. Given $\epsilon > 0$, let $\delta = \epsilon$. Then

$$0 < |s - t| < \delta \implies |s - t| < \epsilon \implies \|\mathbf{f}(s) - \mathbf{f}(t)\| < \epsilon.$$

Consequently, $\lim_{s \to t} \mathbf{f}(s) = \mathbf{f}(t)$, and $\mathbf{f}$ is continuous at t.

Calculator/Computer Exercises

73. After loading the VectorCalculus and plots packages, the trajectory is defined and plotted. We name the plot $Traj$ to display later along with the tangent line.

```
> with(VectorCalculus): with(plots):
  r := <t^2+t,t^2-t>:
  plot( [r[1],r[2],t=-2..2],color=black);
  Traj := %:
```

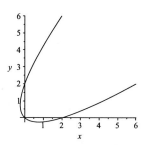

The first entry at the next input prompt defines the vector dr, containing the derivatives of the component functions in r. They are used to define the slope: m, and the tangent line at $t = -1$ (i.e. the point $(0, 2)$). Its plot is then displayed along with the trajectory created above.

```
> dr := diff(r,t):
  m := dr[2]/dr[1]: y := eval(m*(x - r[1]) + r[2],t=-1);
  display( Traj, plot( y, x=-0.5..0.5, color=black);
```

$$y := 3x + 2$$

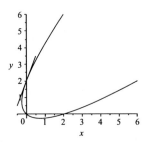

The trajectory and tangent line at $t = -1$.

The slope m is correctly defined because, according to the chain rule, if the trajectory defines y as a function of x, then

$$\frac{dy}{dx} = \frac{dy/dt}{dx/dt} = \frac{r_2'(t)}{r_1'(t)}.$$

75. The trajectory is plotted for $-30 < t < -2$ and $-\frac{1}{2} < t < 30$.

```
> r := <3*t/(1+t^3),3*t^2/(1+t^3)>:
  display(plot( [r[1],r[2],t=-30..-2]),
          plot( [r[1],r[2],t=-0.5..30]),color=black);
  Traj := %:
```

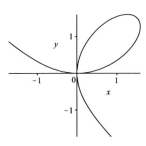

The tangent line is plotted at $t = 2$ (i.e. the point $(2/3, 4/3)$). Its plot is displayed along with the trajectory created above.

```
> dr := diff(r,t):
  m := dr[2]/dr[1]: y := eval(m*(x - r[1]) + r[2],t=2);
  display( Traj, plot( y, x=0..1.3, color=black));
```

$$y := \frac{4}{5}x + \frac{4}{5}$$

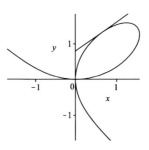

The trajectory and tangent line at $t = 2$.

77. The trajectory is plotted for $-\frac{5}{4} < t < \frac{5}{4}$.

```
> r := <t^3-t,t^2-1>:
  plot( [r[1],r[2],t=-5/4..5/4],color=black);
  Traj := %:
```

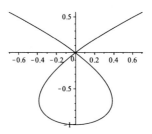

The tangent line is plotted at $t = -\frac{5}{6}$ (i.e. the approximate point $(0.25, -0.31)$).
Its plot is displayed along with the trajectory created above.

```
> dr := diff(r,t):
  m := dr[2]/dr[1]: y := eval(m*(x - r[1]) + r[2],t=-5/6);
  display( Traj, plot( y, x=0..0.5, color=black));
```

$$y := -\frac{20}{13}x + \frac{121}{1404}$$

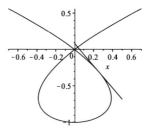

The trajectory and tangent line at $t = -5/6$.

79. The trajectory is plotted for $0 < t < 4\pi$.

```
> r := <t-sin(t),1-cos(t)>:
  plot( [r[1],r[2],t=0..4*Pi],color=black);
```

10.2 Velocity and Acceleration

Problems for Practice

1. $\mathbf{r}(t) = t\,\mathbf{i} + t^2\mathbf{j} + t^3\mathbf{j}$
   $\mathbf{v}(t) = \mathbf{i} + 2t\,\mathbf{j} + 3t^2\mathbf{j}$
   $v(t) = \|\mathbf{v}(t)\| = \sqrt{1 + 4t^2 + 9t^4}$
   $\mathbf{a}(t) = 2\mathbf{j} + 6t\,\mathbf{j}$

3. $\mathbf{r}(t) = (1 - t^2)\,\mathbf{i} + 2t\,\mathbf{j} + (1/(1 + t^2))\,\mathbf{k}$
   $\mathbf{v}(t) = -2t\,\mathbf{i} + 2\,\mathbf{j} - (2t/(1 + t^2)^2)\,\mathbf{k}$
   $v(t) = \|\mathbf{v}(t)\| = \sqrt{4t^2 + 4 + 4t^2/(1 + t^2)^4}$
   $\mathbf{a}(t) = -2\,\mathbf{i} + (6t^2 - 2)/(1 + t^2)^3\,\mathbf{k}$

5. $\mathbf{r}(t) = t\,\mathbf{i} - 5\,\mathbf{j} + e^t\mathbf{j}$
   $\mathbf{v}(t) = \mathbf{i} + e^t\mathbf{j}$
   $v(t) = \|\mathbf{v}(t)\| = \sqrt{1 + e^{2t}}$
   $\mathbf{a}(t) = e^t\mathbf{j}$

7. $\mathbf{r}(t) = -\cos(t^2)\,\mathbf{i} + \sin(t^2)\,\mathbf{j} + t^3\mathbf{j}$
   $\mathbf{v}(t) = 2t\sin(t^2)\,\mathbf{i} + 2t\cos(t^2)\,\mathbf{j} + 3t^2\mathbf{k}$
   $v(t) = \|\mathbf{v}(t)\| = \sqrt{4t^2 + 9t^4}$
   $\mathbf{a}(t) = (4t^2\cos(t^2) + 2\sin(t^2))\,\mathbf{i} - (4t^2\sin(t^2) - 2\cos(t)^2)\,\mathbf{j} + 6t\,\mathbf{k}$

9–11. Since the acceleration is $\mathbf{a}(t) = -32\,\mathbf{k}$, velocity is $\mathbf{v}(t) = -32t\,\mathbf{k} + \mathbf{v}_0$, and position is $\mathbf{r}(t) = -16t^2\mathbf{k} + t\,\mathbf{v}_0 + \mathbf{r}_0$.

9. $\mathbf{r}(t) = -16t^2\mathbf{k} + t\,(3\,\mathbf{i} - 2\,\mathbf{j} + \mathbf{k}) + 2\,\mathbf{i} - 5\,\mathbf{k}$
   $= (3t + 2)\,\mathbf{i} - 2t\,\mathbf{k} - (16t^2 - t + 5)\,\mathbf{k}$

11. $\mathbf{r}(t) = -16t^2\mathbf{k} + t\,(\mathbf{i} + \mathbf{j} + \mathbf{k}) + 3\,\mathbf{i} - 2\,\mathbf{j} - \mathbf{k}$
    $= (t + 3)\,\mathbf{i} + (t - 2)\,\mathbf{j} - (16t^2 - t + 1)\,\mathbf{k}$

13. The position vector at time t is $\mathbf{r}(t) = \langle t, t^2, t^3 \rangle$. The tangent vector at time t is
$$\mathbf{r}'(t) = \langle 1, 2t, 3t^2 \rangle.$$

The point $P = (1, 1, 1)$ corresponds to $t = 1$, so the tangent vector at this point is
$$\mathbf{r}'(1) = \langle 1, 2, 3 \rangle.$$

Consequently, the tangent line to the curve at point P is parametrized by the vector-valued function
$$u \mapsto \mathbf{r}(1) + u\,\mathbf{r}'(1) = \langle 1, 1, 1 \rangle + u\langle 1, 2, 3 \rangle.$$

The parametric equations for the tangent line are
$$x = 1 + u, \quad y = 1 + 2u, \quad z = 1 + 3u.$$

15. The position vector at time t is $\mathbf{r}(t) = \langle e^t, e^{-t}, -e^{-2t} \rangle$. The tangent vector at time t is
$$\mathbf{r}'(t) = \langle e^t, -e^{-t}, 2e^{-2t} \rangle.$$

The point $P = (1, 1, -1)$ corresponds to $t = 0$, so the tangent vector at this point is
$$\mathbf{r}'(0) = \langle 1, -1, 2 \rangle.$$

Consequently, the tangent line to the curve at point P is parametrized by the vector-valued function
$$u \mapsto \mathbf{r}(0) + u\,\mathbf{r}'(0) = \langle 1, 1, -1 \rangle + u\langle 1, -1, 2 \rangle.$$

The parametric equations for the tangent line are
$$x = 1 + u, \quad y = 1 - u, \quad z = -1 + 2u.$$

17. The position vector at time t is $\mathbf{r}(t) = \langle t, -1, e^{t/2} \rangle$. The tangent vector at time t is
$$\mathbf{r}'(t) = \langle 1, 0, (1/2)e^{t/2} \rangle.$$

The point $P = (2, -1, e)$ corresponds to $t = 2$, so the tangent vector at this point is
$$\mathbf{r}'(2) = \langle 1, 0, e/2 \rangle.$$

Consequently, the tangent line to the curve at point P is parametrized by the vector-valued function
$$u \mapsto \mathbf{r}(2) + u\,\mathbf{r}'(2) = \langle 2, -1, e \rangle + u\langle 1, 0, e/2 \rangle.$$

The parametric equations for the tangent line are
$$x = 2 + u, \quad y = -1, \quad z = e + (e/2)u.$$

19. The position vector at time t is $\mathbf{r}(t) = \langle t, -5, 5/t \rangle$. The tangent vector at time t is
$$\mathbf{r}'(t) = \langle 1, 0, -5/t^2 \rangle.$$

The point $P = (5, -5, 1)$ corresponds to $t = 5$, so the tangent vector at this point is
$$\mathbf{r}'(5) = \langle 1, 0, -1/5 \rangle.$$

Consequently, the tangent line to the curve at point P is parametrized by the vector-valued function
$$u \mapsto \mathbf{r}(5) + u\,\mathbf{r}'(5) = \langle 5, -5, 1 \rangle + u\langle 1, 0, -1/5 \rangle.$$

The parametric equations for the tangent line are
$$x = 5 + u, \quad y = -5, \quad z = 1 - (1/5)u.$$

21. The position vector at time t is $\mathbf{r}(t) = \langle (3+t)/(1+t^2), t^2, \sin(\pi t) \rangle$. The tangent vector at time t is

$$\mathbf{r}'(t) = \langle (1 - 6t - t^2)/(1 + t^2)^2, 2t, \pi \cos(\pi t) \rangle.$$

The point $P = (1, 4, 0)$ corresponds to $t = 2$, so the tangent vector at this point is

$$\mathbf{r}'(2) = \langle -3/5, 4, \pi \rangle.$$

Consequently, the tangent line to the curve at point P is parametrized by the vector-valued function

$$u \mapsto \mathbf{r}(2) + u\,\mathbf{r}'(2) = \langle 1, 4, 0 \rangle + u\langle -3/5, 4, \pi \rangle.$$

The parametric equations for the tangent line are

$$x = 1 - (3/5)u, \quad y = 4 + 4u, \quad z = \pi u.$$

23. The trajectory passes through the point $P = (-2, -4, -8)$ when $t = -2$. Since $\mathbf{r}'(t) = \mathbf{i} - 2t\,\mathbf{j} + 3t^2\mathbf{k}$, the tangent vector at that point is

$$\mathbf{r}'(-2) = \mathbf{i} + 4\,\mathbf{j} + 12\mathbf{k}.$$

Consequently, the symmetric equations for the tangent line are

$$x + 2 = \frac{y + 4}{4} = \frac{z + 8}{12}.$$

25. The trajectory passes through the point $P = (5, -5, 1)$ when $t = 5$. Since $\mathbf{r}'(t) = \mathbf{i} - (5/t^2)\mathbf{k}$, the tangent vector at that point is

$$\mathbf{r}'(5) = \mathbf{i} - (1/5)\mathbf{k}.$$

Consequently, the symmetric equations for the tangent line are

$$x - 5 = \frac{z - 1}{-1/5}, y = -5.$$

27. The trajectory passes through the point $P = (1, 4, 1)$ when $t = 2$. Since $\mathbf{r}'(t) = (-2/t^2)\,\mathbf{i} + 2\,\mathbf{j}$, the tangent vector at that point is

$$\mathbf{r}'(2) = -(1/2)\,\mathbf{i} + 2\,\mathbf{j}.$$

Consequently, the symmetric equations for the tangent line are

$$\frac{x - 1}{-1/2} = \frac{y - 4}{2}, z = 1.$$

29. The trajectory passes through the point $P = (1, 0, \sqrt{\pi})$ when $t = \sqrt{\pi}$. Since $\mathbf{r}'(t) = 2t\sin(t^2)\,\mathbf{i} + 2t\cos(t^2)\,\mathbf{j} + \mathbf{k}$, the tangent vector at that point is

$$\mathbf{r}'(\sqrt{\pi}) = -2\sqrt{\pi}\,\mathbf{j} + \mathbf{k}.$$

Consequently, the symmetric equations for the tangent line are

$$x = 1\,, \quad \frac{y}{-2\sqrt{\pi}} = z - \sqrt{\pi}\,.$$

The solution in the text can be obtained by multiplying both sides of the equation by $-2\sqrt{\pi}$.

31. The trajectory passes through the point $P = (1, -4, 1)$ when $t = -2$. Since $\mathbf{r}'(t) = -2t\,\mathbf{i} + 2\,\mathbf{j} - (10t/(1+t^2)^2)\,\mathbf{k}$, the tangent vector at that point is

$$\mathbf{r}'(-2) = 4\,\mathbf{i} + 2\,\mathbf{j} + (4/5)\,\mathbf{k}.$$

Consequently, the symmetric equations for the tangent line are

$$\frac{x-1}{4} = \frac{y+4}{2} = \frac{z-1}{4/5}\,.$$

The solution in the text can be obtained by multiplying all sides of the equations by 4.

Further Theory and Practice

33. We assume that the arrow is shot from the origin and its trajectory is in the yz-plane. Since the acceleration of the arrow is $\mathbf{a}(t) = -32\,\mathbf{k}$, its velocity is $\mathbf{v}(t) = -32t\,\mathbf{k} + \mathbf{v}(0)$, and the position vector of the arrow at time t is $\mathbf{r}(t) = -16t^2\mathbf{k} + t\,\mathbf{v}(0) + \mathbf{r}(0)$.

The arrow is shot with initial height 4 feet implying that $\mathbf{r}(0) = 4\,\mathbf{k}$. Consequently, the position vector is $\mathbf{r}(t) = (-16t^2 + 4)\,\mathbf{k} + t\,\mathbf{v}(0)$.

The initial velocity: $\mathbf{v}(0)$, is a vector of magnitude 100 and direction $\cos(\pi/6)\,\mathbf{j} + \sin(\pi/6)\,\mathbf{k}$. Therefore, $\mathbf{v}(0) = 100(\sqrt{3}/2)\,\mathbf{j} + 100(1/2)\,\mathbf{k} = 50\sqrt{3}\,\mathbf{j} + 50\,\mathbf{k}$. Substitute this into the formula for the position vector to obtain

$$\mathbf{r}(t) = 50\sqrt{3}\,t\,\mathbf{j} + (-16t^2 + 50t + 4)\,\mathbf{k}.$$

The arrow hits the ground when its $\mathbf{k}$-component is 0. That is, when $8t^2 - 25t - 2 = 0$. Using the quadratic formula, this is in

$$t = \frac{25 + \sqrt{689}}{16} \approx 3.2 \text{ seconds}.$$

At this time, $\mathbf{r}(t) = 50\sqrt{3} \cdot \frac{25+\sqrt{689}}{16}\,\mathbf{j}$. Therefore, the arrow travels the horizontal distance $25\sqrt{3}\,(25 + \sqrt{689})/8$ feet. This is approximately 277.4 feet.

The arrow reaches its maximum height when the vertical component of its velocity: $-32t + 50$, is zero. That is, when $t = 25/16$ seconds. Since $\mathbf{r}(25/16) = (625\sqrt{3}/8)\,\mathbf{j} + (689/16)\,\mathbf{k}$, the maximum height is $689/16$ feet. This is approximately 43.1 feet.

35. Since $\mathbf{r}(2) = -20\,\mathbf{i} + 8\,\mathbf{j} + 38\,\mathbf{k}$, the roller coaster leaves the track at $P = (-20, 8, 38)$. Its velocity vector is $\mathbf{v} = \mathbf{r}'(2) = -20\,\mathbf{i} + 12\,\mathbf{j} + 4\,\mathbf{k}$. We can integrate to obtain a new trajectory for the coaster (straight-line) called $\mathbf{c}(t)$. We will restart the clock so $\mathbf{c}(0) = \mathbf{r}(2)$:

$$\mathbf{c}(t) = t\,\mathbf{v} + \mathbf{c}(0)$$
$$= -20t\,\mathbf{i} + 12t\,\mathbf{j} + 4t\,\mathbf{k} + (-20\,\mathbf{i} + 8\,\mathbf{j} + 38\,\mathbf{k})$$
$$= (-20t - 20)\,\mathbf{i} + (12t + 8)\,\mathbf{j} + (4t + 38)\,\mathbf{k}$$

Three seconds later ($t = 5$ on the original clock) the coaster has the position vector $\mathbf{c}(3) = -80\,\mathbf{i} + 44\,\mathbf{j} + 50\,\mathbf{k}$. It will be at the point $(-80, 44, 50)$.

37. Let the acceleration vector be $\mathbf{a}$ so the velocity vector at time t is $\mathbf{v}(t) = \mathbf{a}\,t + \mathbf{v}_0$. Zero initial velocity implies that $\mathbf{v}(t) = \mathbf{a}\,t$. Integrate once more to obtain the position vector at time t:

$$\mathbf{r}(t) = \frac{1}{2}\mathbf{a}\,t^2 + \mathbf{r}_0.$$

The body travels in a straight line path, starting at $\mathbf{r}_0$ and moving in the direction of the acceleration vector.

39. Observe that

$$\frac{d}{dt}(\|\mathbf{r}(t)\|^2) = \frac{d}{dt}(\mathbf{r}(t) \cdot \mathbf{r}(t))$$
$$= \mathbf{r}(t) \cdot \frac{d}{dt}\mathbf{r}(t) + \frac{d}{dt}\mathbf{r}(t) \cdot \mathbf{r}(t)$$
$$= \mathbf{r}(t) \cdot \mathbf{v}(t) + \mathbf{v}(t) \cdot \mathbf{r}(t)$$
$$= 2\,\mathbf{r}(t) \cdot \mathbf{v}(t)$$
$$= 0$$

Consequently $\|\mathbf{r}(t)\|^2$ is constant, and so is its square root $\|\mathbf{r}(t)\|$. In other words, $\|\mathbf{r}(t)\| = R$ for all t, and the trajectory lies on a sphere centered at the origin of radius R.

41. The arrow's position vector at time t is $\mathbf{r}(t) = -16t^2\mathbf{k} + t\,\mathbf{v}(0) + \mathbf{r}(0)$. The initial position is $\mathbf{r}(0) = 500\,\mathbf{k}$, so $\mathbf{r}(t) = (-16t^2 + 500)\,\mathbf{k} + t\,\mathbf{v}(0)$.

Regarding the initial velocity, $\mathbf{v}(0)$, if there were no gust of wind it would be
$$120(\cos(60°)\,\mathbf{i} - \sin(60°)\,\mathbf{k}) = 60\,\mathbf{i} - 60\sqrt{3}\,\mathbf{k}.$$

With the gust of wind the initial velocity vector is $\mathbf{v}(0) = 40\,\mathbf{i} - 60\sqrt{3}\,\mathbf{k}$ and $\mathbf{r}(t) = 40\,t\,\mathbf{i} + (-16t^2 - 60\sqrt{3}\,t + 500)\,\mathbf{k}$.

The arrow hits the ground when its **k**-component is 0. That is, when $4t^2 + 15\sqrt{3}\,t - 125 = 0$. Using the quadratic formula, this is in

$$t_g = \frac{5\sqrt{107} - 15\sqrt{3}}{8} \approx 3.2 \text{ seconds}.$$

At this time, the arrow's velocity is $\mathbf{r}'(t_g) = 40\,\mathbf{i} - 20\sqrt{107}\,\mathbf{k}$ and its position is $\mathbf{r}(t_g) = (25\sqrt{107} - 75\sqrt{3}\,)\,\mathbf{i}$. Horizontally, the arrow travels $25\sqrt{107} - 75\sqrt{3} \approx 128.7$ feet.

43. Observe that triangle $\triangle OP_0P_1$ (O is the origin) is a right triangle with hypotenuse $2r$. Let θ be the angle at P_0. Draw a picture to see that $\|\overrightarrow{P_0P_1}\| = 2r\cos(\theta)$, and the component of the acceleration vector $(-g\,\mathbf{k})$ in the direction of the motion of the particle is $a_m = g\cos(\theta)$.

    Let $d(t)$ be the distance the particle has traveled at time t. Then $d''(t) = a_m$, $d'(t) = a_m t$, and $d(t) = \frac{1}{2}a_m t^2$. The time taken for the particle to reach P_1 is the solution to the equation $d(t) = 2r\cos(\theta)$. That is, $\frac{1}{2}g\cos(\theta)t^2 = 2r\cos(\theta)$. Consequently, $t = 2\sqrt{r/g}$.

45. Differentiate the cross product to obtain

$$\frac{d}{dt}(\mathbf{r}(t) \times \mathbf{r}'(t)) = \mathbf{r}(t) \times \frac{d}{dt}\mathbf{r}'(t) + \frac{d}{dt}\mathbf{r}(t) \times \mathbf{r}'(t)$$
$$= \mathbf{r}(t) \times \mathbf{r}''(t) + \mathbf{r}'(t) \times \mathbf{r}'(t)$$
$$= \mathbf{r}(t) \times \mathbf{r}''(t).$$

Consequently, since $\mathbf{r}(t) \times \mathbf{r}''(t) = 0$, the vector function $t \mapsto \mathbf{r}(t) \times \mathbf{r}'(t)$ differentiates to 0 implying that it is a constant vector.

47. Since the speed is $v(t) = \|\mathbf{v}(t)\|$,

$$v'(t) = \frac{d}{dt}\sqrt{\mathbf{v}(t) \cdot \mathbf{v}(t)}$$
$$= \frac{1}{2\|\mathbf{v}(t)\|}\frac{d}{dt}(\mathbf{v}(t) \cdot \mathbf{v}(t))$$
$$= \frac{1}{\|\mathbf{v}(t)\|}\mathbf{v}(t) \cdot \mathbf{v}'(t)$$
$$= \frac{1}{\|\mathbf{v}(t)\|}\mathbf{v}(t) \cdot \mathbf{a}(t).$$

If the angle is acute, then $\mathbf{v}(t) \cdot \mathbf{a}(t)$ is positive so $v'(t) > 0$, and $v(t)$ is increasing. If the angle is obtuse, then $\mathbf{v}(t) \cdot \mathbf{a}(t)$ is negative, $v'(t) < 0$, and $v(t)$ is decreasing.

49. The tangent line to $\mathcal{C}$ at the point (t, t^2, t^3) is parametrized by the function $u \mapsto \mathbf{r}(t) + u\,\mathbf{r}'(t)$. To find the point P_0 we look for u and t such that $\mathbf{r}(t) + u\,\mathbf{r}'(t) = \langle 0, -1/2, -1/\sqrt{2}\,\rangle$. That is,

$$\langle t, t^2, t^3 \rangle + u\langle 1, 2t, 3t^2 \rangle = \langle 0, -1/2, -1/\sqrt{2}\,\rangle,$$

or

$$\langle t + u, t^2 + 2ut, t^3 + 3ut^2 \rangle = \langle 0, -1/2, -1/\sqrt{2} \rangle.$$

Equate the first components to see that $u = -t$. Substitute this into the equation for the second components to obtain $t^2 - 2t^2 = -1/2$. Therefore, $t = 1/\sqrt{2}$ and $u = -1/\sqrt{2}$. Fortunately, these values are compatible with the third component equation (verify) so the point P_0 has coordinates $(1/\sqrt{2}, 1/2, 1/(2\sqrt{2}))$.

Calculator/Computer Exercises

51. After loading the VectorCalculus package, the position vector **r**, the tangent vector **v**, and the tangent line **T** are defined. Only the formula for **T** is shown. The last entry plots the trajectory and the tangent line.

```
> with(VectorCalculus):
  r := <5+2*cos(t),3+sin(t)>:
  v := diff(r,t):
  T := eval(r,t=Pi/3) + t*eval(v,t=Pi/3);
  plot( [[r[1],r[2],t=0..2*Pi], [T[1],T[2],t=-1..1]],
        color=black);
```

$$T := (6 - \sqrt{3}\,t)e_x + \left(3 + \frac{1}{2}\sqrt{3} + \frac{1}{2}t\right)e_y$$

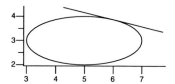

53. See Exercise 51.

```
> r := <tan(t),sec(t)>:
  v := diff(r,t):
  T := eval(r,t=Pi/4) + t*eval(v,t=Pi/4);
  plot( [[r[1],r[2],t=-Pi/3..Pi/3], [T[1],T[2],t=-0.5..0.5]],
        color=black);
```

$$T := (1 - 2t)e_x + (\sqrt{2} + t\sqrt{2})e_y$$

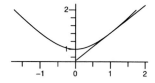

55. See Exercise 51.

```
> r := <exp(t)+exp(-t),exp(t)-exp(-t)>:
  v := diff(r,t):
  T := eval(r,t=2) + t*eval(v,t=2);
  plot( [[r[1],r[2],t=-3..3], [T[1],T[2],t=-1..1]],
        color=black, view=[0..14,-14..14]);
```

$$T := (e^2 + e^{-2} + t(e^2 + e^{-2}))e_x + (e^2 - e^{-2} + t(e^2 + e^{-2}))e_y$$

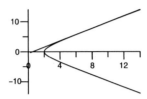

57. See Exercise 51.

```
> r := <2*t/(1+t^2),(1-t^2)/(1+t^2)>;
  v := diff(r,t);
  T := eval(r,t=1/2) + t*eval(v,t=1/2);
  plot( [[r[1],r[2],t=-5..5], [T[1],T[2],t=-1..1]],
        color=black);
```

$$T := \left(\frac{24}{25}t + \frac{4}{5}\right) e_x + \left(-\frac{32}{25}t + \frac{3}{5}\right) e_y$$

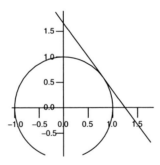

59. The x and y coordinates of a point on the Folium of Descartes are defined
    as functions of t, then the folium is plotted.

```
> with(plots):
  x,y := 3*t/(1+t^3),3*t^2/(1+t^3):
  plot( [x,y,t=-1/2..infinity], color=black);
  Folium := %:
```

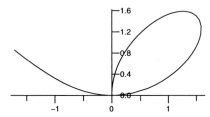

A portion of the Folium of Descartes.

It appears that there are two points on the curve where the tangent line is parallel to the line $y = x$. They can be found by solving the equation $dy/dx = 1$. Recall (Section 10.1, Exercise 73) that $dy/dx = y'(t)/x'(t)$.

```
> diff(y,t)/diff(x,t)=1:
  eqn := simplify(%);
```

$$eqn := \frac{t(-1+t^3)}{-1+2t^3} = 1$$

The solve procedure outputs a sequence of four exact solutions, each one given in terms of a root of a fourth degree polynomial. Only the first solution is shown below.

```
> soln := solve(eqn,t):
  soln[1];
```

$$\text{RootOf}(-2\_Z + \_Z^4 + 1 - 2\_Z^3, index = 1)$$

The first two solutions are real numbers, these are the t values we want to locate the points on the curve. We can get them into a list named Txy by applying the allvalues procedure.

```
> Txy := ['allvalues(soln[j])'$j=1..2];
  evalf[4](Txy);
```

$$Txy := \left[\frac{1}{2} + \frac{1}{2}\sqrt{3} - \frac{1}{2}\sqrt{2}\,3^{1/4}, \frac{1}{2} + \frac{1}{2}\sqrt{3} + \frac{1}{2}\sqrt{2}\,3^{1/4}\right]$$

$$[0.4355, 2.296]$$

The next entry substitutes these t values into the list $[x, y]$ to obtain the points on the curve. Then the points are displayed on the Folium.

```
> subs(t=Txy[1],[x,y]),subs(t=Txy[2],[x,y]):
  TxyPoints := evalf[4](%);
  display( Folium, pointplot( [TxyPoints] ) );
```

$$TxyPoints := [1.206, 0.5256], [0.5259, 1.207]$$

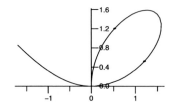

Points on Folium where the tangent
line is parallel to the line $y = x$.

The curve has a horizontal tangent line when $y'(t) = 0$ and $x'(0) \neq 0$. Vertical tangent lines occur when $x'(t) = 0$ and $y'(t) \neq 0$. Clearly there is one horizontal tangent line at the origin $(t = 0)$ and another at the top of the curve.

```
> Th := solve(diff(y,t)=0,t):
  'Horizontal tangents at times',[Th[1],Th[2]];
  ThPoints := subs(t=Th[1],[x,y]),subs(t=Th[2],[x,y]):
  'Horizontal tangents at points',evalf[4](ThPoints);
```

$$Horizontal\ tangents\ at\ times, [0, 2^{1/3}]$$

$$Horizontal\ tangents\ at\ points, [0., 0.], [1.260, 1.587]$$

There appears to be one vertical tangent.[1]

```
> Tv := solve(diff(x,t)=0,t):
  'Vertical tangent at time',Tv[1];
  TvPoint := subs(t=Tv[1],[x,y]):
  'Vertical tangent at point',evalf[4](TvPoint);
```

$$Vertical\ tangent\ at\ time, \frac{1}{2} 2^{2/3}$$

$$Vertical\ tangent\ at\ point, [1.587, 1.260]$$

61. See Exercise 51.

```
> r := <t^2-1,t^4-1,(t-1)*ln(2+t)>:
  v := diff(r,t):
  T1 := eval(r,t=-1) + t*eval(v,t=-1):
  T2 := eval(r,t=1) + t*eval(v,t=1):
  spacecurve( {[r[1],r[2],r[3],t=-6/5..6/5],
              [T1[1],T1[2],T1[3],t=-0.3..0.3],
              [T2[1],T2[2],T2[3],t=-0.3..0.3] },
              color=black, axes=normal, orientation=[-60,70] );
```

---

[1]Two if you want to count the vertical tangent at $t = \infty$.

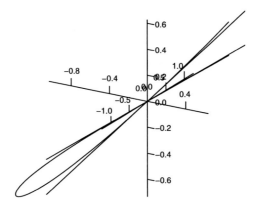

10.3 Tangent Vectors and Arc Length

Problems for Practice

1. Since $\mathbf{r}'(t) = \mathbf{i} - 2t\,\mathbf{j} + 3t^2\mathbf{k}$, the unit tangent vector at $\mathbf{r}(t)$ is

$$\mathbf{T}(t) = \frac{1}{\sqrt{1 + 4t^2 + 9t^4}}(\mathbf{i} - 2t\,\mathbf{j} + 3t^2\mathbf{k})\,.$$

Since $\mathbf{r}'(1) = \mathbf{i} - 2\,\mathbf{j} + 3\,\mathbf{k}$, the tangent line at $(1, -1, 1)$ is

$$x - 1 = \frac{y+1}{-2} = \frac{z-1}{3}\,.$$

3. Since $\mathbf{r}'(t) = \mathbf{i} - (1/2)t^{-1/2}\,\mathbf{k}$, the unit tangent vector at $\mathbf{r}(t)$ is

$$\mathbf{T}(t) = \frac{1}{\sqrt{1 + 1/(4t)}}(\mathbf{i} - (1/2)t^{-1/2}\,\mathbf{k})\,.$$

Since $\mathbf{r}'(4) = \mathbf{i} - (1/4)\,\mathbf{k}$, the tangent line at $(4, 0, -2)$ is

$$x - 4 = \frac{z+2}{-1/4}\,,\ y = 0\,.$$

5. Since $\mathbf{r}'(t) = (e^t + 2t)\,\mathbf{i}$, the unit tangent vector at $\mathbf{r}(t)$ is $\mathbf{T}(t) = \mathbf{i}$. Since $\mathbf{r}'(0) = \mathbf{i}$, the tangent line at $(1, 3, 0)$ is

$$y = 3\,,\ z = 0\,.$$

7. Since $\mathbf{r}'(t) = -\sin(t)\,\mathbf{i} + \cos(t)\,\mathbf{j} + \mathbf{k}$, the unit tangent vector at $\mathbf{r}(t)$ is

$$\mathbf{T}(t) = \frac{1}{\sqrt{2}}(-\sin(t)\,\mathbf{i} + \cos(t)\,\mathbf{j} + \mathbf{k}).$$

Since $\mathbf{r}'(\pi/2) = -\mathbf{i} + \mathbf{k}$, the tangent line at $(0, 1, \pi/2)$ is

$$\frac{x}{-1} = z - \pi/2,\ y = 1.$$

9. Since $\mathbf{r}'(t) = 2t\cos(t^2)\,\mathbf{i} + 2t\sin(t^2)\,\mathbf{j} + 2t\,\mathbf{k}$, $\|\mathbf{r}'(t)\| = \sqrt{8t^2} = 2\sqrt{2}\,t$, $t \geq 0$. Therefore, the arc length calculation can proceed as follows.

$$\int_0^{\sqrt{\pi}} \|\mathbf{r}'(t)\|\,dt = \int_0^{\sqrt{\pi}} 2\sqrt{2}\,t\,dt$$
$$= \sqrt{2}\,t^2 \Big|_0^{\sqrt{\pi}} = \sqrt{2}\,\pi.$$

11. Since $\mathbf{r}'(t) = -3\cos^2(t)\sin(t)\,\mathbf{i} - 3\sin^2(t)\cos(t)\,\mathbf{k}$,

$$\|\mathbf{r}'(t)\| = 3\sqrt{\cos^4(t)\sin^2(t) + \sin^4(t)\cos^2(t)}$$
$$= 3\sqrt{\cos^2(t)\sin^2(t)(\cos^2(t) + \sin^2(t))}$$
$$= 3|\sin(t)\cos(t)|.$$

Therefore, the arc length calculation can proceed as follows.

$$\int_\pi^{3\pi} \|\mathbf{r}'(t)\|\,dt = \int_\pi^{3\pi} 3|\sin(t)\cos(t)|\,dt$$
$$= 3\left(\int_\pi^{3\pi/2} \sin(t)\cos(t)\,dt - \int_{3\pi/2}^{2\pi} \sin(t)\cos(t)\,dt\right.$$
$$\left. + \int_{2\pi}^{5\pi/2} \sin(t)\cos(t)\,dt - \int_{5\pi/2}^{3\pi} \sin(t)\cos(t)\,dt\right)$$
$$= 3\left(\frac{1}{2} + \frac{1}{2} + \frac{1}{2} + \frac{1}{2}\right) = 6.$$

13. Since $\mathbf{r}'(t) = (3/2)\sqrt{1+t}\,\mathbf{i} - (3/2)\sqrt{1-t}\,\mathbf{j} + (3/2)\sqrt{t}\,\mathbf{k}$, $\|\mathbf{r}'(t)\| = (3/2)\sqrt{2+t}$. Therefore, the arc length calculation can proceed as follows.

$$\int_0^1 \|\mathbf{r}'(t)\|\,dt = \int_0^1 (3/2)\sqrt{2+t}\,dt$$
$$= (2+t)^{3/2} \Big|_0^1 = 3\sqrt{3} - 2\sqrt{2}.$$

15. Since $\mathbf{r}'(t) = \mathbf{i} - t\,\mathbf{j} + t^2\mathbf{k}$, $\|\mathbf{r}'(t)\| = \sqrt{1 + t^2 + t^4}$. Therefore, this is not an arc length parametrization.

17. Since $\mathbf{r}'(t) = -(3\sin(t)/5)\,\mathbf{i} + (4\sin(t)/5)\,\mathbf{j} + \cos(t)\,\mathbf{k}$, $\|\mathbf{r}'(t)\| = \sqrt{\sin^2(t) + \cos^2(t)} = 1$. Therefore, this is an arc length parametrization.

19. Because $\mathbf{r}'(t) = 2t\,\mathbf{i} - (1/t)\,\mathbf{j} + 2\,\mathbf{k}$,

$$
\begin{aligned}
\|\mathbf{r}'(t)\| &= \sqrt{4t^2 + 1/t^2 + 4} \\
&= \sqrt{\frac{4t^4 + 4t^2 + 1}{t^2}} \\
&= \sqrt{\frac{(2t^2 + 1)^2}{t^2}} \\
&= \frac{1 + 2t^2}{t}, \ (t > 0).
\end{aligned}
$$

Therefore, since the curve is swept out from $t = 1$ to $t = e$, the arc length calculation can proceed as follows.

$$
\begin{aligned}
\int_1^e \|\mathbf{r}'(t)\|\,dt &= \int_1^e \left(\frac{1}{t} + 2t\right)\,dt \\
&= \ln(t) + t^2 \Big|_1^e = 1 + e^2 - 1 = e^2
\end{aligned}
$$

21. Because $\mathbf{r}'(t) = e^t\mathbf{i} - e^{-t}\mathbf{j} + \sqrt{2}\,\mathbf{k}$,

$$
\begin{aligned}
\|\mathbf{r}'(t)\| &= \sqrt{e^{2t} + e^{-2t} + 2} \\
&= \sqrt{(e^t + e^{-t})^2} \\
&= e^t + e^{-t}.
\end{aligned}
$$

Therefore, since the curve is swept out from $t = 0$ to $t = 1$, the arc length calculation can proceed as follows.

$$
\begin{aligned}
\int_0^1 \|\mathbf{r}'(t)\|\,dt &= \int_0^1 (e^t + e^{-t})\,dt \\
&= (e^t - e^{-t}) \Big|_0^1 = e - e^{-1}
\end{aligned}
$$

23. Since $\mathbf{r}'(t) = \mathbf{i} - 2t\,\mathbf{j} + 3t^2\mathbf{k}$, the unit tangent vector at $\mathbf{r}(t)$ is

$$
\begin{aligned}
\mathbf{T}(t) &= \frac{1}{\sqrt{1 + 4t^2 + 9t^4}}(\mathbf{i} - 2t\,\mathbf{j} + 3t^2\mathbf{k}) \\
&= \frac{1}{\sqrt{1 + 4t^2 + 9t^4}}\,\mathbf{i} - \frac{2t}{\sqrt{1 + 4t^2 + 9t^4}}\,\mathbf{j} + \frac{3t^2}{\sqrt{1 + 4t^2 + 9t^4}}\,\mathbf{k}.
\end{aligned}
$$

Differentiation and simplification yields the vector

$$\mathbf{T}'(t) = \frac{1}{(1 + 4t^2 + 9t^4)^{3/2}}(-2t(9t^2 + 2)\,\mathbf{i} + 2(9t^4 - 1)\,\mathbf{j} + 6t(2t^2 + 1)\mathbf{k})$$

which has the same direction as the unit normal. Substitute $t = 1$ and multiply by the scalar $14^{3/2}/2$ to obtain the vector

$$(14^{3/2}/2)\,\mathbf{T}'(1) = -11\,\mathbf{i} + 8\,\mathbf{j} + 9\,\mathbf{k}\,,$$

which is also parallel to the unit normal. Therefore, the normal line at $(1, -1, 1)$ has symmetric equations

$$\frac{x-1}{-11} = \frac{y+1}{8} = \frac{z-1}{9}\,.$$

25. Since $\mathbf{r}'(t) = -\sin(t)\,\mathbf{i} + \cos(t)\,\mathbf{j} + \mathbf{k}$, the unit tangent vector at $\mathbf{r}(t)$ is

$$\mathbf{T}(t) = \frac{1}{\sqrt{2}}(-\sin(t)\,\mathbf{i} + \cos(t)\,\mathbf{j} + \mathbf{k})\,.$$

Differentiation yields the vector

$$\mathbf{T}'(t) = \frac{1}{\sqrt{2}}(-\cos(t)\,\mathbf{i} - \sin(t)\,\mathbf{j})\,,$$

which has the same direction as the unit normal. Substitute $t = \pi/2$ and multiply by the scalar $\sqrt{2}$ to obtain the vector

$$\sqrt{2}\,\mathbf{T}'(\pi/2) = -\mathbf{j}\,,$$

which is, in fact, the unit normal. Therefore, the normal line at $(0, 1, \pi/2)$ has symmetric equations

$$x = 0\,,\ z = \pi/2\,.$$

27. Since $\mathbf{r}'(t) = e^t\mathbf{i} - e^{-t}\mathbf{j} - 2t\,\mathbf{k}$, the unit tangent vector at $\mathbf{r}(t)$ is

$$\mathbf{T}(t) = \frac{1}{\sqrt{e^{2t} + e^{-2t} + 4t^2}}(e^t\mathbf{i} - e^{-t}\mathbf{j} - 2t\,\mathbf{k})$$

$$= \frac{e^t}{\sqrt{e^{2t} + e^{-2t} + 4t^2}}\,\mathbf{i} - \frac{e^{-t}}{\sqrt{e^{2t} + e^{-2t} + 4t^2}}\,\mathbf{j} - \frac{2t}{\sqrt{e^{2t} + e^{-2t} + 4t^2}}\,\mathbf{k}\,.$$

Differentiation and simplification yields the vector

$$\mathbf{T}'(t) = \frac{2}{(e^{2t} + e^{-2t} + 4t^2)^{3/2}}(e^t(e^{-2t} - 2t + 2t^2)\,\mathbf{i}$$

$$+ e^{-t}(e^{2t} + 2t + 2t^2)\,\mathbf{j} + ((t-1)e^{2t} - (t+1)e^{-2t})\,\mathbf{k})\,,$$

which has the same direction as the unit normal. Substitute $t = 0$ and multiply by the scalar $\sqrt{2}$ to obtain the vector

$$\sqrt{2}\,\mathbf{T}'(0) = \mathbf{i} + \mathbf{j} - 2\,\mathbf{k}\,,$$

which is also parallel to the unit normal. Therefore, the normal line at $(1, 1, 0)$ has symmetric equations

$$x - 1 = y - 1 = \frac{z}{-2}\,.$$

29. Since $\mathbf{r}'(t) = \frac{1}{2}t^{-1/2}\,\mathbf{i} - 4t^{-2}\,\mathbf{j} - 2t^{-3/2}\,\mathbf{k} = \frac{1}{2}t^{-2}(t^{3/2}\,\mathbf{i} - 8\,\mathbf{j} - 4t^{1/2}\,\mathbf{k})$, the unit tangent vector at $\mathbf{r}(t)$ is

$$\mathbf{T}(t) = \frac{1}{\sqrt{t^3 + 16t + 64}}\,(t^{3/2}\,\mathbf{i} - 8\,\mathbf{j} - 4t^{1/2}\,\mathbf{k})$$

$$= \frac{t^{3/2}}{\sqrt{t^3 + 16t + 64}}\,\mathbf{i} - \frac{8}{\sqrt{t^3 + 16t + 64}}\,\mathbf{j} - \frac{4t^{1/2}}{\sqrt{t^3 + 16t + 64}}\,\mathbf{k}\,.$$

Differentiation and simplification yields the vector

$$\mathbf{T}'(t) = \frac{4}{(t^3 + 16t + 64)^{3/2}}\,((4t^{3/2} + 24t^{1/2})\,\mathbf{i} + (3t^2 + 16)\,\mathbf{j} + (t^{5/2} - 32t^{-1/2})\,\mathbf{k})\,,$$

which has the same direction as the unit normal. Substitute $t = 4$ and delete the scalar coefficient to obtain the vector $80\,\mathbf{i} + 64\,\mathbf{j} + 16\,\mathbf{k}$. This vector is also parallel to the unit normal, as is the vector $5\,\mathbf{i} + 4\,\mathbf{j} + \mathbf{k}$. Therefore, the normal line at $(2, 1, 2)$ has symmetric equations

$$\frac{x - 2}{5} = \frac{y - 1}{4} = z - 2\,.$$

31. Since $\mathbf{r}'(t) = \mathbf{i} - 2t\,\mathbf{j} + 3t^2\mathbf{k}$, the tangent vector at the point $P_0 = (1, -1, 1)$ is $\langle 1, -2, 3 \rangle$. This vector is a normal vector for the normal plane so its equation is of the form $x - 2y + 3z = D$. Substitute the coordinates of P_0 to see that $D = 6$. The normal plane's equation is $x - 2y + 3z = 6$.

33. Since $\mathbf{r}'(t) = \mathbf{i} - (2/t^2)\,\mathbf{j} + 2t\,\mathbf{k}$, the tangent vector at the point $P_0 = (2, 1, 4)$ is $\langle 1, -1/2, 4 \rangle$. This vector is a normal vector for the normal plane so its equation is of the form $x - y/2 + 4z = D$. Substitute the coordinates of P_0 to see that $D = 35/2$. The normal plane's equation is $x - y/2 + 4z = 35/2$.

35. Since $\mathbf{r}'(t) = e^t\mathbf{i} - e^{-t}\mathbf{j} - 2t\,\mathbf{k}$, the tangent vector at the point $P_0 = (1, 1, 0)$ is $\langle 1, -1, 0 \rangle$. This vector is a normal vector for the normal plane so its equation is of the form $x - y = D$. Substitute the coordinates of P_0 to see that $D = 0$. The normal plane's equation is $x - y = 0$.

37. Since $\mathbf{r}'(t) = 1/(2\sqrt{t})\,\mathbf{i} - (4/t^2)\,\mathbf{j} + 2t\,\mathbf{k}$, the tangent vector at the point $P_0 = (2, 1, 1)$ is $\langle 1/4, -1/4, 8 \rangle$. This vector is a normal vector for the normal plane so its equation is of the form $x - y + 32z = D$. Substitute the coordinates of P_0 to see that $D = 33$. The normal plane's equation is $x - y + 32z = 33$.

Further Theory and Practice

39. Since $\mathbf{r}'(t) = \mathbf{i} + \mathbf{j} + (3/2)t^{1/2}\mathbf{k}$, $\|\mathbf{r}'(t)\| = \sqrt{2 + 9t/4}$. Therefore, the arc length function, starting at $t = 0$, can be calculated as follows.

$$\sigma(t) = \int_0^t \|\mathbf{r}'(\tau)\| \, d\tau = \int_0^t \sqrt{2 + 9\tau/4} \, d\tau$$

$$= \frac{8}{27}(2 + 9\tau/4)^{3/2}\Big|_0^t$$

$$= \frac{1}{27}(8 + 9t)^{3/2} - \frac{16}{27}\sqrt{2}$$

Solve the equation $\sigma(t) = s$ for t to obtain the formula for the inverse function:

$$\sigma^{-1}(s) = \frac{1}{9}(27s + 16\sqrt{2})^{2/3} - \frac{8}{9}.$$

The arc length parametrization is

$$\mathbf{p}(s) = \mathbf{r}(\sigma^{-1}(s)) = \sigma^{-1}(s)\,\mathbf{i} + \sigma^{-1}(s)\,\mathbf{j} + (\sigma^{-1}(s))^{3/2}\,\mathbf{k}.$$

41. Since $\mathbf{r}'(t) = e^t\mathbf{i} - e^{-t}\mathbf{j} + \sqrt{2}\mathbf{k}$, $\|\mathbf{r}'(t)\| = \sqrt{e^{2t} + e^{-2t} + 2}$. This simplifies to $\|\mathbf{r}'(t)\| = \sqrt{(e^t + e^{-t})^2} = e^t + e^{-t}$, so the arc length function, starting at $t = 0$, can be calculated as follows.

$$\sigma(t) = \int_0^t \|\mathbf{r}'(\tau)\| \, d\tau = \int_0^t e^\tau + e^{-\tau} \, d\tau$$

$$= e^\tau - e^{-\tau}\Big|_0^t$$

$$= e^t - e^{-t}$$

Solve the equation $\sigma(t) = s$ for t to obtain the formula for the inverse function:[2]

$$\sigma^{-1}(s) = \ln\left(\frac{s + \sqrt{s^2 + 4}}{2}\right).$$

The arc length parametrization is

$$\mathbf{p}(s) = \mathbf{r}(\sigma^{-1}(s))$$

$$= \left(\frac{s + \sqrt{s^2 + 4}}{2}\right)\mathbf{i} + \left(\frac{2}{s + \sqrt{s^2 + 4}}\right)\mathbf{j}$$

$$+ \sqrt{2}\ln\left(\frac{s + \sqrt{s^2 + 4}}{2}\right)\mathbf{k}.$$

---

[2]Rearrange $e^t - e^{-t} = s$ to $e^{2t} - se^t - 1 = 0$ and use the quadratic formula to solve for e^t.

43. Observe that $\mathbf{r}(\pi/3) = \langle 6, 3 + \sqrt{3}/2 \rangle$. Since $\mathbf{r}'(t) = -2\sin(t)\,\mathbf{i} + \cos(t)\,\mathbf{j}$, $\mathbf{r}'(\pi/3) = \langle -\sqrt{3}, 1/2 \rangle$. Therefore, the symmetric equation for the tangent line is
$$\frac{x-6}{-\sqrt{3}} = \frac{y-3-\sqrt{3}/2}{1/2}.$$

Because this is a planar curve the principal unit normal vector is in the same plane. A normal vector can be obtained by simply interchanging the two components of $\mathbf{r}'(\pi/3)$ and then negating the first component. (This rotates the tangent vector 90 degrees in the clockwise direction.) Therefore, the normal line has the symmetric equation
$$\frac{x-6}{-1/2} = \frac{y-3-\sqrt{3}/2}{-\sqrt{3}}.$$

45. Observe that $\mathbf{r}(\pi/4) = \langle 1, \sqrt{2} \rangle$. Since $\mathbf{r}'(t) = \sec^2(t)\,\mathbf{i} + \sec(t)\tan(t)\,\mathbf{j}$, $\mathbf{r}'(\pi/4) = \langle 2, \sqrt{2} \rangle$. Therefore, the symmetric equation for the tangent line is
$$\frac{x-1}{2} = \frac{y-\sqrt{2}}{\sqrt{2}}.$$

Using this we see that the normal line has the following symmetric equation (see Exercise 43).
$$\frac{x-1}{-\sqrt{2}} = \frac{y-\sqrt{2}}{2}.$$

47. Observe that $\mathbf{r}(0) = \langle 2, 0 \rangle$. Since $\mathbf{r}'(t) = (e^t - e^{-t})\,\mathbf{i} + (e^t + e^{-t})\,\mathbf{j}$, $\mathbf{r}'(0) = \langle 0, 2 \rangle$. Therefore, the symmetric equation for the tangent line is
$$x = 2.$$

Using this we see that the normal line has the symmetric equation
$$y = 0.$$

49. Observe that $\mathbf{r}(1/2) = \langle 4/5, 3/5 \rangle$. Since $\mathbf{r}'(t) = 2(1 - t^2)/(t^2 + 1)^2\,\mathbf{i} - 4t/(t^2 + 1)^2\,\mathbf{j}$, $\mathbf{r}'(1/2) = \langle 24/25, -32/25 \rangle$. Using the fact that $\mathbf{r}'(1/2)$ is parallel to $\langle 3, -4 \rangle$, the symmetric equation for the tangent line is
$$\frac{x-4/5}{3} = \frac{y-3/5}{-4}.$$

In view of this, the normal line has the following symmetric equation (see Exercise 43).
$$\frac{x-4/5}{4} = \frac{y-3/5}{3}$$

51. Since $\mathbf{r}'(t) = -\sin(t)\,\mathbf{i} + 2\cos(2t)\,\mathbf{j} + \mathbf{k}$, the unit tangent vector at $\mathbf{r}(t)$ is

$$\mathbf{T}(t) = \frac{1}{\sqrt{\sin^2(t) + 4\cos^2(2t) + 1}}\left(-\sin(t)\,\mathbf{i} + 2\cos(2t)\,\mathbf{j} + \mathbf{k}\right).$$

Differentiate, using the product rule, to obtain

$$\mathbf{T}'(t) = \frac{1}{\sqrt{\sin^2(t) + 4\cos^2(2t) + 1}}\left(-\cos(t)\,\mathbf{i} - 4\sin(2t)\,\mathbf{j}\right)$$

$$-\frac{1}{2} \cdot \frac{2\sin(t)\cos(t) - 16\cos(2t)\sin(2t)}{(\sin^2(t) + 4\cos^2(2t) + 1)^{3/2}}\left(-\sin(t)\,\mathbf{i} + 2\cos(2t)\,\mathbf{j} + \mathbf{k}\right).$$

Substitute $t = \pi/4$ to get

$$\mathbf{T}(\pi/4) = \langle -\sqrt{3}/3, 0, \sqrt{6}/3 \rangle \quad \text{and} \quad \mathbf{T}'(\pi/4) = -\frac{\sqrt{6}}{9}\langle \sqrt{2}, 12, 1 \rangle.$$

Therefore,

$$\mathbf{N}(\pi/4) = -\frac{1}{\sqrt{147}}\langle \sqrt{2}, 12, 1 \rangle = -\frac{1}{7\sqrt{3}}\langle \sqrt{2}, 12, 1 \rangle$$

and

$$\mathbf{B}(\pi/4) = \mathbf{T}(\pi/4) \times \mathbf{N}(\pi/4)$$

$$= -\frac{1}{7\sqrt{3}} \det\left(\begin{bmatrix} \mathbf{i} & \mathbf{j} & \mathbf{k} \\ -\frac{\sqrt{3}}{3} & 0 & \frac{\sqrt{6}}{3} \\ \sqrt{2} & 12 & 1 \end{bmatrix}\right)$$

$$= -\frac{1}{7\sqrt{3}}\langle -4\sqrt{6}, \sqrt{3}, -4\sqrt{3} \rangle$$

$$= \langle \frac{4}{7}\sqrt{2}, -\frac{1}{7}, \frac{4}{7} \rangle.$$

53. Since $\mathbf{r}'(t) = \sec^2(t)\,\mathbf{i} + \sec(t)\tan(t)\,\mathbf{j} + \mathbf{k}$, the unit tangent vector at $\mathbf{r}(t)$ is

$$\mathbf{T}(t) = \frac{1}{\sqrt{\sec^4(t) + \sec^2(t)\tan^2(t) + 1}}\left(\sec^2(t)\,\mathbf{i} + \sec(t)\tan(t)\,\mathbf{j} + \mathbf{k}\right).$$

Simplify the radical by converting to tangents then differentiate, using the product rule, to obtain

$$\mathbf{T}'(t) = \frac{1}{\sqrt{2\tan^4(t) + 3\tan^2(t) + 2}}\left(2\sec^2(t)\tan(t)\,\mathbf{i} + (\sec^3(t) + \sec(t)\tan^2(t))\,\mathbf{j}\right)$$

$$-\frac{1}{2} \cdot \frac{8\tan^3(t)\sec^2(t) + 6\tan(t)\sec^2(t)}{(2\tan^4(t) + 3\tan^2(t) + 2)^{3/2}}\left(\sec^2(t)\,\mathbf{i} + \sec(t)\tan(t)\,\mathbf{j} + \mathbf{k}\right).$$

Substitute $t = \pi/4$ to get

$$\mathbf{T}(\pi/4) = (1/\sqrt{7})\,\langle 2, \sqrt{2}, 1\rangle \quad \text{and} \quad \mathbf{T}'(\pi/4) = (1/\sqrt{7})\,\langle 0, \sqrt{2}, -2\rangle\,.$$

Therefore,

$$\mathbf{N}(\pi/4) = (1/\sqrt{6})\,\langle 0, \sqrt{2}, -2\rangle$$

and

$$\begin{aligned}
\mathbf{B}(\pi/4) &= \mathbf{T}(\pi/4) \times \mathbf{N}(\pi/4) \\
&= \frac{1}{\sqrt{42}}\det\left(\begin{bmatrix} \mathbf{i} & \mathbf{j} & \mathbf{k} \\ 2 & \sqrt{2} & 1 \\ 0 & \sqrt{2} & -2 \end{bmatrix}\right) \\
&= \frac{1}{\sqrt{42}}\langle -3\sqrt{2}, 4, 2\sqrt{2}\rangle \\
&= \frac{1}{\sqrt{21}}\langle -3, 2\sqrt{2}, 2\rangle\,.
\end{aligned}$$

55. Since $\mathbf{r}'(t) = e^t\,\mathbf{i} - e^{-t}\,\mathbf{j} + (e^t - e^{-t})\,\mathbf{k}$, the unit tangent vector at $\mathbf{r}(t)$ is

$$\mathbf{T}(t) = \frac{1}{\sqrt{e^{2t} + e^{-2t} + (e^t - e^{-t})^2}}\,(e^t\,\mathbf{i} - e^{-t}\,\mathbf{j} + (e^t - e^{-t})\,\mathbf{k})\,.$$

Simplify the radical then differentiate, using the product rule, to obtain

$$\begin{aligned}
\mathbf{T}'(t) &= \frac{1}{\sqrt{2e^{2t} + 2e^{-2t} - 2}}\,(e^t\,\mathbf{i} + e^{-t}\,\mathbf{j} + (e^t + e^{-t})\,\mathbf{k}) \\
&\quad - \frac{1}{2}\cdot\frac{4e^{2t} - 4e^{-2t}}{(2e^{2t} + 2e^{-2t} - 2)^{3/2}}\,(e^t\,\mathbf{i} - e^{-t}\,\mathbf{j} + (e^t - e^{-t})\,\mathbf{k})\,.
\end{aligned}$$

Substitute $t = \ln(2)$ to get

$$\mathbf{T}(\ln(2)) = (\sqrt{2/13})\,\langle 2, -1/2, 3/2\rangle = (1/\sqrt{26})\,\langle 4, -1, 3\rangle$$

and

$$\mathbf{T}'(\ln(2)) = (2/13)^{3/2}\,\langle -2, 7, 5\rangle\,.$$

Therefore,

$$\mathbf{N}(\ln(2)) = (1/\sqrt{78})\,\langle -2, 7, 5\rangle$$

and

$$\mathbf{B}(\ln(2)) = \mathbf{T}(\ln(2)) \times \mathbf{N}(\ln(2))$$

$$= \frac{1}{\sqrt{26 \cdot 78}} \det \left(\begin{bmatrix} \mathbf{i} & \mathbf{j} & \mathbf{k} \\ 4 & -1 & 3 \\ -2 & 7 & 5 \end{bmatrix} \right)$$

$$= \frac{1}{26\sqrt{3}} \langle -26, -26, 26 \rangle$$

$$= \frac{1}{\sqrt{3}} \langle -1, -1, 1 \rangle .$$

57. Let the curve consist of straight line segments that zig-zag back and forth from the left side of the square $[0,1] \times [0,1]$ to the right side, never crossing itself. If there are N segments (N a positive integer), then the length of the curve will be greater than N.

A similar effect can be accomplished smoothly using

$$x(t) = \sin^2 \left(\frac{N \pi t}{2} \right) , \ y(t) = t ,$$

on the interval $0 \le t \le 1$. If N is a positive integer, then the length of the curve will be greater than N. The figure below shows the trajectory for $N = 5$.

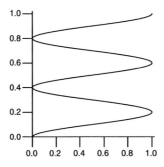

Calculator/Computer Exercises

59-61. Both of these Exercises call for Simpson Rule integral approximations. The following entry defines a procedure named $Simp$ which takes $f, a, b,$ and n as input and outputs the $2n$ Simpson approximation for $\int_a^b f(x)\,dx$:

$$\int_a^b f(x)\,dx \approx \frac{h}{3}(f(x_0) + 4f(x_1) + 2f(x_2) + \cdots + 4f(x_{2n-1}) + f(x_{2n})) .$$

```
> Simp := proc(f,a,b,n)
  local m,h;
    m := evalf(2*n);
    h := evalf((b-a)/m);
    h/3*(f(a)+f(b)+4*add(f(a+(2*k-1)*h),k=1..n)
                    +2*add(f(a+2*k*h),k=1..n-1));
    evalf(%);
  end proc:
```

The procedure is tested on $\int_0^3 e^x \cos(2x)\,dx$. The exact 10 digit value can be compared to the Simpson Rule approximations for $n = 10, 20, 30$.

```
> f := x -> exp(x)*cos(2*x):
  Int(f(x),x=0..3) = evalf(int(f(x),x=0..3));
  'Simp(f,0,3,10*n)'$n=1..3;
```

$$\int_0^3 e^x \cos(2x)\,dx = 1.412223026$$

$$1.411621745, 1.412185632, 1.412215647$$

59. Since $\mathbf{r}'(t) = 2t\,\mathbf{i} - 4t^3\,\mathbf{j} - t^{-2}\,\mathbf{k}$, $\|\mathbf{r}'(t)\| = \sqrt{4t^2 + 16t^6 + t^{-4}}$. Therefore, the arc length can calculated with the following integral.

$$\int_1^4 \|\mathbf{r}'(t)\|\,dt = \int_1^4 \sqrt{4t^2 + 16t^6 + t^{-4}}\,dt$$

```
> f := t -> sqrt(4*t^2 + 16*t^6 + t^(-4));
  'Simp(f,1,4,10*n)'$n=1..3;
```

$$f := t \to \sqrt{4t^2 + 16t^6 + t^{-4}}$$

$$255.7052549, 255.7051694, 255.7051644$$

61. Since $\mathbf{r}'(t) = -\sin(t)\,\mathbf{i} + 2\cos(2t)\,\mathbf{j} + 1/(2t^{1/2})\,\mathbf{k}$,

$$\|\mathbf{r}'(t)\| = \sqrt{\sin^2(t) + 4\cos^2(2t) + 1/(4t)}.$$

Therefore, the arc length can calculated with the following integral.

$$\int_1^4 \|\mathbf{r}'(t)\|\,dt = \int_1^4 \sqrt{\sin^2(t) + 4\cos^2(2t) + 1/(4t)}\,dt$$

```
> f := t -> sqrt(sin(t)^2 + 4*cos(2*t)^2 + 1/(4*t));
  'Simp(f,1,4,10*n)'$n=1..3;
```

$$f := t \to \sqrt{\sin(t)^2 + 4\cos(2t)^2 + \frac{1}{4t}}$$

$$4.693824165, 4.694911296, 4.694923752$$

10.4 Curvature

Problems for Practice

1. Since $\mathbf{r}'(s) = \cos(s)\,\mathbf{i} + \sin(s)\,\mathbf{j}$, $\|\mathbf{r}'(s)\| = \sqrt{\cos^2(s) + \sin^2(s)} = 1$. This verifies that $\mathbf{r}$ is an arc length parametrization.

   (a) $\mathbf{T}(s) = \mathbf{r}'(s) = \cos(s)\,\mathbf{i} + \sin(s)\,\mathbf{j}$.

   (b) Since $\mathbf{T}'(s) = -\sin(s)\,\mathbf{i} + \cos(s)\,\mathbf{j}$, and this is a unit vector, $\mathbf{N}(s) = -\sin(s)\,\mathbf{i} + \cos(s)\,\mathbf{j}$.

   (c) $\kappa(s) = \|\mathbf{T}'(s)\| = 1$.

3. Since $\mathbf{r}'(s) = -(1/\sqrt{2})\,\sin(s/\sqrt{2})\,\mathbf{i} - (1/\sqrt{2})\,\cos(s/\sqrt{2})\,\mathbf{j} + (1/\sqrt{2})\,\mathbf{k}$,

$$\|\mathbf{r}'(s)\| = (1/\sqrt{2})\,\sqrt{\sin^2(s\sqrt{2}) + \cos^2(s\sqrt{2}) + 1} = 1.$$

   This verifies that $\mathbf{r}$ is an arc length parametrization.

   (a) $\mathbf{T}(s) = \mathbf{r}'(s) = -(1/\sqrt{2})\,\sin(s/\sqrt{2})\,\mathbf{i} - (1/\sqrt{2})\,\cos(s/\sqrt{2})\,\mathbf{j} + (1/\sqrt{2})\,\mathbf{k}$.

   (b) Since $\mathbf{T}'(s) = -(1/2)\,\cos(s/\sqrt{2})\,\mathbf{i} + (1/2)\,\sin(s/\sqrt{2})\,\mathbf{j}$, and $\|\mathbf{T}'(s)\| = 1/2$, $\mathbf{N}(s) = 2\mathbf{T}'(s) = -\cos(s/\sqrt{2})\,\mathbf{i} + \sin(s/\sqrt{2})\,\mathbf{j}$.

   (c) $\kappa(s) = \|\mathbf{T}'(s)\| = 1/2$.

5. Since $\mathbf{r}'(s) = (3\cos(s)/5)\,\mathbf{i} - \sin(s)\,\mathbf{j} + (4\cos(s)/5)\,\mathbf{k}$,

$$\|\mathbf{r}'(s)\| = \sqrt{9\cos^2(s)/25 + \sin^2(s) + 16\cos^2(s)/25} = 1.$$

   This verifies that $\mathbf{r}$ is an arc length parametrization.

   (a) $\mathbf{T}(s) = \mathbf{r}'(s) = (3\cos(s)/5)\,\mathbf{i} - \sin(s)\,\mathbf{j} + (4\cos(s)/5)\,\mathbf{k}$.

   (b) Since $\mathbf{T}'(s) = (-3\sin(s)/5)\,\mathbf{i} - \cos(s)\,\mathbf{j} - (4\sin(s)/5)\,\mathbf{k}$, and $\|\mathbf{T}'(s)\| = 1$ (the length calculation is just like the one for $\|\mathbf{r}'(s)\|$), $\mathbf{N}(s) = \mathbf{T}'(s) = (-3\sin(s)/5)\,\mathbf{i} - \cos(s)\,\mathbf{j} - (4\sin(s)/5)\,\mathbf{k}$.

   (c) $\kappa(s) = \|\mathbf{T}'(s)\| = 1$.

7. Look back at Exercise 3 to see that $\rho(s) = 1/\kappa(s) = 2$. Therefore, the center of the osculating circle has the position vector

$$\mathbf{r}(s) + 2\mathbf{N}(s) = -\cos(s/\sqrt{2})\,\mathbf{i} + \sin(s/\sqrt{2})\,\mathbf{j} + (s/\sqrt{2})\,\mathbf{k}.$$

9. Look back at Exercise 5 to see that, $\rho(s) = 1/\kappa(s) = 1$. Therefore, the center of the osculating circle has the position vector

$$\mathbf{r}(s) + \mathbf{N}(s) = \langle 0, 0, 0 \rangle.$$

11. Use the first formula in Theorem 1, as it applies to a planar curve (see Theorem 2). Since $\mathbf{r}'(t) = \langle e^t, -2e^{2t} \rangle$ and $\mathbf{r}''(t) = \langle e^t, -4e^{2t} \rangle$,

$$\kappa_{\mathbf{r}}(t) = \frac{|x'(t)y''(t) - y'(t)x''(t)|}{(x'(t)^2 + y'(t)^2)^{3/2}} = \frac{|e^t \cdot (-4e^{2t}) - (-2e^{2t}) \cdot e^t|}{(e^{2t} + 4e^{4t})^{3/2}}$$

$$= \frac{2e^{3t}}{(e^{2t} + 4e^{4t})^{3/2}}.$$

Factor the denominator and simplify to obtain $\kappa_{\mathbf{r}}(t) = 1/(1 + 4e^{2t})^{3/2}$.

13. Use the first formula in Theorem 1, as it applies to a planar curve (see Theorem 2). Since $\mathbf{r}'(t) = \langle 1/t, 1/t^2 \rangle$ and $\mathbf{r}''(t) = \langle -1/t^2, -2/t^3 \rangle$,

$$\kappa_{\mathbf{r}}(t) = \frac{|x'(t)y''(t) - y'(t)x''(t)|}{(x'(t)^2 + y'(t)^2)^{3/2}} = \frac{|(1/t) \cdot (-2/t^3) - (1/t^2) \cdot (-1/t^2)|}{((1/t)^2 + (1/t^2)^2)^{3/2}}$$

$$= \frac{1/t^4}{(1/t^2 + 1/t^4)^{3/2}}.$$

Write the denominator in the form $((t^2 + 1)t^{-4})^{3/2}$ and simplify to obtain $\kappa_{\mathbf{r}}(t) = t^2/(t^2 + 1)^{3/2}$.

15. Use the first formula in Theorem 1, as it applies to a planar curve (see Theorem 2). Since $\mathbf{r}'(t) = \langle \sec^2(t), \sec(t)\tan(t) \rangle$ and

$$\mathbf{r}''(t) = \langle 2\sec^2(t)\tan(t), \sec^3(t) + \sec(t)\tan^2(t) \rangle,$$

$$\kappa_{\mathbf{r}}(t) = \frac{|x'(t)y''(t) - y'(t)x''(t)|}{(x'(t)^2 + y'(t)^2)^{3/2}}$$

$$= \frac{|\sec^2(t)(\sec^3(t) + \sec(t)\tan^2(t)) - \sec(t)\tan(t) \cdot 2\sec^2(t)\tan(t)|}{(\sec^4(t) + \sec^2(t)\tan^2(t))^{3/2}}.$$

Factor $|\sec^3(t)|$ out of the denominator and move it into the numerator to obtain

$$\kappa_{\mathbf{r}}(t) = \frac{1}{(\sec^2(t) + \tan^2(t))^{3/2}}.$$

The answer in the text can be obtained by converting this expression to sines and cosines.

17. Use formula (10.4.13). Since $f'(x) = (1/2)x^{-1/2}$ and $f''(x) = (-1/4)x^{-3/2}$,

$$\kappa(x) = \frac{|f''(x)|}{(1 + f'(x)^2)^{3/2}} = \frac{|(-1/4)x^{-3/2}|}{(1 + (1/4)x^{-1})^{3/2}}.$$

Since x is positive, $x^{-3/2}$ can be moved into the denominator as $x^{3/2}$ yielding

$$\kappa(x) = \frac{1/4}{(x + (1/4))^{3/2}}.$$

Multiply the numerator and the denominator by 8 to obtain the formula in the text.

19. Use formula (10.4.13). Since $f'(x) = \tan(x)$ and $f''(x) = \sec^2(x)$,

$$\kappa(x) = \frac{|f''(x)|}{(1 + f'(x)^2)^{3/2}} = \frac{|\sec^2(x)|}{(1 + \tan^2(x))^{3/2}}.$$

Use the identity $1 + \tan^2(x) = \sec^2(x)$ to obtain $\kappa(x) = |\cos(x)|$.

21. Use formula (10.4.13). Since $f'(x) = x^2/2 - 1/(2x^2)$ and $f''(x) = x + 1/x^3$,

$$\kappa(x) = \frac{|f''(x)|}{(1 + f'(x)^2)^{3/2}} = \frac{\left|x + \frac{1}{x^3}\right|}{\left(1 + \left(\frac{x^2}{2} - \frac{1}{2x^2}\right)^2\right)^{3/2}}$$

$$= \frac{x^4 + 1}{|x^3|\left(\frac{x^4}{4} + \frac{1}{2} + \frac{1}{4x^4}\right)^{3/2}} = \frac{x^4 + 1}{|x^3|\left(\frac{x^2}{2} + \frac{1}{2x^2}\right)^3}$$

$$= \frac{x^4 + 1}{|x^3| \cdot \frac{(x^4 + 1)^3}{8x^6}} = \frac{8|x^3|}{(x^4 + 1)^2}.$$

23. Since $\mathbf{r}'(t) = \langle 2t, 2t, 2\rangle$ and $\mathbf{r}''(t) = \langle 2, 2, 0\rangle$, the curvature is

$$\kappa_{\mathbf{r}}(t) = \frac{\|\mathbf{r}'(t) \times \mathbf{r}''(t)\|}{\|\mathbf{r}'(t)\|^3} = \frac{1}{(8t^2 + 4)^{3/2}} \left\| \det\left(\begin{bmatrix} \mathbf{i} & \mathbf{j} & \mathbf{k} \\ 2t & 2t & 2 \\ 2 & 2 & 0 \end{bmatrix} \right) \right\|$$

$$= \frac{\|\langle -4, 4, 0\rangle\|}{(8t^2 + 4)^{3/2}} = \frac{4\sqrt{2}}{(8t^2 + 4)^{3/2}} = \frac{\sqrt{2}}{2(2t^2 + 1)^{3/2}}.$$

The radius of curvature is $\rho(t) = \sqrt{2}\,(2t^2 + 1)^{3/2}$.

The unit tangent vector is $\mathbf{T}(t) = \frac{1}{\sqrt{2t^2 + 1}}\langle t, t, 1\rangle$. Differentiation yields

$$\mathbf{T}'(t) = \frac{1}{(2t^2 + 1)^{3/2}}\langle 1, 1, -2t\rangle.$$

Consequently, the principal unit normal is $\mathbf{N}(t) = \frac{1}{\sqrt{4t^2 + 2}}\langle 1, 1, -2t\rangle$, and the center of curvature has the position vector

$$\mathbf{r}(t) + \rho(t)\mathbf{N}(t) = \langle 3t^2 + 1, 3t^2 + 1, 1 - 4t^3\rangle.$$

25. Since $\mathbf{r}'(t) = \langle e^t, 1, e^t\rangle$ and $\mathbf{r}''(t) = \langle e^t, 0, e^t\rangle$, the curvature is

$$\kappa_{\mathbf{r}}(t) = \frac{\|\mathbf{r}'(t) \times \mathbf{r}''(t)\|}{\|\mathbf{r}'(t)\|^3} = \frac{1}{(2e^{2t} + 1)^{3/2}} \left\| \det\left(\begin{bmatrix} \mathbf{i} & \mathbf{j} & \mathbf{k} \\ e^t & 1 & e^t \\ e^t & 0 & e^t \end{bmatrix} \right) \right\|$$

$$= \frac{\|\langle e^t, 0, -e^t\rangle\|}{(2e^{2t} + 1)^{3/2}} = \frac{\sqrt{2}\,e^t}{(2e^{2t} + 1)^{3/2}}.$$

The radius of curvature is $\rho(t) = \frac{\sqrt{2}}{2}e^{-t}(2e^{2t}+1)^{3/2}$.

The unit tangent vector is $\mathbf{T}(t) = \frac{1}{\sqrt{2e^{2t}+1}}\langle e^t, 1, e^t\rangle$. Differentiation yields

$$\mathbf{T}'(t) = \frac{1}{(2e^{2t}+1)^{3/2}}\langle e^t, -2e^{2t}, e^t\rangle = \frac{e^t}{(2e^{2t}+1)^{3/2}}\langle 1, -2e^t, 1\rangle.$$

Consequently, the principal unit normal is $\mathbf{N}(t) = \frac{1}{\sqrt{4e^{2t}+2}}\langle 1, -2e^t, 1\rangle$, and the center of curvature has the position vector

$$\mathbf{r}(t) + \rho(t)\mathbf{N}(t) = \left\langle \frac{1}{2}(4e^t + e^{-t}), t-1-2e^{2t}, \frac{1}{2}(4e^t+e^{-t})\right\rangle.$$

27. Since $\mathbf{r}'(t) = \langle 4t^3, 1, 2\rangle$ and $\mathbf{r}''(t) = \langle 12t^2, 0, 0\rangle$, the curvature is

$$\kappa_{\mathbf{r}}(t) = \frac{\|\mathbf{r}'(t) \times \mathbf{r}''(t)\|}{\|\mathbf{r}'(t)\|^3} = \frac{1}{(16t^6+5)^{3/2}}\left\| \det\left(\begin{bmatrix} \mathbf{i} & \mathbf{j} & \mathbf{k} \\ 4t^3 & 1 & 2 \\ 12t^2 & 0 & 0 \end{bmatrix}\right)\right\|$$

$$= \frac{\|\langle 0, 24t^2, -12t^2\rangle\|}{(16t^6+5)^{3/2}} = \frac{12\sqrt{5}\,t^2}{(16t^6+5)^{3/2}}.$$

The radius of curvature is $\rho(t) = \frac{(16t^6+5)^{3/2}}{12\sqrt{5}\,t^2} = \frac{\sqrt{5}\,(16t^6+5)^{3/2}}{60t^2}$.

The unit tangent vector is $\mathbf{T}(t) = \frac{1}{\sqrt{16t^6+5}}\langle 4t^3, 1, 2\rangle$. Differentiation yields

$$\mathbf{T}'(t) = \frac{12t^2}{(16t^6+5)^{3/2}}\langle 5, -4t^3, -8t^3\rangle.$$

Consequently, the principal unit normal is $\mathbf{N}(t) = \frac{1}{\sqrt{25+80t^6}}\langle 5, -4t^3, -8t^3\rangle$, and the center of curvature has the position vector

$$\mathbf{r}(t) + \rho(t)\mathbf{N}(t) = \left\langle \frac{28t^6+5}{12t^2}, \frac{2t(5-8t^6)}{15}, \frac{4t(5-8t^6)}{15}\right\rangle.$$

29. The osculating circle at $\mathbf{r}(t)$ has normal vector $\mathbf{r}'(t) \times \mathbf{r}''(t)$. Since $\mathbf{r}'(t) = \langle e^t, 1, 2e^{2t}\rangle$ and $\mathbf{r}''(t) = \langle e^t, 0, 4e^{2t}\rangle$, and $t=0$, the plane that contains the point $(1, 0, 1)$ has a normal vector

$$\mathbf{r}'(0) \times \mathbf{r}''(0) = \det\left(\begin{bmatrix} \mathbf{i} & \mathbf{j} & \mathbf{k} \\ 1 & 1 & 2 \\ 1 & 0 & 4 \end{bmatrix}\right) = \langle 4, -2, -1\rangle.$$

Its equation is $4x - 2y - z = 3$.

31. The osculating circle at $\mathbf{r}(t)$ has normal vector $\mathbf{r}'(t) \times \mathbf{r}''(t)$. Since $\mathbf{r}'(t) = \langle 2t+1, 2t-1, 2t\rangle$ and $\mathbf{r}''(t) = \langle 2, 2, 2\rangle$, and $t=-2$, the plane that contains

the point $(2, 6, 4)$ has a normal vector

$$\mathbf{r}'(-2) \times \mathbf{r}''(-2) = \det\left(\begin{bmatrix} \mathbf{i} & \mathbf{j} & \mathbf{k} \\ -3 & -5 & -4 \\ 2 & 2 & 2 \end{bmatrix}\right) = \langle -2, -2, 4 \rangle.$$

Its equation is $-2x - 2y + 4z = 0$.

Further Theory and Practice

33. Using $\mathbf{r}'(t) = \langle -2t, 6t^2 \rangle$ and $\mathbf{r}''(t) = \langle -2, 12t \rangle$ the unit tangent vector simplifies to the formula appearing below. The principal unit normal on the right is obtained by differentiating $\mathbf{T}(t)$ and then normalizing (i.e. converting to a unit vector).

$$\mathbf{T}(t) = \left\langle \frac{-t}{\sqrt{t^2 + 9t^4}}, \frac{3t^2}{\sqrt{t^2 + 9t^4}} \right\rangle \quad \text{and} \quad \mathbf{N}(t) = \left\langle \frac{3t^2}{\sqrt{t^2 + 9t^4}}, \frac{t}{\sqrt{t^2 + 9t^4}} \right\rangle.$$

Using (10.4.12), the curvature is $(3/2)t^2(t^2 + 9t^4)^{-3/2}$ and the radius of curvature is $\rho(t) = (2/3)t^{-2}(t^2 + 9t^4)^{3/2}$. Consequently, the trajectory of the evolute simplifies to the formula shown below.

$$\mathbf{r}(t) + \rho(t)\mathbf{N}(t) = \langle 18t^4 + t^2 + 2, 8t^3 + 2t/3 \rangle$$

35. Using $\mathbf{r}'(t) = \langle 1, e^t \rangle$ and $\mathbf{r}''(t) = \langle 0, e^t \rangle$ the unit tangent vector simplifies to the formula appearing below. The principal unit normal on the right is obtained by differentiating $\mathbf{T}(t)$ and then normalizing (i.e. converting to a unit vector).

$$\mathbf{T}(t) = \left\langle \frac{1}{\sqrt{1 + e^{2t}}}, \frac{e^t}{\sqrt{1 + e^{2t}}} \right\rangle \quad \text{and} \quad \mathbf{N}(t) = \left\langle \frac{-e^t}{\sqrt{1 + e^{2t}}}, \frac{1}{\sqrt{1 + e^{2t}}} \right\rangle.$$

Using (10.4.12), the curvature is $e^t(1 + e^{2t})^{-3/2}$ and the radius of curvature is $\rho(t) = e^{-t}(1 + e^{2t})^{3/2}$. Consequently, the trajectory of the evolute simplifies to the formula shown below.

$$\mathbf{r}(t) + \rho(t)\mathbf{N}(t) = \langle t - 1 - e^{2t}, 2e^t + e^{-t} \rangle$$

37. Using $\mathbf{r}'(t) = \langle 1 - \cos(t), \sin(t) \rangle$ and $\mathbf{r}''(t) = \langle \sin(t), \cos(t) \rangle$ the unit tangent vector simplifies to the formula appearing below. The principal unit normal on the right is obtained by differentiating $\mathbf{T}(t)$ and then normalizing (i.e. converting to a unit vector).

$$\mathbf{T}(t) = \left\langle \frac{1 - \cos(t)}{\sqrt{2 - 2\cos(t)}}, \frac{\sin(t)}{\sqrt{2 - 2\cos(t)}} \right\rangle$$

and

$$\mathbf{N}(t) = \left\langle \frac{\sin(t)}{\sqrt{2 - 2\cos(t)}}, \frac{\cos(t) - 1}{\sqrt{2 - 2\cos(t)}} \right\rangle.$$

Using (10.4.12), the curvature is $(1-\cos(t))(2-2\cos(t))^{-3/2}$ and the radius of curvature is $\rho(t) = (1 - \cos(t))^{-1}(2 - 2\cos(t))^{3/2}$. The trajectory of the evolute simplifies to the formula shown below.

$$\mathbf{r}(t) + \rho(t)\mathbf{N}(t) = \langle t + \sin(t), \cos(t) - 1 \rangle$$

39. The first and second derivatives are $\mathbf{r}'(t) = \langle 1, \tan(t) \rangle$ and $\mathbf{r}''(t) = \langle 0, \sec^2(t) \rangle$. Assuming $\cos(t) > 0$, the unit tangent vector is $\mathbf{T}(t) = \langle \cos(t), \sin(t) \rangle$ and the principal unit normal is $\mathbf{N}(t) = \langle -\sin(t), \cos(t) \rangle$. Using (10.4.12) the curvature simplifies to $\kappa(t) = \cos(t)$ and the radius of curvature is $\rho(t) = 1/\cos(t)$. Therefore, when $-\frac{\pi}{2} < t < \frac{\pi}{2}$, the trajectory of the evolute is

$$\mathbf{r}(t) + \rho(t)\mathbf{N}(t) = \langle t - \tan(t), \ln(\sec(t)) + 1 \rangle .$$

41. Using (10.4.13), the curvature is $\kappa(x) = 2/(1 + 4x^2)^{3/2}$. This is clearly maximum at $x = 0$, $f(0) = 0$.

43. Observe that the graph of f is symmetric with respect to the y axis. Using (10.4.13), the curvature simplifies to $\kappa(x) = 2|x|/(x^2 + 4)^{3/2}$. Since $\kappa'(x) = 4(2 - x^2)/(x^2 + 4)^{5/2}$ when $x > 0$, the maximum curvature is at $x = \pm\sqrt{2}$; $f(\pm\sqrt{2}) = 2\ln(2)$.

45. If $x/a = \cos(t)$ and $y/b = \sin(t)$, then $x^2/a^2 + y^2/b^2 = 1$. Thus the ellipse can be parametrized using $\mathbf{r}(t) = \langle a\cos(t), b\sin(t) \rangle$. Use (10.4.12) to obtain the curvature as a function of t,

$$\kappa_{\mathbf{r}}(t) = \frac{|ab|}{(a^2 \sin^2(t) + b^2 \cos^2(t))^{3/2}} .$$

Since $\mathbf{r}(\pi/4) = (a/\sqrt{2}, b/\sqrt{2})$, the curvature at this point is $\kappa_{\mathbf{r}}(\pi/4) = |ab|/(a^2/2 + b^2/2)^{3/2} = 2\sqrt{2}\,|ab|/(a^2 + b^2)^{3/2}$.

47. If $x^{1/3} = \cos(t)$ and $y^{1/3} = \sin(t)$, then $x^{2/3} + y^{2/3} = 1$. Thus the curve can be parametrized using $\mathbf{r}(t) = \langle \cos^3(t), \sin^3(t) \rangle$. Use (10.4.12) to obtain the curvature as a function of t. The formula simplifies to

$$\kappa_{\mathbf{r}}(t) = \frac{1/3}{|\sin(t)\cos(t)|} .$$

Since $\mathbf{r}(\pi/3) = (1/8, 3\sqrt{3}/8)$, the curvature at this point is $\kappa_{\mathbf{r}}(\pi/3) = 4\sqrt{3}/9$.

49. Since the curvature is 0 at every point, $\mathbf{r}'(t) \times \mathbf{r}''(t) = \mathbf{0}$, and there is a real-valued function λ such that $\mathbf{r}''(t) = \lambda(t)\mathbf{r}'(t)$. Therefore, $x''(t) = \lambda(t)x'(t)$ and

$$\frac{x''(t)}{x'(t)} = \lambda(t) .$$

Integrate this equation over the interval $[0, t]$ to obtain

$$\ln \left(\frac{x'(t)}{x'(0)} \right) = \Lambda(t)$$

where $\Lambda(t) = \int_0^t \lambda(t) \, dt$. (We are assuming that $x'(t) > 0$.) This implies that $x'(t) = x'(0)e^{\Lambda(t)}$.

Similarly, $y'(t) = y'(0)e^{\Lambda(t)}$ and $z'(t) = z'(0)e^{\Lambda(t)}$. In other words,

$$\mathbf{r}'(t) = e^{\Lambda(t)}\mathbf{r}'(0).$$

Integrate once more, over the interval $[0, t]$, to obtain

$$\mathbf{r}(t) = \mathbf{r}(0) + \phi(t)\mathbf{r}'(0),$$

where $\phi(t) = \int_0^t e^{\Lambda(t)} \, dt$. This is a straight line trajectory.

51. Let $v(t) = \|\mathbf{r}'(t)\| (= ds/dt)$, so $\mathbf{r}'(t) = v(t)\mathbf{T}(t)$. Using the fact that $\mathbf{T}'(t) = \frac{d}{ds}(\mathbf{T}(s))\frac{ds}{dt} = v(t)\kappa(t)\mathbf{N}(t)$,

$$\begin{aligned}
\mathbf{r}''(t) &= v'(t)\mathbf{T}(t) + v(t)\mathbf{T}'(t) \\
&= v'(t)\mathbf{T}(t) + v(t)^2\kappa(t)\mathbf{N}(t),
\end{aligned}$$

so

$$\begin{aligned}
\mathbf{r}'(t) \times \mathbf{r}''(t) &= v(t)\mathbf{T}(t) \times v(t)^2\kappa(t)\mathbf{N}(t) \\
&= v(t)^3\kappa(t)\mathbf{B}(t).
\end{aligned}$$

In addition,

$$\begin{aligned}
\mathbf{r}'''(t) = v''(t)\mathbf{T}(t) &+ v'(t)v(t)\kappa(t)\mathbf{N}(t) \\
&+ \frac{d}{dt}(v^2(t)\kappa(t))\mathbf{N}(t) + v(t)^2\kappa(t)\mathbf{N}'(t).
\end{aligned} \tag{1}$$

Consequently, because $\mathbf{B}(t)$ is perpendicular to both $\mathbf{T}(t)$ and $\mathbf{N}(t)$,

$$(\mathbf{r}'(t) \times \mathbf{r}''(t)) \cdot \mathbf{r}'''(t) = v(t)^5\kappa(t)^2\mathbf{B}(t) \cdot \mathbf{N}'(t). \tag{2}$$

Using the second Frenet Formula,

$$\begin{aligned}
\mathbf{N}'(t) &= \frac{d}{ds}(\mathbf{N}(s))\frac{ds}{dt} \\
&= v(t)(-\kappa(t)\mathbf{T}(t) + \tau(t)\mathbf{B}(t)).
\end{aligned}$$

Substitute this into (2) to obtain

$$(\mathbf{r}'(t) \times \mathbf{r}''(t)) \cdot \mathbf{r}'''(t) = v(t)^6\kappa(t)^2\tau(t).$$

Therefore, using the curvature formula (10.4.2),

$$\tau(t) = \frac{(\mathbf{r}'(t) \times \mathbf{r}''(t)) \cdot \mathbf{r}'''(t)}{v(t)^6 \kappa(t)^2}$$

$$= \frac{(\mathbf{r}'(t) \times \mathbf{r}''(t)) \cdot \mathbf{r}'''(t)}{\|\mathbf{r}'(t) \times \mathbf{r}''(t)\|^2}.$$

53. Suppose that the torsion is 0. Then $\mathbf{B}'(s) = \mathbf{0}$ so $\frac{d}{ds}(\mathbf{T}(s) \times \mathbf{N}(s)) = \mathbf{0}$ and $\mathbf{T}(s) \times \mathbf{N}(s) = \mathbf{c}$, a constant vector. Therefore, if the trajectory passes through a point P_0, then the plane containing P_0 having normal vector $\mathbf{c}$ contains the vectors $\mathbf{T}(s)$ for all s. It follows that the motion of the particle must be entirely in this plane.

If the curve is planar, then the vectors $\mathbf{r}'(t)$, $\mathbf{r}''(t)$, and $\mathbf{r}'''(t)$ all lie in the same plane, the osculating plane. Therefore, $(\mathbf{r}'(t) \times \mathbf{r}''(t)) \cdot \mathbf{r}'''(t) = 0$ and, in view of the torsion formula in Exercise 51, $\tau(s) = 0$.

55. According to Theorem 3a in Section 9.4,

$$\|\mathbf{r}'(t) \times \mathbf{r}''(t)\| = \sqrt{(\|\mathbf{r}'(t)\| \cdot \|\mathbf{r}''(t)\|)^2 - (\mathbf{r}'(t) \cdot \mathbf{r}''(t))^2}.$$

Substitute into (10.4.2).

Calculator/Computer Exercises

57. The Maple code is mostly self-explanatory. The assumption that t is real permits the simplify procedure to simplify the formulas considerably. Only the formulas for the trajectory and the evolute are displayed.

```
> assume(t::real):
  r := <2-t^2,2*t^3,0>;
  T := Normalize(diff(r,t)):
  N := simplify(Normalize(diff(T,t))):
  kappa := Norm(diff(r,t) &x diff(r,t,t))/Norm(diff(r,t))^3:
  evol := simplify(r + 1/kappa*N);
  plot( [[r[1],r[2],t=-1..1],[evol[1],evol[2],t=-0.6..0.6]],
        color=black, scaling=constrained, thickness=[2,1],
        view=[0..5,-2..2]);
```

$$r := (2 - t^2)\, e_x + 2t^3\, e_y$$

$$evol := (2 + t^2 + 18t^4)\, e_x + \left(8t^3 + \frac{2}{3}t\right) e_y$$

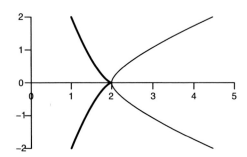

The trajectory is the thicker curve.

59. Only the formulas for the trajectory and the evolute are displayed.

```
> assume(t::real):
  r := <t,exp(t),0>;
  T := Normalize(diff(r,t)):
  N := simplify(Normalize(diff(T,t))):
  kappa := Norm(diff(r,t) &x diff(r,t,t))/Norm(diff(r,t))^3:
  evol := simplify(r + 1/kappa*N);
  plot( [[r[1],r[2],t=-5..2],[evol[1],evol[2],t=-5..2]],
        color=black, scaling=constrained, thickness=[2,1],
        view=[-12..5,-1..12]);
```

$$r := (t)\, e_x + (e^t)\, e_y$$

$$evol := (t - 1 - e^{2t})\, e_x + (2e^t + e^{-t}) e_y$$

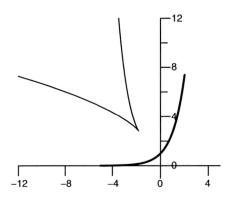

The trajectory is the thicker curve.

61. Only the formulas for the trajectory and the evolute are displayed.

```
> assume(t::real):
  r := <t-sin(t),1-cos(t),0>;
  T := Normalize(diff(r,t)):
```

```
N := simplify(Normalize(diff(T,t))):
kappa := Norm(diff(r,t) &x diff(r,t,t))/Norm(diff(r,t))^3:
evol := simplify(r + 1/kappa*N);
plot( [[r[1],r[2],t=-7..7],[evol[1],evol[2],t=-7..7]],
        color=black, scaling=constrained, thickness=[2,1],
        view=[-7..7,-2.5..2.5]);
```

$$r := (t - \sin(t))\, e_x + (1 - \cos(t))\, e_y$$

$$evol := (t + \sin(t))\, e_x + (\cos(t) - 1)e_y$$

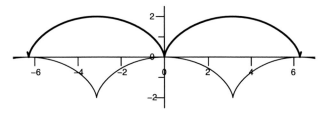

The trajectory is the thicker curve.

63. Only the formulas for the trajectory and the evolute are displayed.

```
> with(plots): assume(t::real):
  r := <cos(t),sin(t),t>;
  T := Normalize(diff(r,t)):
  N := simplify(Normalize(diff(T,t))):
  kappa := Norm(diff(r,t) &x diff(r,t,t))/Norm(diff(r,t))^3:
  evol := simplify(r + 1/kappa*N);
  display( spacecurve( r, t=0..2*Pi,thickness=3),
            spacecurve( evol, t=0..2*Pi, thickness=1),
              color=black, scaling=constrained,
              axes=normal, orientation=[-60,70]);
```

$$r := (\cos(t))\, e_x + (\sin(t))\, e_y + (t)\, e_z$$

$$evol := (-\cos(t))\, e_x + (-\sin(t))\, e_y + (t)\, e_z$$

The trajectory is the thicker curve.

65. Only the formulas for the trajectory and the evolute are displayed.

```
> with(plots): assume(t::real):
  r := <cos(t),sin(t),t^2>;
  T := Normalize(diff(r,t)):
  N := simplify(Normalize(diff(T,t))):
  kappa := Norm(diff(r,t) &x diff(r,t,t))/Norm(diff(r,t))^3:
  evol := simplify(r + 1/kappa*N);
  display( spacecurve( r, t=-Pi/2..Pi/2,thickness=3),
           spacecurve( evol, t=-Pi/2..Pi/2, thickness=1),
             color=black, scaling=constrained,
             axes=normal, orientation=[-60,70]);
```

$$r := (\cos(t))\, e_x + (\sin(t))\, e_y + (t^2)\, e_z$$

$$
\begin{aligned}
evol := {} & \frac{4(-t^2\cos(t) + \cos(t) + 4t^3\sin(t) - 4t^4\cos(t) + t\sin(t))}{4t^2 + 5}\, e_x \\
& + \frac{4(-t^2\sin(t) + \sin(t) - 4t^3\cos(t) - 4t^4\sin(t) - t\cos(t))}{4t^2 + 5}\, e_y \\
& + \left(\frac{4t^4 + 13t^2 + 2}{4t^2 + 5}\right)\, e_z \quad .
\end{aligned}
$$

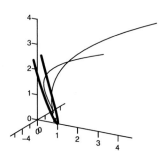

The trajectory is the thicker curve.

10.5   Applications of Vector-Valued Functions to Motion

Problems for Practice

1. Since $\mathbf{r}'(t) = \langle 1, 0, -2t\rangle$ and $\mathbf{r}''(t) = \langle 0, 0, -2\rangle$, $v(t) = \|\mathbf{r}'(t)\| = \sqrt{1 + 4t^2}$

and $a_T = v'(t) = 4t/\sqrt{1+4t^2}$. The normal component of acceleration is

$$a_N = \sqrt{\|\mathbf{r}''(t)\|^2 - a_T^2} = \sqrt{4 - \frac{16t^2}{1+4t^2}} = \frac{2}{\sqrt{1+4t^2}}.$$

The unit tangent and principal unit normal are

$$\mathbf{T}(t) = \frac{1}{\sqrt{1+4t^2}}\langle 1, 0, -2t\rangle \quad \text{and} \quad \mathbf{N}(t) = \frac{1}{\sqrt{1+4t^2}}\langle -2t, 0, -1\rangle.$$

(The normal can be found easily by differentiating $\mathbf{T}(t)$.)

Since $a_N(t) = v(t)^2 \kappa_\mathbf{r}(t)$, the curvature is

$$\kappa_\mathbf{r}(t) = \frac{a_N(t)}{v(t)^2} = \frac{2}{(1+4t^2)^{3/2}}.$$

3. Since $\mathbf{r}'(t) = \langle 1-2t, 1+2t\rangle$ and $\mathbf{r}''(t) = \langle -2, 2\rangle$, $v(t) = \|\mathbf{r}'(t)\| = \sqrt{2+8t^2}$ and $a_T = v'(t) = 8t/\sqrt{2+8t^2}$. The normal component of acceleration is

$$a_N = \sqrt{\|\mathbf{r}''(t)\|^2 - a_T^2} = \sqrt{8 - \frac{64t^2}{2+8t^2}} = \frac{4}{\sqrt{2+8t^2}}.$$

The unit tangent and principal unit normal are

$$\mathbf{T}(t) = \frac{1}{\sqrt{2+8t^2}}\langle 1-2t, 1+2t\rangle \quad \text{and} \quad \mathbf{N}(t) = \frac{1}{\sqrt{2+8t^2}}\langle -1-2t, 1-2t\rangle.$$

(The normal can be found easily by differentiating $\mathbf{T}(t)$.)

Since $a_N(t) = v(t)^2 \kappa_\mathbf{r}(t)$, the curvature is

$$\kappa_\mathbf{r}(t) = \frac{a_N(t)}{v(t)^2} = \frac{4}{(2+8t^2)^{3/2}}.$$

5. Since $\mathbf{r}'(t) = \langle 1-\cos(t), \sin(t)\rangle$ and $\mathbf{r}''(t) = \langle \sin(t), \cos(t)\rangle$, $v(t) = \|\mathbf{r}'(t)\| = \sqrt{2-2\cos(t)}$ and $a_T = v'(t) = \sin(t)/\sqrt{2-2\cos(t)}$. The normal component of acceleration is

$$a_N = \sqrt{\|\mathbf{r}''(t)\|^2 - a_T^2} = \sqrt{1 - \frac{\sin^2(t)}{2-2\cos(t)}} = \frac{1}{2}\sqrt{2-2\cos(t)}.$$

(Note that $\sin^2(t) = (1-\cos(t))(1+\cos(t))$.)

The unit tangent and principal unit normal are

$$\mathbf{T}(t) = \frac{1}{\sqrt{2-2\cos(t)}}\langle 1-\cos(t), \sin(t)\rangle$$

and
$$\mathbf{N}(t) = \frac{1}{\sqrt{2 - 2\cos(t)}} \langle \sin(t), -1 + \cos(t) \rangle.$$

(The normal can be found easily by differentiating $\mathbf{T}(t)$.)

Since $a_N(t) = v(t)^2 \kappa_{\mathbf{r}}(t)$, the curvature is
$$\kappa_{\mathbf{r}}(t) = \frac{a_N(t)}{v(t)^2} = \frac{1/2}{\sqrt{2 - 2\cos(t)}}.$$

7. Since $\mathbf{r}'(t) = \langle (3/2)t^{1/2}, (1/2)t^{-1/2} \rangle = (1/2)t^{-1/2}\langle 3t, 1 \rangle$ and
$$\mathbf{r}''(t) = \langle (3/4)t^{-1/2}, (-1/4)t^{-3/2} \rangle = (1/4)t^{-3/2}\langle 3t, -1 \rangle,$$
$$v(t) = \|\mathbf{r}'(t)\| = (1/2)t^{-1/2}\sqrt{9t^2 + 1} = (1/2)\sqrt{9t + t^{-1}} \text{ and}$$
$$a_T = v'(t) = \frac{1}{4} \cdot \frac{9 - t^{-2}}{\sqrt{9t + t^{-1}}} = \frac{1}{4} \cdot \frac{9t^2 - 1}{t^{3/2}\sqrt{9t^2 + 1}}.$$

The normal component of acceleration is
$$a_N = \sqrt{\|\mathbf{r}''(t)\|^2 - a_T^2} = \frac{1}{4t^{3/2}}\sqrt{9t^2 + 1 - \frac{(9t^2 - 1)^2}{9t^2 + 1}}$$
$$= \frac{3/2}{\sqrt{t}\sqrt{9t^2 + 1}}.$$

(Note that $9t^2 + 1 - (9t^2 - 1)^2/(9t^2 + 1) = 36t^2/(9t^2 + 1)$.)

The unit tangent and principal unit normal are
$$\mathbf{T}(t) = \frac{1}{\sqrt{9t^2 + 1}}\langle 3t, 1 \rangle \quad \text{and} \quad \mathbf{N}(t) = \frac{1}{\sqrt{9t^2 + 1}}\langle 1, -3t \rangle.$$

(The normal can be found easily by differentiating $\mathbf{T}(t)$.)

Since $a_N(t) = v(t)^2 \kappa_{\mathbf{r}}(t)$, the curvature is
$$\kappa_{\mathbf{r}}(t) = \frac{a_N(t)}{v(t)^2} = \frac{6\sqrt{t}}{(9t^2 + 1)^{3/2}}.$$

9. Since $\mathbf{r}'(t) = \langle 1, 2t, 3t^2 \rangle$ and $\mathbf{r}''(t) = \langle 0, 2, 6t \rangle$, $v(t) = \|\mathbf{r}'(t)\| = \sqrt{1 + 4t^2 + 9t^4}$ and
$$a_T = v'(t) = \frac{4t + 18t^3}{\sqrt{1 + 4t^2 + 9t^4}}.$$

The normal component of acceleration is
$$a_N = \sqrt{\|\mathbf{r}''(t)\|^2 - a_T^2} = \sqrt{4 + 36t^2 - \frac{(4t + 18t^3)^2}{1 + 4t^2 + 9t^4}}$$
$$= \frac{2\sqrt{1 + 9t^2 + 9t^4}}{\sqrt{1 + 4t^2 + 9t^4}}.$$

The unit tangent and principal unit normal are

$$\mathbf{T}(t) = \frac{1}{\sqrt{1 + 4t^2 + 9t^4}} \langle 1, 2t, 3t^2 \rangle$$

and

$$\mathbf{N}(t) = \frac{1}{\sqrt{81t^8 + 117t^6 + 54t^4 + 13t^2 + 1}} \langle -t(2 + 9t^2), 1 - 9t^4, 3t(1 + 2t^2) \rangle .$$

(The normal can be found by differentiating $\mathbf{T}(t)$.)

Since $a_N(t) = v(t)^2 \kappa_{\mathbf{r}}(t)$, the curvature is

$$\kappa_{\mathbf{r}}(t) = \frac{a_N(t)}{v(t)^2} = \frac{2\sqrt{1 + 9t^2 + 9t^4}}{(1 + 4t^2 + 9t^4)^{3/2}} .$$

11. Since $\mathbf{r}'(t) = \langle (3/2)\sqrt{1 + t}, -(3/2)\sqrt{1 - t}, 1 \rangle$ and

$$\mathbf{r}''(t) = \langle (3/4)(1 + t)^{-1/2}, (3/4)(1 - t)^{-1/2}, 0 \rangle ,$$

$v(t) = \|\mathbf{r}'(t)\| = \sqrt{22}/2$ and $a_T = v'(t) = 0$.

The normal component of acceleration is

$$a_N = \sqrt{\|\mathbf{r}''(t)\|^2 - a_T^2} = \|\mathbf{r}''(t)\| = \frac{3}{4}\sqrt{\frac{2}{1 - t^2}} .$$

The unit tangent and principal unit normal are

$$\mathbf{T}(t) = \frac{3}{\sqrt{22}} \langle \sqrt{1 + t}, -\sqrt{1 - t}, 2/3 \rangle$$

and

$$\mathbf{N}(t) = \frac{\sqrt{2}}{2} \langle \sqrt{1 - t}, \sqrt{1 + t}, 0 \rangle .$$

(The normal can be found by differentiating $\mathbf{T}(t)$.)

Since $a_N(t) = v(t)^2 \kappa_{\mathbf{r}}(t)$, the curvature is

$$\kappa_{\mathbf{r}}(t) = \frac{a_N(t)}{v(t)^2} = \frac{3}{22}\sqrt{\frac{2}{1 - t^2}} .$$

13. Since $\mathbf{r}'(t) = \langle e^t, -e^{-t}, \sqrt{2} \rangle$ and $\mathbf{r}''(t) = \langle e^t, e^{-t}, 0 \rangle$, $v(t) = \|\mathbf{r}'(t)\| = \sqrt{e^{2t} + e^{-2t} + 2} = e^t + e^{-t}$ and $a_T = v'(t) = e^t - e^{-t}$.

The normal component of acceleration is

$$a_N = \sqrt{\|\mathbf{r}''(t)\|^2 - a_T^2} = \sqrt{e^{2t} + e^{-2t} - (e^t - e^{-t})^2} = \sqrt{2}.$$

The unit tangent and principal unit normal are

$$\mathbf{T}(t) = \frac{1}{e^t + e^{-t}}\langle e^t, -e^{-t}, \sqrt{2}\rangle \quad \text{and} \quad \mathbf{N}(t) = \frac{1}{e^t + e^{-t}}\langle \sqrt{2}, \sqrt{2}, e^{-t} - e^t\rangle.$$

(The normal can be found by differentiating $\mathbf{T}(t)$.)

Since $a_N(t) = v(t)^2 \kappa_{\mathbf{r}}(t)$, the curvature is

$$\kappa_{\mathbf{r}}(t) = \frac{a_N(t)}{v(t)^2} = \frac{\sqrt{2}}{(e^t + e^{-t})^2}.$$

15. Since $\mathbf{r}'(t) = \langle -\sin(t), \cos(t), 2t\rangle$ and $\mathbf{r}''(t) = \langle -\cos(t), -\sin(t), 2\rangle$,

$$v(t) = \|\mathbf{r}'(t)\| = \sqrt{1 + 4t^2} \quad \text{and} \quad a_T = v'(t) = 4t/\sqrt{1 + 4t^2}.$$

The normal component of acceleration is

$$a_N = \sqrt{\|\mathbf{r}''(t)\|^2 - a_T^2} = \sqrt{5 - \frac{16t^2}{1 + 4t^2}} = \frac{\sqrt{5 + 4t^2}}{\sqrt{1 + 4t^2}}.$$

The unit tangent tangent is

$$\mathbf{T}(t) = \frac{1}{\sqrt{1 + 4t^2}}\langle -\sin(t), \cos(t), 2t\rangle$$

and the formula for the principal unit normal simplifies to

$$\frac{-1}{\sqrt{1 + 4t^2}\sqrt{5 + 4t^2}}\langle 4t^2\cos(t) - 4t\sin(t) + \cos(t), 4t^2\sin(t) + 4t\cos(t) + \sin(t), -2\rangle.$$

This can be obtained by normalizing, then simplifying, the derivatiave of $\mathbf{T}(t)$. However, it is easier to solve the equation

$$\mathbf{a}(t) = a_T\mathbf{T}(t) + a_N\mathbf{N}(t)$$

for $\mathbf{N}(t)$. That is, use $\mathbf{N}(t) = a_N^{-1}(\mathbf{a}(t) - a_T\mathbf{T}(t))$. See the Insight following equation (10.5.8).

Since $a_N(t) = v(t)^2 \kappa_{\mathbf{r}}(t)$, the curvature is

$$\kappa_{\mathbf{r}}(t) = \frac{a_N(t)}{v(t)^2} = \frac{\sqrt{5 + 4t^2}}{(1 + 4t^2)^{3/2}}.$$

17. Since $\mathbf{r}'(t) = \langle 1, 2t, 2\rangle$ and $\mathbf{r}''(t) = \mathbf{a}(t) = \langle 0, 2, 0\rangle$,

$$v(t) = \|\mathbf{r}'(t)\| = \sqrt{4t^2 + 5} \quad \text{and} \quad a_T = v'(t) = \frac{4t}{\sqrt{4t^2 + 5}}.$$

The normal component of acceleration is

$$a_N = \sqrt{\|\mathbf{a}(t)\|^2 - a_T^2} = \sqrt{4 - \frac{16t^2}{4t^2 + 5}} = \frac{2\sqrt{5}}{\sqrt{4t^2 + 5}}.$$

19. Since $\mathbf{r}'(t) = \langle e^t, -e^{-t}, 1 \rangle$ and $\mathbf{r}''(t) = \mathbf{a}(t) = \langle e^t, e^{-t}, 0 \rangle$,

$$v(t) = \|\mathbf{r}'(t)\| = \sqrt{e^{2t} + e^{-2t} + 1} \quad \text{and} \quad a_T = v'(t) = \frac{e^{2t} - e^{-2t}}{\sqrt{e^{2t} + e^{-2t} + 1}}.$$

The normal component of acceleration is

$$
\begin{aligned}
a_N = \sqrt{\|\mathbf{a}(t)\|^2 - a_T^2} &= \sqrt{e^{2t} + e^{-2t} - \frac{(e^{2t} - e^{-2t})^2}{e^{2t} + e^{-2t} + 1}} \\
&= \frac{\sqrt{e^{2t} + e^{-2t} + 4}}{\sqrt{e^{2t} + e^{-2t} + 1}}.
\end{aligned}
$$

21. Since $a = 3$ and $b = 2$, the equation of the ellipse is $x^2/2^2 + y^2/3^2 = 1$.

23. Since $c = 4$ and $b = 3$, $a^2 = 9 + 16 = 25$. Therefore, the equation of the ellipse is $x^2/25 + y^2/9 = 1$.

25. The center of the ellipse is at $(1, 2)$, and $c = 12$. Since the eccentricity is $12/13$, $a = 13c/12 = 13$ so $b^2 = a^2 - c^2 = 25$. Therefore, the equation of the ellipse is $(x - 1)^2/5^2 + (y - 2)^2/13^2 = 1$.

27. Since $c = 8$ and the eccentricity is 0.8, $a = c/0.8 = 10$. Therefore, $b^2 = a^2 - c^2 = 36$ and, since the center is at $(1,2)$, the equation of the ellipse is $(x - 1)^2/6^2 + (y - 2)^2/10^2 = 1$.

29. Let the vertex that is nearest to the directrix be $(x_0, 0)$. Since the directrix is at $(6, 0)$ and the nearest focus is $(2, 0)$, $(x_0 - 2)/(6 - x_0) = e = 1/2$ and $x_0 = 10/3$.

    Let the center of the ellipse be $(z_0, 0)$. The fact that $c/a = e$ implies that $(2 - z_0)/(10/3 - z_0) = 1/2$ and $z_0 = 2/3$. Thus the center of the ellipse is $(2/3, 0)$, $c = 4/3$, and $a = 8/3$. It follows that $b^2 = a^2 - c^2 = 16/3$ so the equation of the ellipse is $(x - 2/3)^2/(8/3)^2 + y^2/(4/\sqrt{3})^2 = 1$.

Further Theory and Practice

31. They are perpendicular. This is because constant speed implies that $v'(t) = a_T = 0$.

33. Constant speed implies that $a_T = 0$ so the acceleration is

$$
\begin{aligned}
\mathbf{a}(t) = a_N \mathbf{N}(t) &= v(t)^2 \kappa_{\mathbf{r}}(t) \mathbf{N}(t) \\
&= \frac{v(t)^2}{\rho(t)} \mathbf{N}(t).
\end{aligned}
$$

If the motion is circular, radius r, then $\rho(t) = r$ and the force acting on the body is $\mathbf{F}(t) = m\mathbf{a}(t) = (mv/r^2)\mathbf{N}(t)$. This vector has magnitude mv^2/r.

35. Let T be the time it takes for the satellite to complete one orbit. Then, according to Kepler's Third Law,

$$\frac{T^2}{(1850/2)^3} = \frac{27.322^2}{384400^3} .$$

Solve for T to obtain $T = 0.003225$ days.

37. Let a be the semi-major axis of the satellite's orbit. The orbit is swept out in $T = 15 \cdot 60^2$ seconds. According to Kepler's Third Law,

$$\frac{(15 \cdot 60^2)^2}{a^3} = \frac{4\pi^2}{(6.637 \times 10^{-8}) \cdot (5.976 \times 10^{27})} .$$

Solve for a and double it to obtain a major axis of 6.166×10^9 centimeters.

39. Let $\mathbf{r}_0$ and $\mathbf{v}_0$ be the position vector and the velocity vector of the satellite at the moment that it is 4310 miles from the Earth's center. If the satellite were to continue to move in a straight line in the direction of $\mathbf{v}_0$, then the area swept out by its position vector in t hours is

$$A(t) = \frac{1}{2} \|\mathbf{r}_0 \times t\mathbf{v}_0\| = \frac{t}{2} \|\mathbf{r}_0\| \cdot \|\mathbf{v}_0\| \sin(\pi/3) .$$

Therefore, $A'(t) = 1/2 \cdot 4310 \cdot 900 \cdot \sqrt{3}/2 = 969750 \sqrt{3}$ mi^2/hr.

41. Let a denote Mercury's semi-major axis. According to Kepler's Third Law,

$$\frac{87.97^2}{a^3} = \frac{365.256^2}{149597887^3} .$$

Therefore, $a = 5.791 \times 10^7$ kilometers. Since $e = c/a$, the distance from the center of Mercury's orbit to the sun is $c = a \cdot e = 1.191 \times 10^7$ kilometers. Consequently, the perihelion distance from Mercury to the sun is $a - c = 4.600 \times 10^7$ kilometers.

43. Let $d = 816,041,455$. Observe that $a + c = d$ where a is its semimajor axis and c is its semifocal distance. Since $c = ae$, $a + ae = d$ and $a = d/(1+e)$. Thus

$$a = \frac{d}{1+e} = \frac{816041455}{1.04839} = 778375847 \text{ km.}$$

Using Kepler's Third Law, where time is measured in Earth years, the period T of Jupiter's orbit satisfies the equation

$$\frac{T^2}{a^3} = \frac{1}{149597887^3} .$$

Solve for T to find that Jupiter's period is 11.87 Earth years.

45. Let $\mathbf{r} = \langle x, y \rangle$ be the position vector for the planet. Then $r = \|\mathbf{r}\|$ and $-(1/r)\mathbf{r}$ is a unit vector pointing from the planet to the sun. According the Newton's Law of Gravitation, the force on the planet is

$$\mathbf{F} = \left(\frac{GMm}{r^2} \right)(-1/r)\mathbf{r} = -\frac{GMm}{r^3}\langle x, y \rangle.$$

Since $\mathbf{F} = m\mathbf{a}$, the gravitational acceleration is

$$\mathbf{a} = -\frac{GM}{r^3}\langle x, y \rangle.$$

In the x- and y-directions, $\ddot{x} = -(GM/r^3)x$ and $\ddot{y} = -(GM/r^3)y$.

Using $x = r\cos(\theta)$, the x equation can be written as $\ddot{x} = -(GM/r^3)r\cos(\theta)$ or $\ddot{x} = -(GM/r^2)\cos(\theta)$. Since $r^2\dot{\theta} = C$ this can be expressed in the form

$$\frac{\ddot{x}}{\dot{\theta}} = -\frac{GM}{C}\cos(\theta).$$

Apply the Chain Rule to the left side of this equation to obtain

$$\frac{\ddot{x}}{\dot{\theta}} = \frac{d\dot{x}/dt}{d\theta/dt} = \frac{d\dot{x}}{dt}\frac{dt}{d\theta} = \frac{d\dot{x}}{d\theta}.$$

Therefore,

$$\frac{d\dot{x}}{d\theta} = -\frac{GM}{C}\cos(\theta).$$

A similar calculation can be used to obtain the equation in the y-direction.

47. Substitute and differentiate to obtain

$$-r\sin(\theta)\dot{\theta} + \dot{r}\cos(\theta) = -\frac{GM}{C}\sin(\theta) + a$$

and

$$r\cos(\theta)\dot{\theta} + \dot{r}\sin(\theta) = \frac{GM}{C}\cos(\theta) + b.$$

Multiply the first equation by $-\sin(\theta)$, the second equation by $\cos(\theta)$, and add them to obtain the first equation displayed in Exercise 47.

The second equation follows from the first. Use the fact that $r^2\dot{\theta} = C$, but in the form $r\dot{\theta} = C/r$, to obtain

$$\frac{C}{r} = \frac{GM}{C} - a\sin(\theta) + b\cos(\theta).$$

Now solve for r.

49. The first two equations in this exercise are derived in the first paragraph of the solution to Exercise 45. The third equation follows from the first via the two identities

$$\frac{d}{dt}(\dot{x})^2 = 2\dot{x}\frac{d\dot{x}}{dt} \quad \text{and} \quad \frac{d}{dt}x^2 = 2x\dot{x}.$$

To see how, multiply both sides of the first equation by $2\dot{x}$. Similarly, the fourth equation follows from the second one. Multiply both sides by $2\dot{y}$.

51. Since $\|\mathbf{a}(t)\| = \sqrt{a_T(t)^2 + a_N(t)^2}$, the fact that $\|\mathbf{a}(t_0)\| = a_N(t_0)$ means that $a_T(t_0) = 0$. Therefore, because

$$\frac{d}{dt}(\|\mathbf{a}(t)\|) = \frac{a_T(t)a'_T(t) + a_N(t)a'_N(t)}{\sqrt{a_T(t)^2 + a_N(t)^2}},$$

$$\frac{d}{dt}(\|\mathbf{a}(t)\|)\Big|_{t=t_0} = \frac{a_N(t_0)a'_N(t_0)}{\sqrt{a_N(t_0)^2}} = a'_N(t_0).$$

Thus the tangent lines to the graphs of $t \mapsto \|\mathbf{a}(t)\|$ and $t \mapsto a_N(t)$ have the same slope at t_0.

Calculator/Computer Exercises

53. After loading the VectorCalculus package and declaring t to be a real variable, the trajectory is defined as **r**. The velocity, **v**, and acceleration, **a**, are defined, followed by the speed, s. The tangential and normal components of acceleration are defined as aT and aN respectively. They are then plotted along with the norm of **a**.

```
> with(VectorCalculus): assume(t::real):
> r := <t^3,exp(t),t^2>: v := diff(r,t): a := diff(v,t):
  s := Norm(v):
> aT := diff(s,t):
  aN := simplify(sqrt(Norm(a)^2 - aT^2)):
> plot( [Norm(a),aT,aN], t=-1..1, -1..5, color=black,
        linestyle=[SOLID,DASH,DOT]);
```

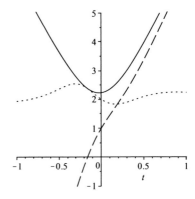

The solid curve is $y = \|\mathbf{a}(t)\|$, the dashed curve is $y = a_T(t)$, and the dotted curve is $y = a_N(t)$.

According to the following calculation, $a_T(t) = a_N(t)$ when $t = 0.1875$.

```
> fsolve( aT = aN, t=0..0.5);
```

$$0.1874852934$$

The next calculation shows that the graph of $t \mapsto \|\mathbf{a}(t)\|$ touches the graph of $t \mapsto a_N(t)$ when $t = -.1618$. At such a point of contact, $a_T(t) = 0$. See the figure above.

```
> fsolve( Norm(a) = aN, t=-0.5..0);
```

$$-.1618138732$$

55. See Exercise 53.

```
> r := <cos(t),sin(2*t),cos(3*t)+sin(5*t)>: v := diff(r,t):
  a := diff(v,t): s := Norm(v):
> aT := diff(s,t):
  aN := simplify(sqrt(Norm(a)^2 - aT^2)):
> plot( [Norm(a),aT,aN], t=1/4..1, -30..35, color=black,
        linestyle=[SOLID,DASH,DOT]);
```

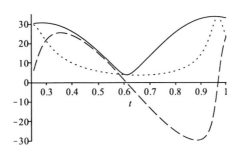

The particle is slowing down when $a_T(t) < 0$. According to the following calculation this is when $0.608 < t < 0.967$.

```
> fsolve( aT=0, t=0.4..0.8), fsolve( aT=0, t=0.8..1.0);
```

$$0.6080645791, 0.9673008157$$

The next calculation shows that both components of acceleration are decreasing on the interval $0.354 < t < 0.655$.

```
> fsolve( diff(aT,t)=0, t=0.2..0.4),
  fsolve( diff(aN,t)=0, t=0.6..0.8);
```

$$0.3544789980, 0.6549207414$$

57. See Exercise 53.

```
> r := <exp(sin(t)),exp(cos(t)),ln(2+cos(t))>: v := diff(r,t):
  a := diff(v,t): s := Norm(v):
> aT := diff(s,t):
```

```
aN := simplify(sqrt(Norm(a)^2 - aT^2)):
> plot( [Norm(a),aT,aN], t=0..2*Pi, -7/4..3, color=black,
          linestyle=[SOLID,DASH,DOT]);
```

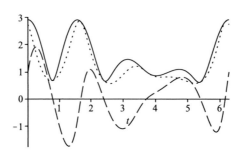

According to the following calculation the maximum value of normal acceleration is $a_N = 2.866$ at $t = 1.638$.

```
> L,C,R := 0,fsolve( diff(aN,t)=0, t=1..2),
            fsolve( diff(aN,t)=0,t=6..2*Pi);
  seq(aN, t=[L,C,R]): evalf(%);
```

$$L, C, R := 0, 1.637823794, 6.199169178$$

$$2.738643314, 2.865622973, 2.860238531$$

The next calculation shows that the absolute maximum value of the magnitude of the acceleretion vector is $A = 2.915$ at $t = 0$ and $t = 2\pi$.

```
> L,C,R := 0,fsolve( diff(Norm(a),t)=0, t=1..2),2*Pi;
  seq(Norm(a), t=[L,C,R]): evalf(%);
```

$$L, C, R := 0, 1.568882519, 2\pi$$

$$2.915504623, 2.907176589, 2.915504623$$

59. The trajectory is defined and plotted.

```
> r := <1/(1+t^2),t/(1+t^2),t^3>:
  spacecurve( r, t=-0.5..0.5, color=black, axes=normal,
              orientation=[40,60]);
```

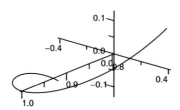

The next entries calculate the components of acceleration and plots the
trajectory $t \mapsto \langle a_T(t), a_N(t) \rangle$.

```
> v := diff(r,t): a := diff(v,t): s := Norm(v):
  aT := diff(s,t): aN := sqrt(Norm(a)^2 - aT^2):
> plot( [aT,aN, t=-0.5..0.5], color=black,
        view=[-0.3..0.3,1.9..2.7],
        labels=["aT","aN"]); Traj := %:
```

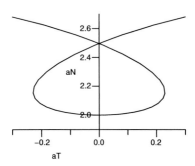

Judging from the values of a_T (on the horizontal axis), the loop, which
starts at the upper left, indicates that the particle is initially slowing down,
then it speeds up and slows down in a symmetric fashion, and finally begins
to speed up again. The values of a_N, read off the vertical axis, correspond
to a particle that is always bending in the same direction.

The tangent lines to the loop are vertical at those t-values where $a_T'(t) = 0$.
The next entries find these t values then display the vertical tangent lines
on the plot of the trajectory.

```
> a,b := fsolve(diff(aT,t)=0,t=-1..0),
         fsolve(diff(aT,t)=0,t=0..1);
  plots[display]( Traj,
    plot( [subs(t=a,aT),t,t=2.0..2.3], color=black),
    plot( [subs(t=b,aT),t,t=2.0..2.3], color=black));
```

$$a, b := -.1707961835, .1707961835$$

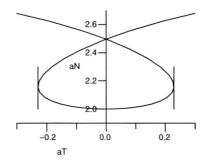

Chapter 11

Functions of Several Variables

11.1 Functions of Several Variables

Problems for Practice

1. Using the volume formula: (area of base) × (height), $V(h,r) = \pi r^2 h$. The domain is all ordered pairs (h,r) with both entries positive.

3. Using the volume formula: $(1/3)$(area of base)×(height), $V(h,r) = \pi r^2 h/3$. The domain is all ordered pairs (h,r) with both entries positive.

5. The integral provides the area: $A(a,b) = \int_a^b x^2 \, dx$. The domain is all ordered pairs (a,b) with $a < b$. The formula can be simplified to $A(a,b) = (b^3 - a^3)/3$.

7. The formula for mass at time t when the half-life is τ can be expressed as $m_0 \cdot 2^{-T/\tau}$, with m_0 denoting the mass at time 0. Therefore,

$$m(M, \tau, T) = M \cdot 2^{-T/\tau}.$$

In base e this is $m(M, \tau, T) = M \cdot e^{-(T/\tau)\ln(2)}$.

9. $(2f + g)(5,2) = 2f(5,2) + g(5,2) = 2e^3 + 25 - 8 = 2e^3 + 17$

11. $(h - k/3)(1/2, 3, 0) = h(1/2, 3, 0) - k(1/2, 3, 0)/3$
$$= (1/2) \cdot 9 - (3/(1/2))/3 = 5/2$$

13. $(\phi \circ f)(3,4) = \sqrt{f(3,4)} = \sqrt{36/(9+16)} = 6/5$

15. $(\phi \circ (2f + 11g))(1,1) = \sqrt{2f(1,1) + 11g(1,1)} = \sqrt{2 \cdot 3/2 + 11 \cdot 3} = 6$

17. $((2\phi) \circ (f \cdot g))(3,4) = 2\sqrt{f(3,4) \cdot g(3,4)} = 2\sqrt{(36/25) \cdot 17} = 12\sqrt{17}/5$

19. Level sets are vertical lines,

$$3x = a,$$

or

$$x = a/3$$

for

$$a = -6, -4, -2, 0, 2, 4, 6.$$

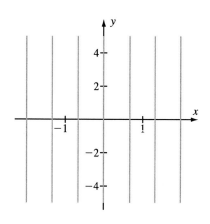

21. Level sets are skew lines,

$$3x - 2y = a,$$

or

$$y = \frac{3}{2}x - \frac{1}{2}a$$

for

$$a = -6, -4, -2, 0, 2, 4, 6.$$

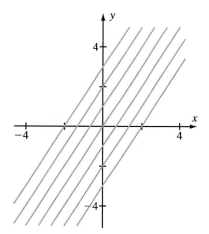

23. Level sets are parabolas opening to the right,

$$1 + x - y^2 = a,$$

or

$$x = y^2 + a - 1$$

for

$$a = -6, -4, -2, 0, 2, 4, 6.$$

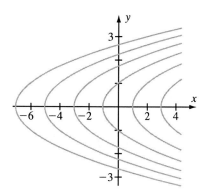

25. Level sets are ellipses centered at the origin,

$$1 - \frac{x^2}{9} - y^2 = a \,,$$

or

$$\frac{x^2}{9} + y^2 = 1 - a$$

for

$$a = -6, -4, -2, 0 \,.$$

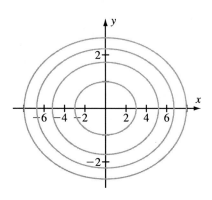

27. Level sets are hyperbolas centered at the origin,

$$x^2 - y^2 = a \,,$$

for

$$a = -6, -4, -2, 0, 2, 4, 6 \,.$$

The value $a = 0$ yields the crossing lines, $y = \pm x$. These are the asymptotes for the hyperbolas.

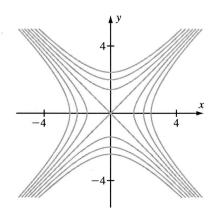

29. Level sets are vertical lines,

$$1 + x = a \,,$$

or

$$x = a - 1$$

for

$$a = -6, -4, -2, 0, 2, 4, 6 \,.$$

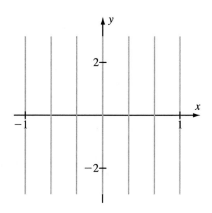

Lift the level set to the appropriate level, $z = a$, and the surface is a plane.

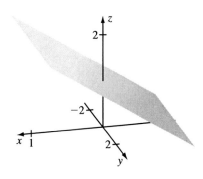

31. Level sets are circles,

$$x^2 + y^2 - 1 = a \,,$$

or

$$x^2 + y^2 = a + 1 \,.$$

Several are sketched on the right.

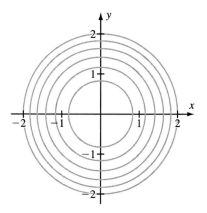

Lift the level set to height $z = a$ and the surface can be seen by adding a sketch of the slice in the yz-plane which is a parabola:

$$z = y^2 - 1 \,.$$

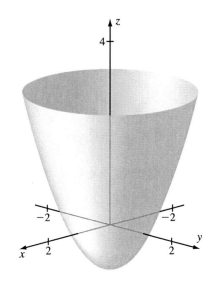

33. Level sets are circles,

$$\sqrt{x^2 + y^2} = a \,,$$

or

$$x^2 + y^2 = a^2 \,.$$

Several are sketched on the right.

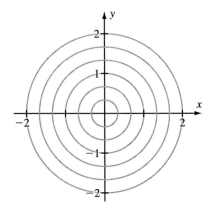

Lift the level set to height $z = a$ and the surface can be seen by adding a sketch of the slice in the yz-plane which forms a $\vee$:

$$z = |y| \,.$$

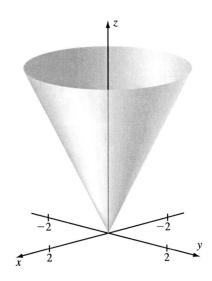

35. Level sets are circles,

$$\sqrt{1 - x^2 - y^2} = a \,,$$

or

$$x^2 + y^2 = 1 - a^2 \,.$$

Several are sketched on the right.

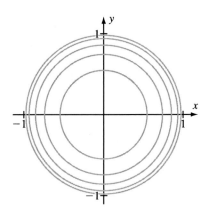

Lift the level set to height $z = a$ and the surface can be seen by adding a sketch of the slice in the yz-plane which is a semi-circle:

$$z = \sqrt{1 - y^2}\,.$$

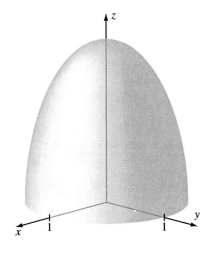

37. Level sets are circles,

$$y^2 + 2x + x^2 = a\,,$$

or

$$(x + 1)^2 + y^2 = a + 1\,.$$

Several are sketched on the right.

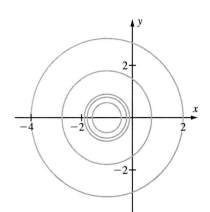

Lift the level set to height $z = a$ and the surface can be seen by adding a sketch of the slice in the xz-plane which is a parabola:

$$z = x^2 + 2x\,.$$

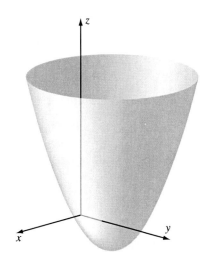

39. Level sets are parallel planes,

$$x + y + z = a\,,$$

for $a = -8, -4, 0, 4, 8$.

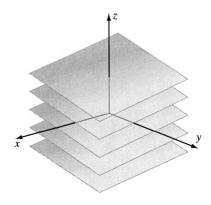

41. Level sets are planes parallel to the xy-plane,

$$12 - z^2 = a\,,$$

or

$$z = \pm\sqrt{12 - a}$$

for $a = -8, -4, 0, 4, 8$.

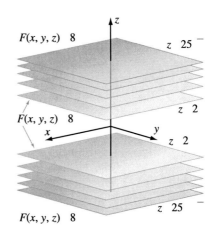

43. Level sets are concentric spheres,

$$x^2 + 2x + y^2 + 4y + z^2 + 6z = a\,,$$

or

$$(x+1)^2 + (y+2)^2 + (z+3)^2 = a+14$$

for $a = -8, -4, 0, 4, 8$. Only two of them are shown.

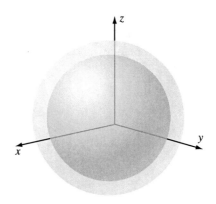

Further Theory and Practice

45. $f(f(x,y), f(x,y)) = f(x,y)f(x,y)/\sqrt{2f(x,y)^2}$. This simplifies as follows.

$$f(f(x,y), f(x,y)) = \frac{f(x,y)^2}{\sqrt{2}\,|f(x,y)|} = \frac{\sqrt{2}}{2} \cdot |f(x,y)|$$

$$= \frac{\sqrt{2}}{2} \cdot \frac{|xy|}{\sqrt{x^2 + y^2}}$$

47. Each level set for f_1 is a line: $x - y = a$, with a any real number. Each level set for f_2 consists of all points (x,y) such that $(x - y)^2 = a$, with $a \geq 0$. If $a = 0$ this is one line: $x = y$. However, if $a > 0$, then the level set consists of two lines, $x - y = \sqrt{a}$ and $x - y = -\sqrt{a}$.

49. The tangent line at the point $(c, h+e^c)$ has slope e^c. Therefore, its equation is $y = h + e^c + e^c(x - c)$. To obtain the x-intercept, set $y = 0$:

$$y = 0 \implies 0 = h + e^c + e^c(x - c) \implies x - c = -e^{-c}(h + e^c),$$

and solve for x. Therefore, the x-intercept is $a(h,c) = c - 1 - he^{-c}$.

51. The boundary of the set consists of the points (x, y) such that $f(x,y) = g(x,y)$:

$$x^2 + y^2 = 4 - 2x^2 - 2y^2,$$

or

$$3x^2 + 3y^2 = 4.$$

This is a circle, centered at the origin with radius $2/\sqrt{3}$. The set S consists of the points that lie outside of this circle.

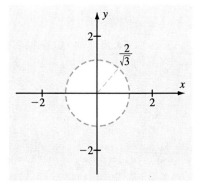

53. The boundary curves for the set S are level sets of f:

$$x^2 - y + 4 = a,$$

or

$$y = x^2 + 4 - a.$$

These are parabolas opening upward. One of them, $f(x,y) = 2$, is the parabola $y = x^2 + 2$ with vertex $(0, 2)$. It is in S. The other boundary, $f(x,y) = 4$, is not in S. It is the parabola $y = x^2$.

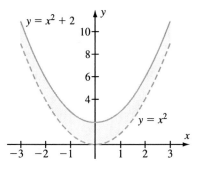

55. Given a number c with $|c| \leq 1$, f's level set $f(x,y) = c$ is the family of parabolas $y = x^2 + \gamma$ where $\sin(\gamma) = c$. There are an infinite number of

such parabolas since the possibilities for γ are not only $\arcsin(c) + 2n\pi$ but also $-\arcsin(c) + (2n+1)\pi$, n any integer.

57. Given a number c with $|c| \leq 1$, f's level set $f(x,y) = c$ is the family of circles $x^2 + y^2 = \gamma$ where $\sin(\gamma) = c$ and $\gamma \geq 0$. There are an infinite number of such circles since the possibilities for γ are either $\arcsin(c) + 2n\pi$, n an integer such that $\arcsin(c) + 2n\pi \geq 0$ (i.e. $n \geq -\arcsin(c)/2\pi$) or $-\arcsin(c)+(2n+1)\pi$, n an integer such that $-\arcsin(c)+(2n+1)\pi \geq 0$ (i.e. $n \geq (\arcsin(c) - \pi)/2\pi$).

59. The level set $f(x,y) = 0$ is the x-axis. If $c > 0$, then the level set $f(x,y) = c$ is all (x,y) such that $y^2/(1 + x^2) = c$ or $y^2 = c + cx^2$. Equivalently, $y^2 - cx^2 = c$. This is a hyperbola, centered at the origin, with vertices $(0, \sqrt{c})$ and $(0, -\sqrt{c})$.

61. We recognize the infinite sum $\sum_{j=0}^{\infty}(x/y)^j$ as a geometric series with first term 1 and ratio x/y, provided $x \neq 0$.[1] The series converges to $1/(1-x/y)$ when $|x/y| < 1$. Therefore, $f(x,y) = y/(y - x)$, and its domain is all ordered pairs (x,y) such that $0 < |x| < |y|$.

63. The simplest such function is $f(x,y) = x^2 + y^3$.

65. A typical level curve has the equation $x^2 - 2xy + 2y^2 = c$ or $(x-y)^2 + y^2 = c$. If $c = 0$, then $x = y = 0$ and the level set is the single point $(0,0)$. Parametrize this with $x = 0$, $y = 0$ for all t. If $c > 0$ write the equation in the form

$$\left(\frac{x - y}{\sqrt{c}}\right)^2 + \left(\frac{y}{\sqrt{c}}\right)^2 = 1.$$

This suggests that the curve can be parametrized using the fact that for each point (x,y) on the curve there is a unique number t in the interval $[0, 2\pi)$ such that $(x - y)/\sqrt{c} = \cos(t)$ and $y/\sqrt{c} = \sin(t)$. Therefore, one parametrization is

$$x = \sqrt{c}\,(\cos(t) + \sin(t))$$
$$y = \sqrt{c}\,\sin(t)$$

Calculator/Computer Exercises

67. The plots package is loaded for access to the contourplot procedure.

```
> with(plots):
  contourplot( x^2 + x*y + y^3, x=-1.2..1, y=-1..1, color=black,
            contours = [k/4 $ k=-3..3], grid=[80,80] );
```

---

[1] If $x = 0$, then the first term in the series is 0^0 which is undefined.

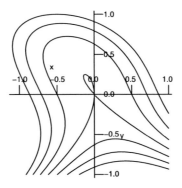

69. The plots package is loaded for access to the contourplot procedure.

```
> with(plots):
  contourplot( 2*x*y^2 + x^4*y, x = -2..2, y=-2..2, color=black,
               contours = [k/4 $ k=-3..3], grid=[80,80] );
```

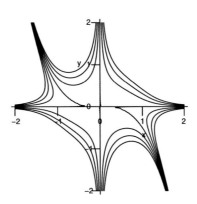

11.2   Cylinders and Quadratic Surfaces

Problems for Practice

1. Since z is missing all slices perpendicular to the z-axis are the same, the line

   $$x + y = 5 .$$

   The surface is a plane.

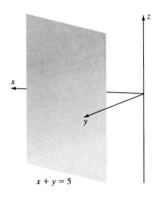

3. Since y is missing all slices perpendicular to the y-axis are the same, the parabola

   $$4x^2 + z = 4 .$$

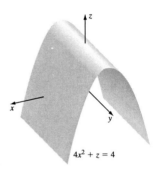

5. Since y is missing all slices perpendicular to the y-axis are the same, the parabola

   $$x^2 + z = -4 .$$

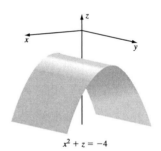

7. Since z is missing all slices perpendicular to the z-axis are the same, the parabola

   $$x^2 + y = 4 .$$

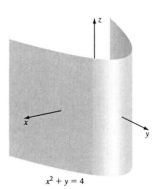

9. Since z is missing all slices perpendicular to the z-axis are the same, the hyperbola

$$x^2 - 2y^2 = 1.$$

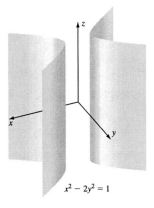

11. Since x is missing all slices perpendicular to the x-axis are the same, the hyperbola

$$y^2 - 4z^2 = 1.$$

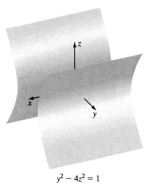

13. Since z is missing all slices perpendicular to the z-axis are the same, the graph of

$$x - 2xy = 1,$$

or

$$x = \frac{1}{1 - 2y}.$$

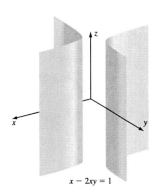

15. This is a hyperboloid of one sheet. Write the equation in the form

$$x^2 + 9z^2 = 4y^2 + 36$$

to see that slices in planes perpendicular to the y-axis are ellipses having the same eccentricity. The slice in the xy-plane is the hyperbola $x^2 - 4y^2 = 36$.

This is not the graph of a function.

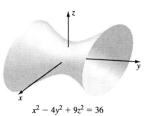

17. This is a circular paraboloid. Write the equation in the form
$$y = x^2 + z^2$$
to see that slices in planes perpendicular to the y-axis are circles. The slice in the xy-plane is the parbola $y = x^2$.

This is not the graph of a function $z = f(x, y)$. It is the graph of a function $y = g(x, z)$.

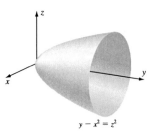

$y - x^2 = z^2$

19. This is an ellipsoid. Slices in all planes that are perpendicular to an axis are ellipses.

This is not the graph of a function.

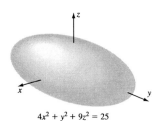

$4x^2 + y^2 + 9z^2 = 25$

21. This is a circular paraboloid. Since $z = 4 - x^2 - y^2$, the slices in planes perpendicular to the z-axis are circles. The slice in the xz-plane is the parbola $z = 4 - x^2$.

This is the graph of a function.

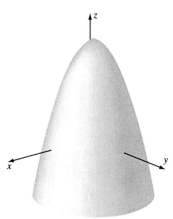

$x^2 + y^2 + z = 4$

23. This is an elliptic paraboloid. Slices in planes perpendicular to the z-axis are ellipses. The slice in the xz-plane is the parbola

$$z = 4x^2 .$$

This is the graph of a function.

$4x^2 + y^2 = z$

25. This is a hyperboloid of two sheets. Write the equation in the form

$$y^2 - 1 = x^2 + z^2$$

to see that for $|y| > 1$ slices in planes perpendicular to the y-axis are circles. The slice in the xy-plane is the hyperbola $y^2 - x^2 = 1$.

This is not the graph of a function.

27. This is an elliptic cone. Slices in planes perpendicular to the x-axis are ellipses having the same eccentricity. The slice in the xy-plane is crossing lines $y = \pm x$.

This is not the graph of a function.

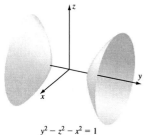

$y^2 + 8z^2 = x^2$

29. This is a hyperbolic paraboloid. Write the equation in the form

$$x = y^2 - z^2$$

to see that slices in planes perpendicular to the x-axis are hyperbolas and slices in planes perpendicular to the y- and z-axis are parabolas.

This is not the graph of a function $z = f(x, y)$. It is the graph of a function $x = g(y, z)$.

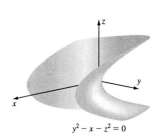

$y^2 - x - z^2 = 0$

Further Theory and Practice

31. This is false. For example, the surface determined by the equation $xyz = 1$ is not a quadric surface yet every level curve of the surface is the graph of a quadratic equation.

33. Consider the two dimensional slice obtained by setting $y = 0$: $z = f(x)$. Fix a point (x, z) on the graph of f with $x > 0$ and let β be the angle between the incoming line and the tangent line at this point. Observe that if α is the angle between the incoming line and the normal to the tangent line, then $\alpha + \beta = \pi/2$. Moreover, α is also the angle between the tangent line and the positive x-axis (verify). Thus $\tan(\alpha) = f'(x)$ and $\tan(\beta) = 1/f'(x)$. Draw a picture to confirm that if β is less than $\pi/4$, then the point $(0, p)$ where the outgoing line intersects the vertical axis is below the point $(0, z)$. Examination of the right triangle with vertices $(0, p), (0, z), (x, z)$ reveals that

$$\tan(\alpha - \beta) = \frac{z - p}{x}.$$

Use the identity $\tan(\alpha - \beta) = (\tan(\alpha) - \tan(\beta))/(1 + \tan(\alpha)\tan(\beta))$ to see that

$$\frac{f'(x) - 1/f'(x)}{2} = \frac{f(x) - p}{x}.$$

Substitute $g(x) = (f(x) - p)/x$ and simplify[2] to obtain the equation

$$(xg'(x))^2 = 1 + g(x)^2.$$

Since x and $g'(x)$ are positive this can be written in the form

$$\frac{g'(x)}{\sqrt{1 + g(x)^2}} = \frac{1}{x}.$$

Integrate with respect to x,

$$\ln(g(x) + \sqrt{1 + g(x)^2}) = \ln(x) + c,$$

replace c with $\ln(C)$, and exponentiate both sides to obtain the equation

$$g(x) + \sqrt{1 + g(x)^2} = Cx.$$

Consequently, $g(x) = (C^2x^2 - 1)/(2Cx)$ and

$$f(x) = \frac{1}{2}Cx^2 - \frac{1}{2C} + p.$$

Thus, for example, if $f(0) = 0$, then $C = 1/2p$ and $f(x) = x^2/4p$, a parabola with vertex at $(0, 0)$ and focus at $(0, p)$.

35. The intersection is the set of points (x, y, z) satisfying the equations

$$z = x^2/2 - y^2 \quad \text{and} \quad z = 1 - x^2/2 - 5y^2/4.$$

Therefore, the projection in the xy-plane consists of the ordered pairs (x, y) such that

$$x^2/2 - y^2 = 1 - x^2/2 - 5y^2/4.$$

---

[2]Since $f(x) = xg(x) + p$, $f'(x) = xg'(x) + g(x)$.

Equivalently,

$$x^2 + y^2/4 = 1\,.$$

This is an ellipse, centered at the origin.

Calculator/Computer Exercises

37. The plots package is loaded to have access to the implicitplot3d procedure.

```
> with(plots):
  implicitplot3d( x^2 + 2*x*y + 2*y^2 - z^2 = 0,
                  x=-1..1, y=-1..1, z=-1..1,
                  style=patchnogrid, axes=normal,
                  orientation=[45,70], grid=[40,40,40],
                  lightmodel=light1 );
```

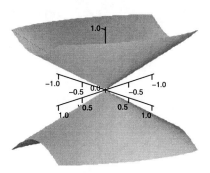

39. The plots package is loaded to have access to the implicitplot3d procedure.

```
> with(plots):
  implicitplot3d( x^2 + 2*x*y + 2*y^2 - z^2 = -1,
                  x=-5..5, y=-5..5, z=-5..5,
                  style=patchnogrid, axes=normal,
                  orientation=[45,70], grid=[40,40,40],
                  lightmodel=light1 );
```

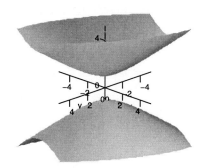

41. The plots package is loaded to have access to the spacecurve and display procedures. The cone is plotted first.

```
> with(plots):
  plot3d( 2*sqrt(x^2 + y^2), x=-3..3, y=-3..3, view=0..6, axes=normal,
          orientation=[45,70], style=patchnogrid); Cone := %:
```

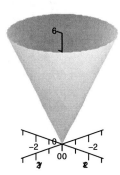

Now the plane.

```
> plot3d( sqrt(2)*(1-x), x=-6*sqrt(2)+1..1, y=-3..3, view=0..6,
          axes=normal, orientation=[45,70], style=patchnogrid,
          color=green);
  Plane := %:
```

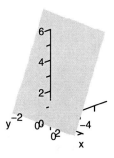

Finally, the curve of intersection. This curve projects in the xy-plane to the points (x, y) such that $2(1 - x)^2 = 4(x^2 + y^2)$ or $2x^2 + 4x + 4y^2 = 2$. This is an ellipse $2(x + 1)^2 + 4y^2 = 4$ or, in standard form,

$$\frac{(x + 1)^2}{2} + y^2 = 1.$$

Therefore, the xy-projection of the curve of intersection can be parametrized with $x = \sqrt{2}\cos(\theta) - 1$, $y = \sin(\theta)$, $0 \le \theta \le 2\pi$. The points on the projection are be lifted up to the plane by adding the equation

$$z = \sqrt{2}(1 - x) = 2\sqrt{2} - 2\cos(\theta).$$

See the following display.

```
> display( Cone, Plane,
  spacecurve( [sqrt(2)*cos(theta)-1,sin(theta),2*sqrt(2)-2*cos(theta)],
          theta=0..2*Pi, color=black, thickness=2),
  orientation=[150,60] );
```

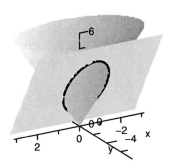

11.3 Limits and Continuity

Problems for Practice

1. The function $f(x,y) = x(2xy-3)^2$ is the product of two functions, $(x,y) \mapsto x$ and $(x,y) \mapsto (2xy-3)^2$ that are continuous at all points. Therefore, f is continuous at all points and

$$\lim_{(x,y)\to(2,3)} f(x,y) = f(2,3) = 162\,.$$

3. The function $f(x,y) = x^2y - x$ is continuous at all points and $x \mapsto \sqrt{x}$ is continuous at $f(4,2)$. Therefore,

$$\lim_{(x,y)\to(4,2)} \sqrt{f(x,y)} = \sqrt{f(4,2)} = \sqrt{28} = 2\sqrt{7}\,.$$

5. The function $f(x,y) = y^2 + xy$ is continuous at all points and $x \mapsto \ln x$ is continuous at $f(-1,4)$. Therefore,

$$\lim_{(x,y)\to(-1,4)} \ln(f(x,y)) = \ln(f(-1,4)) = \ln(12)\,.$$

7. Both $(x,y) \mapsto \arccos(x-y)$ and $(x,y) \mapsto e^{x-y}$ are continuous at $(1/4,1/4)$. Moreover, $e^{1/4-1/4} = 1$ so their quotient is also continuous at $(1/4,1/4)$. Therefore,

$$\lim_{(x,y)\to(1/4,1/4)} \frac{\arccos(x-y)}{e^{x-y}} = \frac{\arccos(0)}{e^0} = \frac{\pi}{2}\,.$$

9. The function $f(x,y) = \sin(y\cos(x)) + \cos(x + \cos(y))$ is continuous at all points (why?). Therefore,

$$\lim_{(x,y)\to(0,\pi/2)} f(x,y) = f(0,\pi/2) = 1 + 1 = 2\,.$$

11. Factor the numerator and the denominator, and cancel common factors, to obtain

$$\lim_{(x,y)\to(1,2)} \frac{(x-1)(y-2)(y+2)}{(y-2)(x-1)(x+1)} = \lim_{(x,y)\to(1,2)} \frac{y+2}{x+1} = 2\,.$$

13. Factor the numerator, and cancel common factors, to obtain

$$\lim_{(x,y)\to(6,3)} \frac{(x-2y)(x+y)}{x-2y} = \lim_{(x,y)\to(6,3)} (x+y) = 9\,.$$

15. Approach along the x-axis and the limiting value is 0. Approach along the line $y = x$ and the limiting value is 1.

17. Approach along the line $y = x$, $x > 0$, and the limiting value is 1. Approach along the line $y = x$, $x < 0$, and the limiting value is -1.

19. Approach along the x-axis and the limit is 1. Approach along the y-axis and the limit is -1.

Further Theory and Practice

21. The function is continuous at $(0,0)$ because

$$\lim_{(x,y)\to(0,0)} \frac{\sin(x^2+y^2)}{x^2+y^2} = \lim_{t\to 0^+} \frac{\sin(t)}{t} = 1 = f(0,0)\,.$$

23. The function is not continuous at $(0,0)$ because $\lim_{(x,y)\to(0,0)} f(x,y)$ does not exist. Observe that if (x,y) approaches $(0,0)$ along the x-axis, then the limiting value is -2 and if (x,y) approaches $(0,0)$ along the y-axis, then the limiting value is 1.

25. Let $f(x,y) = \phi(x) + \psi(y)$. We show that f is continuous at a point (x_0, y_0) by verifying that $\lim_{(x,y)\to(x_0,y_0)} f(x,y) = f(x_0, y_0)$. Let $\epsilon > 0$. Since ϕ is continuous at x_0 and ψ is continuous at y_0 there is a positive number δ_1 and a positive number δ_2 such that

$$0 < |x - x_0| < \delta_1 \implies |\phi(x) - \phi(x_0)| < \epsilon/2$$

and

$$0 < |y - y_0| < \delta_2 \implies |\psi(y) - \psi(y_0)| < \epsilon/2\,.$$

Let δ be the smaller of the two numbers δ_1 and δ_2. Then[3]

$$0 < d((x,y),(x_0,y_0)) < \delta \implies 0 < |x - x_0| < \delta_1 \quad \text{and} \quad 0 < |y - y_0| < \delta_2$$

so, for all (x,y) in the punctured disk $D_*((x_0,y_0),\delta)$,

$$\begin{aligned}
|f(x,y) - f(x_0,y_0)| &= |\phi(x) + \psi(y) - (\phi(x_0) + \psi(y_0))| \\
&= |\phi(x) - \phi(x_0) + \psi(y) - \psi(y_0)| \\
&\leq |\phi(x) - \phi(x_0)| + |\psi(y) - \psi(y_0)| \\
&< \epsilon/2 + \epsilon/2 = \epsilon.
\end{aligned}$$

27. It is not true. Consider, for example, the function f defined as

$$f(x,y) = \begin{cases} x & , \quad y = 0 \\ y & , \quad x = 0 \\ 1 & , \quad \text{otherwise}. \end{cases}$$

Then $\phi(x) = f(x,0) = x$ and $\psi(y) = f(0,y) = y$ are both continuous at 0. However, f is not continuous at $(0,0)$ because $\lim_{(x,y)\to(0,0)} f(x,y)$ does not exist.

29. The limit is 0. This is because, by l'Hôpital's Rule, $\lim_{t\to 0^+} t \ln(t) = 0$ (verify). Therefore,

$$\begin{aligned}
0 \leq |f(x,y)| &= x^2 |\ln(x^2 + y^2)| \\
&\leq (x^2 + y^2)|\ln(x^2 + y^2)| \xrightarrow[(x,y)\to(0,0)]{} 0.
\end{aligned}$$

31. First of all, $\lim_{y\to 0} f(0,y) = \lim_{y\to 0} 0 = 0$. Moreover, given any slope m, $f(x,mx) = m^2 x^3/(x^2 + m^4 x^4) = m^2 x/(1 + m^4 x^2)$ so

$$\lim_{x\to 0} f(x,mx) = \lim_{x\to 0} \frac{m^2 x}{1 + m^4 x^2} = 0$$

also. Thus $f(x,y)$ has limiting value 0 along every straight line path towards $(0,0)$.

If a point (x,y) approaches $(0,0)$ along the path $x = y^2$, then the values of f are $f(y^2,y) = y^4/(y^4 + y^4) = 1/2$ and the limiting value is $1/2$. Therefore, $f(x,y)$ does not have a limit as (x,y) approaches $(0,0)$.

Calculator/Computer Exercises

33. The plot appears below.

```
> plot3d( x*y/(x^2 + y^2), x=-0.6..0.6, y=-0.6..0.6, axes=normal,
          style=patchnogrid, grid=[40,40],
          view=[-0.8..0.8,-0.8..0.8,-0.6..0.6],
          orientation=[25,70], lightmodel=light1);
```

---

[3]See the Insight on page 884.

3. $\phi(x) = f(x,-1) = 4$ and $\psi(y) = f(2,y) = 7 - 3y^4$. Therefore, $\phi'(x) = 0$, $\psi'(y) = -12y^3$, and

$$\phi'(2) = 0 \quad \text{and} \quad \psi'(-1) = 12.$$

$\frac{\partial f}{\partial x}(x,y) = 0$ and $\frac{\partial f}{\partial y}(x,y) = -12y^3$. Therefore,

$$\frac{\partial f}{\partial x}(2,-1) = 0 \quad \text{and} \quad \frac{\partial f}{\partial y}(2,-1) = 12.$$

5. $\phi(x) = f(x,1/2) = \cos(x/4)$ and $\psi(y) = f(2,y) = \cos(\pi y^2)$ so $\phi'(x) = -\sin(x/4)/4$ and $\psi'(y) = -2\pi y \sin(\pi y^2)$. Therefore,

$$\phi'(\pi) = -\sqrt{2}/8 \quad \text{and} \quad \psi'(1/2) = -\pi\sqrt{2}/2.$$

$\frac{\partial f}{\partial x}(x,y) = -y^2 \sin(xy^2)$ and $\frac{\partial f}{\partial y}(x,y) = -2xy \sin(xy^2$. Therefore,

$$\frac{\partial f}{\partial x}(\pi,1/2) = -\sqrt{2}/8 \quad \text{and} \quad \frac{\partial f}{\partial y}(\pi,1/2) = -\pi\sqrt{2}/2.$$

7. $\phi(x) = f(x,e) = x \ln(ex)$ and $\psi(y) = f(1,y) = \ln(y)$ so $\phi'(x) = 1 + \ln(ex)$ and $\psi'(y) = 1/y$. Therefore,

$$\phi'(1) = 2 \quad \text{and} \quad \psi'(e) = \frac{1}{e}.$$

$\frac{\partial f}{\partial x}(x,y) = 1 + \ln(xy)$ and $\frac{\partial f}{\partial y}(x,y) = x/y$. Therefore,

$$\frac{\partial f}{\partial x}(1,e) = 2 \quad \text{and} \quad \frac{\partial f}{\partial y}(1,e) = \frac{1}{e}.$$

9. $\phi(x) = f(x,2) = x^2$ and $\psi(y) = f(2,y) = 2^y$ so $\phi'(x) = 2x$ and $\psi'(y) = 2^y \ln(y)$. Therefore,

$$\phi'(2) = 4 \quad \text{and} \quad \psi'(2) = 4\ln(2).$$

$\frac{\partial f}{\partial x}(x,y) = yx^{y-1}$ and $\frac{\partial f}{\partial y}(x,y) = x^y \ln(x)$. Therefore,

$$\frac{\partial f}{\partial x}(2,2) = 4 \quad \text{and} \quad \frac{\partial f}{\partial y}(2,2) = 4\ln(2).$$

11. $f_x(x,y) = 2xy - y$, $\qquad f_{xx}(x,y) = 2y$, $\qquad f_{xy}(x,y) = 2x - 1$;

$\quad f_y(x,y) = x^2 - x + 5y^4$, $\quad f_{yy}(x,y) = 20y^3$, $\qquad f_{yx}(x,y) = 2x - 1$.

13. $f_x(x,y) = 2x - 1/y$, $\qquad f_{xx}(x,y) = 2$, $\qquad f_{xy} = 1/y^2$;

$\quad f_y(x,y) = x/y^2$, $\qquad\qquad f_{yy} = -2x/y^3$, $\qquad f_{yx} = 1/y^2$.

15. $f_x(x,y) = \dfrac{\cos(x)}{\cos(y)}$, $f_{xx}(x,y) = -\dfrac{\sin(x)}{\cos(y)}$, $f_{xy}(x,y) = \dfrac{\cos(x)\sin(y)}{\cos^2(y)}$;

$f_y(x,y) = \dfrac{\sin(x)\sin(y)}{\cos^2(y)}$, $f_{yy}(x,y) = \dfrac{\sin(x)(\sin^2(y)+1)}{\cos^3(y)}$,

$f_{yx}(x,y) = \dfrac{\cos(x)\sin(y)}{\cos^2(y)}$.

17. $f_x(x,y) = \dfrac{1}{x-y}$, $f_{xx}(x,y) = -\dfrac{1}{(x-y)^2}$, $f_{xy}(x,y) = \dfrac{1}{(x-y)^2}$;

$f_y(x,y) = \dfrac{x-2y}{xy-x^2}$, $f_{yy}(x,y) = \dfrac{2xy-2y^2-x^2}{(xy-x^2)^2}$, $f_{yx}(x,y) = \dfrac{1}{(x-y)^2}$.

19. $f_x(x,y) = -\sin(x)\sin(y)$, $f_{xx}(x,y) = -\cos(x)\sin(y)$, $f_{xy}(x,y) = -\sin(x)\cos(y)$;

$f_y(x,y) = \cos(x)\cos(y)$, $f_{yy}(x,y) = -\cos(y)\sin(x)$, $f_{yx}(x,y) = -\sin(x)\cos(y)$.

21. $D_y f(x,y) = x\sec(y)\tan(y)$, $D_{yy} f(x,y) = x\sec(y)(2\tan^2(y)+1)$,

$D_{yx} f(x,y) = \sec(y)\tan(y)$;

$f_1(x,y) = \sec(y)$ $f_{12}(x,y) = \sec(y)\tan(y)$, $f_{11}(x,y) = 0$.

23. $D_y f(x,y) = 2\cos(x^3+2y)$, $D_{yy} f(x,y) = -4\sin(x^3+2y)$,

$D_{yx} f(x,y) = -6x^2\sin(x^3+2y)$;

$f_1(x,y) = 3x^2\cos(x^3+2y)$, $f_{11}(x,y) = 6x\cos(x^3+2y) - 9x^4\sin(x^3+2y)$,

$f_{12}(x,y) = -6x^2\sin(x^3+2y)$.

25. $D_y f(x,y) = 5(\sin(x)-\cos(y))^4\sin(y)$,

$D_{yy} f(x,y) = 20(\sin(x)-\cos(y))^3\sin^2(y) + 5(\sin(x)-\cos(y))^4\cos(y)$,

$D_{yx} f(x,y) = 20(\sin(x)-\cos(y))^3\sin(y)\cos(x)$;

$f_1(x,y) = 5(\sin(x)-\cos(y))^4\cos(x)$,

$f_{11}(x,y) = 20(\sin(x)-\cos(y))^3\cos^2(x) - 5(\sin(x)-\cos(y))^4\sin(x)$,

$f_{12}(x,y) = 20(\sin(x)-\cos(y))^3\sin(y)\cos(x)$.

27. $D_y f(x,y) = 12(x+3y)^3$, $D_{yy} f(x,y) = 108(x+3y)^2$,

$D_{yx} f(x,y) = 36(x+3y)^2$

$f_1(x,y) = 4(x+3y)^3$, $f_{11}(x,y) = 12(x+3y)^2$

$f_{12}(x,y) = 36(x+3y)^2$

29. $D_y f(x,y) = \dfrac{2y^3}{\sqrt{1+3x^2+y^4}}$, $D_{yy} f(x,y) = \dfrac{2y^2(3+9x^2+y^4)}{(1+3x^2+y^4)^{3/2}}$,

$D_{yx} f(x,y) = -\dfrac{6xy^3}{(1+3x^2+y^4)^{3/2}}$;

$$f_1(x,y) = \frac{3x}{\sqrt{1+3x^2+y^4}}, \qquad f_{11}(x,y) = -\frac{3(1+y^4)}{(1+3x^2+y^4)^{3/2}},$$

$$f_{12}(x,y) = -\frac{6xy^3}{(1+3x^2+y^4)^{3/2}}.$$

31–39. See text.

41. Each column in the following table displays the successive partial derivatives for the third partial that appears at the top: first (f_*), second (f_{**}), and third (f_{***}).

	f_{xyz}	f_{yxz}	f_{xzy}	f_{zxy}	f_{yzx}	f_{zyx}
f_*	y^2z^3	$2xyz^3$	y^2z^3	$3xy^2z^3$	$2xyz^3$	$3xy^2z^2$
f_{**}	$2yz^3$	$2yz^3$	$3y^2z^2$	$3y^2z^2$	$6xyz^2$	$6xyz^2$
f_{***}	$6yz^2$	$6yz^2$	$6yz^2$	$6yz^2$	$6yz^2$	$6yz^2$

43. Each column in the following table displays the successive partial derivatives for the third partial that appears at the top: first (f_*), second (f_{**}), and third (f_{***}).

	f_{xyz}	f_{yxz}	f_{xzy}	f_{zxy}	f_{yzx}	f_{zyx}
f_*	$-\frac{3+y\cos(z))}{x^2}$	$\frac{\cos(z)}{x}$	$-\frac{3+y\cos(z))}{x^2}$	$-\frac{y\sin(z)}{x}$	$\frac{\cos(z)}{x}$	$-\frac{y\sin(z)}{x}$
f_{**}	$-\frac{\cos(z)}{x^2}$	$-\frac{\cos(z)}{x^2}$	$\frac{y\sin(z)}{x^2}$	$\frac{y\sin(z)}{x^2}$	$-\frac{\sin(z)}{x}$	$-\frac{\sin(z)}{x}$
f_{***}	$\frac{\sin(z)}{x^2}$	$\frac{\sin(z)}{x^2}$	$\frac{\sin(z)}{x^2}$	$\frac{\sin(z)}{x^2}$	$\frac{\sin(z)}{x^2}$	$\frac{\sin(z)}{x^2}$

Further Theory and Practice

45. Calculate the partials for $f(x,y) = \frac{x-y}{x+y}$.

$$f_x(x,y) = \frac{2y}{(x+y)^2} \quad, \quad f_{xx}(x,y) = -\frac{4y}{(x+y)^3} \quad, \quad f_{xy}(x,y) = \frac{2(x-y)}{(x+y)^3}$$

$$f_y(x,y) = -\frac{2x}{(x+y)^2} \quad, \quad f_{yy}(x,y) = \frac{4x}{(x+y)^3} \quad, \quad f_{yx}(x,y) = \frac{2(x-y)}{(x+y)^3}$$

47. Calculate the partials for $f(x,y) = \tan(x^2y)$.

$$f_x(x,y) = 2xy\sec^2(x^2y) \quad \text{and} \quad f_y(x,y) = x^2\sec^2(x^2y)$$

$$f_{xx}(x,y) = 2y\sec^2(x^2y)(4x^2y\tan(x^2y)+1) \,, \ f_{yy}(x,y) = 2x^4\sec^2(x^2y)\tan(x^2y)$$

$$f_{yx}(x,y) = 2x\sec^2(x^2y)(2x^2y\tan(x^2y)+1) = f_{xy}(x,y)$$

49. Calculate the partials for $f(x, y) = \arctan(x + 2y)$.

$$f_x(x, y) = \frac{1}{1 + (x + 2y)^2} \quad \text{and} \quad f_y(x, y) = \frac{2}{1 + (x + 2y)^2}$$

$$f_{xx}(x, y) = -\frac{2(x + 2y)}{(1 + (x + 2y)^2)^2}, \quad f_{yy}(x, y) = -\frac{8(x + 2y)}{(1 + (x + 2y)^2)^2}$$

$$f_{yx}(x, y) = -\frac{4(x + 2y)}{(1 + (x + 2y)^2)^2} = f_{xy}(x, y)$$

51. The curve $\mathcal{C}_1$ is parametrized with the position vector $\mathbf{r}(y) = \langle 2, y, f(2, y) \rangle$. Its tangent vector at $\mathbf{r}(y)$ is $\mathbf{r}'(y) = \langle 0, 1, f_y(2, y) \rangle$. Because $\mathbf{r}(1) = \langle 2, 1, f(2, 1) \rangle = \langle 2, 1, 4 \rangle$ and $\mathbf{r}'(1) = \langle 0, 1, f_y(2, 1) \rangle = \langle 0, 1, -4 \rangle$, the tangent line to $\mathcal{C}_1$ at $(2, 1, 4)$ has the parametrization $u \mapsto \langle 2, 1, 4 \rangle + u\langle 0, 1, -4 \rangle$. Therefore, ℓ_1 meets the xy-plane when $u = 1$ at the point $(2, 2, 0)$.

The curve $\mathcal{C}_2$ is parametrized with the position vector $\mathbf{r}(x) = \langle x, 1, f(x, 1) \rangle$. Its tangent vector at $\mathbf{r}(x)$ is $\mathbf{r}'(x) = \langle 1, 0, f_x(x, 1) \rangle$. Because $\mathbf{r}'(2) = \langle 1, 0, f_x(2, 1) \rangle = \langle 1, 0, 10 \rangle$, the tangent line to $\mathcal{C}_2$ at $(2, 1, 4)$ has parametrization $u \mapsto \langle 2, 1, 4 \rangle + u\langle 1, 0, 10 \rangle$. Therefore, ℓ_2 meets the xy-plane when $u = -2/5$ at the point $(8/5, 1, 0)$.

53. $\mathcal{C}_1$ is parametrized with the position vector $\mathbf{r}(y) = \langle 2, y, f(2, y) \rangle$. Its tangent vector is $\mathbf{r}'(y) = \langle 0, 1, f_y(2, y) \rangle$. Because $\mathbf{r}(-2) = \langle 2, -2, f(2, -2) \rangle = \langle 2, -2, 12 \rangle$ and $\mathbf{r}'(-2) = \langle 0, 1, f_y(2, -2) \rangle = \langle 0, 1, -8/3 \rangle$, the tangent line to $\mathcal{C}_1$ at $(2, -2, 12)$ has the parametrization $u \mapsto \langle 2, -2, 12 \rangle + u\langle 0, 1, -8/3 \rangle$. Therefore, ℓ_1 meets the xy-plane when $u = 9/2$ at the point $(2, 5/2, 0)$.

$\mathcal{C}_2$ is parametrized with the position vector $\mathbf{r}(x) = \langle x, -2, f(x, -2) \rangle$. Its tangent vector is $\mathbf{r}'(x) = \langle 1, 0, f_x(x, -2) \rangle$. Because $\mathbf{r}'(2) = \langle 1, 0, f_x(2, -2) \rangle = \langle 1, 0, 40/3 \rangle$, the tangent line to $\mathcal{C}_2$ at $(2, -2, 12)$ has parametrization $u \mapsto \langle 2, -2, 12 \rangle + u\langle 1, 0, 40/3 \rangle$. Therefore, ℓ_2 meets the xy-plane when $u = -9/10$ at the point $(11/10, -2, 0)$.

55. $V = \dfrac{1}{3}\pi r^2 h$,

rate of change of volume with repect to radius: $\quad \dfrac{\partial V}{\partial r} = \dfrac{2}{3}\pi r h$,

rate of change of volume with repect to height: $\quad \dfrac{\partial V}{\partial h} = \dfrac{1}{3}\pi r^2$.

57. Since $P = kT/V$, $\partial P/\partial V = -kT/V^2$. Since $V = kT/P$, $\partial V/\partial T = k/P$. Because $T = PV/k$, $\partial T/\partial P = V/k$. Consequently,

$$\frac{\partial P}{\partial V} \cdot \frac{\partial V}{\partial T} \cdot \frac{\partial T}{\partial P} = -\frac{kT}{V^2} \cdot \frac{k}{P} \cdot \frac{V}{k} = -\frac{kT}{PV} = -1.$$

59. Since $u(x, y) = \ln(e^x + e^y)$,

$$u_x(x, y) = \frac{e^x}{e^x + e^y} \quad \text{and} \quad u_y(x, y) = \frac{e^y}{e^x + e^y}.$$

Clearly, $u_x(x, y) + u_y(x, y) = 1$.

61. (a) Harmonic because $u_{xx} + u_{yy} = 2 - 2 = 0$.

    (b) Harmonic because $u_{xx} + u_{yy} = 0 + 0 = 0$.

    (c) Harmonic because $u_{xx} + u_{yy} = e^x \cos(y) - e^x \cos(y) = 0$.

    (d) Not harmonic because $u_{xx} + u_{yy} = 2 + 2$.

    (e) Not harmonic because $u_{xx} + u_{yy} = 6x$.

    (f) Harmonic because $u_{xx} + u_{yy} = e^{-x} \sin(y) - e^{-x} \sin(y) = 0$.

63. Observe that

$$\frac{\partial P}{\partial x}(x, y) = -\frac{2xy}{\pi(x^2 + y^2)^2} \quad \text{so} \quad \frac{\partial^2 P}{\partial x^2}(x, y) = \frac{2y(3x^2 - y^2)}{\pi(x^2 + y^2)^3}$$

and

$$\frac{\partial P}{\partial y}(x, y) = \frac{x^2 - y^2}{\pi(x^2 + y^2)^2} \quad \text{so} \quad \frac{\partial^2 P}{\partial y^2}(x, y) = -\frac{2y(3x^2 - y^2)}{\pi(x^2 + y^2)^3}.$$

Clearly, $\dfrac{\partial^2 P}{\partial x^2}(x, y) + \dfrac{\partial^2 P}{\partial x^2}(x, y) = 0$.

65. Since $u_x = 3Ax^2 + 2Bxy + Cy^2$ and $u_y = Bx^2 + 2Cxy + 3Ey^2$,

$$u_{xx} + u_{yy} = 6Ax + 2By + 2Cx + 6Ey$$
$$= (6A + 2C)x + (2B + 6E)y.$$

Therefore, u is harmonic if and only if $B = -3E$ and $C = -3A$.

67. Observe that $\partial v_n / \partial x = 2nx(x^2 + y^2)^{n-1}$ and $\partial v_n / \partial y = 2ny(x^2 + y^2)^{n-1}$. Therefore,

$$\frac{\partial^2 v_n}{\partial x^2} = 4n(n - 1)x^2(x^2 + y^2)^{n-2} + 2n(x^2 + y^2)^{n-1}$$

and

$$\frac{\partial^2 v_n}{\partial y^2} = 4n(n - 1)y^2(x^2 + y^2)^{n-2} + 2n(x^2 + y^2)^{n-1}.$$

Consequently,

$$\left(\frac{\partial^2}{\partial x^2} + \frac{\partial^2}{\partial y^2}\right) v_n(x,y) = 4n(n-1)(x^2+y^2)^{n-1} + 4n(x^2+y^2)^{n-1}$$

$$= 4n^2(x^2+y^2)^{n-1}$$

$$= \frac{4n^2}{x^2+y^2}\left(x^2+y^2\right)^n$$

$$= \frac{4n^2}{x^2+y^2}\, v_n(x,y)\,.$$

Thus $\left(\frac{\partial^2}{\partial x^2} + \frac{\partial^2}{\partial y^2}\right) v_n(x,y) = 4n^2(x^2+y^2)^{n-1}$ and $v_n(x,y)$ is harmonic if, and only if, $n = 0$.

69. Observe that

$$\frac{\partial u_n}{\partial t} = -n^2\alpha^2 e^{-n^2\alpha^2 t}\sin(nx) = -n^2\alpha^2 u_n(x,t)$$

and

$$\frac{\partial^2 u_n}{\partial x^2} = -n^2 e^{-n^2\alpha^2 t}\sin(nx) = -n^2 u_n(x,t)\,.$$

Therefore, $\dfrac{\partial u_n}{\partial t} = \alpha^2 \dfrac{\partial^2 u_n}{\partial x^2}$.

71. Since $f_x(x,y) = 2xy$, f must be of the form $f(x,y) = x^2 y + g(y)$. This implies that $f_y(x,y) = x^2 + g'(y)$. Since we require that $f_y(x,y) = x^2 + 1$ it must be the case that $g'(y) = 1$ and $g(y) = y + C$. Conclusion:

$$f(x,y) = x^2 y + y + C\,.$$

73. Since $f_x(x,y) = 2x$, f must be of the form $f(x,y) = x^2 + g(y)$. This implies that $f_y(x,y) = g'(y)$. Since we require that $f_y(x,y) = 1 - 2y$ it must be the case that $g'(y) = 1 - 2y$ and $g(y) = y - y^2 + C$. Conclusion:

$$f(x,y) = x^2 + y - y^2 + C\,.$$

75. It is not difficult to verify[4] that if f and g are functions of a single variable x, then

$$\frac{d^n}{dx^n}(f(x)g(x)) = \sum_{k=0}^{n} \binom{n}{k} \frac{d^k}{dx^k}(f(x)) \frac{d^{n-k}}{dx^{n-k}}(g(x))\,,$$

---

[4]Write out the formula for $n = 1, 2, 3$, then construct an inductive argument.

where $\binom{n}{k}$ denotes the binomial coefficient $\binom{n}{k} = n!/k!(n-k)!$. Applying this to the problem at hand, we have

$$
\begin{aligned}
\frac{\partial^{n+m}}{\partial x^n \partial y^m} e^{xy} &= \frac{\partial^n}{\partial x^n}\left(\frac{\partial^m}{\partial y^m} e^{xy}\right) \\
&= \frac{\partial^n}{\partial x^n}(x^m e^{xy}) \\
&= \sum_{k=0}^{n}\binom{n}{k}\frac{\partial^k}{\partial x^k}(x^m)\frac{\partial^{n-k}}{\partial x^{n-k}}(e^{xy}) \\
&= \sum_{k=0}^{n}\binom{n}{k}m\cdot(m-1)\cdots(m-k+1)x^{m-k}y^{n-k}e^{xy} \\
&= e^{xy}\sum_{k=0}^{n}\binom{n}{k}\frac{m!}{(m-k)!}x^{m-k}y^{n-k} \\
&= e^{xy}\sum_{k=0}^{n}k!\binom{n}{k}\binom{m}{k}x^{m-k}y^{n-k},
\end{aligned}
$$

and the result follows.

77. Observe that $f(x,0) = 0$ for all x and $f(0,y) = 0$ for all y. It follows that $f_x(0,0) = 0 = f_y(0,0)$. However, f is not continuous at $(0,0)$ because $\lim_{x\to(0,0)} f(x,y)$ does not exist. Indeed, the limiting value of $f(x,y)$ is 0 if (x,y) approaches $(0,0)$ along either coordinate axis, but the limiting value is $1/2$ if (x,y) approaches $(0,0)$ along the line $y = x$.

Calculator/Computer Exercises

79, 81. The first entry defines the function f, the coordinates x0 and y0, and the step sizes h and k (which we take to be 0.01). The transformation rule for f is displayed.

```
> f := (x,y) -> (sqrt(x) + exp(y-x))/sqrt(3 + x^2 + y^4);
  x0,y0 := 4.0,3.0: h,k := 0.01,0.01:
```

$$
f := (x,y) \to \frac{\sqrt{x} + e^{y-x}}{\sqrt{3 + x^2 + y^4}}
$$

Approximate values for $f_x(x_0,y_0)$ and $f_y(x_0,y_0)$ are obtained by adapting the approximate derivative formula

$$
g'(x_0) \approx \frac{g(x_0+h) - g(x_0-h)}{2h}.
$$

```
> fx = (f(x0+h,y0) - f(x0-h,y0))/(2*h);
  fy = (f(x0,y0+k) - f(x0,y0-k))/(2*k);
```

$$
fx = -0.02126011500
$$

$$fy = -0.09107770000$$

Approximate values for $f_{xx}(x_0, y_0)$ and $f_{yy}(x_0, y_0)$ are obtained by adapting the approximate second derivative formula

$$g''(x_0) \approx \frac{g(x_0 + h) - 2g(x0) + g(x_0 - h)}{h^2} \, .$$

```
> fxx = (f(x0+h,y0) - 2*f(x0,y0) + f(x0-h,y0))/h^2;
  fyy = (f(x0,y0+k) - 2*f(x0,y0) + f(x0,y0-k))/k^2;
```

$$fxx = 0.03337500000$$

$$fyy = 0.07633400000$$

The last entry checks the first and second x-partials.

```
> Check;
  'D[1](f)(x0,y0)' = D[1](f)(x0,y0);
  'D[1,1](f)(x0,y0)' = D[1,1](f)(x0,y0);
```

Check

$$\mathrm{D}_1(f)(x0, y0) = -0.02125946188$$

$$\mathrm{D}_{1,1}(f)(x0, y0) = 0.03337468234$$

11.5 Differentiability and the Chain Rule

Problems for Practice

1. (a) (Chain Rule)

$$\begin{aligned}
\frac{dz}{ds} &= \frac{\partial z}{\partial x} \cdot \frac{dx}{ds} + \frac{dz}{\partial y} \cdot \frac{dy}{ds} \\
&= 2xy \cdot 3s^2 + (x^2 - 3y^2) \cdot (-s^{-2}) \\
&= 2(s^3 - 1)s^{-1} \cdot 3s^2 - \frac{(s^3 - 1)^2 - 3(s^{-1})^2}{s^2} \\
&= 5s^4 - 4s - s^{-2} + 3s^{-4}
\end{aligned}$$

   (b) (Subsitution) $z = f(s^3 - 1, s^{-1}) = (s^3 - 1)^2 s^{-1} - (s^{-1})^3$
   $$= s^5 - 2s^2 + s^{-1} - s^{-3} \, .$$

   Therefore, $dz/ds = 5s^4 - 4s - s^{-2} + 3s^{-4} \, .$

3. (a) (Chain Rule)

$$\frac{dz}{ds} = \frac{\partial z}{\partial x} \cdot \frac{dx}{ds} + \frac{dz}{\partial y} \cdot \frac{dy}{ds}$$

$$= \frac{y^2 - x^2}{(x^2 + y^2)^2} \cdot (-\sin(s)) - \frac{2xy}{(x^2 + y^2)^2} \cdot \cos(s)$$

$$= (\sin^2(s) - \cos^2(s)) \cdot (-\sin(s)) - 2\cos(s)\sin(s) \cdot \cos(s)$$

$$= -\sin^3(s) - \cos^2(s)\sin(s)$$

$$= -\sin(s)(\sin^2(s) + \cos^2(s)) = -\sin(s)$$

(b) (Subsitution) $z = f(\cos(s), \sin(s)) = \dfrac{\cos(s)}{\cos^2(s) + \sin^2(s)} = \cos(s)$.

Therefore, $dz/ds = -\sin(s)$.

5. (a) (Chain Rule)

$$\frac{dz}{ds} = \frac{\partial z}{\partial x} \cdot \frac{dx}{ds} + \frac{dz}{\partial y} \cdot \frac{dy}{ds}$$

$$= \frac{3x^2}{x^3 + y} \cdot \frac{1}{2\sqrt{s}} + \frac{1}{x^3 + y} \cdot \frac{3\sqrt{s}}{2}$$

$$= \frac{3s}{s^{3/2} + s^{3/2}} \cdot \frac{1}{2\sqrt{s}} + \frac{1}{s^{3/2} + s^{3/2}} \cdot \frac{3\sqrt{s}}{2}$$

$$= \frac{3\sqrt{s}}{2} \cdot \frac{1}{s\sqrt{s}} = \frac{3}{2s}$$

(b) (Substitution) $z = f(s^{1/2}, s^{3/2}) = \ln(s^{3/2} + s^{3/2}) = \ln(2s^{3/2}) = \ln(2) + (3/2)\ln(s)$. Therefore, $dz/ds = (3/2) \cdot (1/s) = 3/2s$.

7. (a) (Chain Rule)

$$\frac{dz}{ds} = \frac{\partial z}{\partial x} \cdot \frac{dx}{ds} + \frac{dz}{\partial y} \cdot \frac{dy}{ds}$$

$$= 2e^{2x-3y} \cdot (-\sin(s)) - 3e^{2x-3y} \cdot \cos(s)$$

$$= -(2\sin(s) + 3\cos(s))e^{2\cos(s)-3\sin(s)}$$

(b) (Subsitution) $z = f(\cos(s), \sin(s)) = e^{2\cos(s)-3\sin(s)}$.

Therefore, $dz/ds = -(2\sin(s) + 3\cos(s))e^{2\cos(s)-3\sin(s)}$.

9. (a) (Chain Rule)

$$\frac{dz}{ds} = \frac{\partial z}{\partial x} \cdot \frac{dx}{ds} + \frac{dz}{dy} \cdot \frac{dy}{ds}$$

$$= -\frac{2y}{(x-y)^2} \cdot (-2s\sin(s^2)) + \frac{2x}{(x-y)^2} \cdot 2s\cos(s^2)$$

$$= \frac{4s(\sin^2(s^2) + \cos^2(s^2))}{(\cos^2(s) - \sin^2(s))^2}$$

$$= \frac{4s}{(\cos^2(s) - \sin^2(s))^2}$$

(b) (Subsitution) $z = f(\cos(s^2), \sin(s^2)) = \dfrac{\cos(s^2) + \sin(s^2)}{\cos(s^2) - \sin(s^2)}$.

Calculating dz/ds using the quotient rule starts with

$$\frac{(\cos(s^2) - \sin(s^2))(-2s\sin(s^2) + 2s\cos(s^2)) - (\cos(s^2) + \sin(s^2))(-2s\sin(s)^2 - 2s\cos(s^2))}{(\cos(s^2) - \sin(s^2))^2}$$

which, upon expansion of the numerator, can be simplified to $\dfrac{4s}{(\cos^2(s) - \sin^2(s))^2}$.
Verify!

11. Partial derivative calculations using the Chain Rule

$$\frac{\partial z}{\partial s} = \frac{\partial z}{\partial x} \cdot \frac{\partial x}{\partial s} + \frac{\partial z}{\partial y} \cdot \frac{\partial y}{\partial s}$$

$$= 2x \cdot 3t\cos(3st) - 7y^6 \cdot (-3t\sin(3st))$$

$$= 2\sin(3st) \cdot 3t\cos(3st) - 7\cos^6(3st) \cdot (-3t\sin(3st))$$

$$= 3t\sin(3st)\cos(3st)(2 + 7\cos^5(3st))$$

A similar calculation will show that $\partial z/\partial t = 3s\sin(3st)\cos(3st)(2 + 7\cos^5(3st))$.

13. Partial derivative calculations using the Chain Rule

$$\frac{\partial z}{\partial s} = \frac{\partial z}{\partial x} \cdot \frac{\partial x}{\partial s} + \frac{\partial z}{\partial y} \cdot \frac{\partial y}{\partial s}$$

$$= 3ye^{3xy} \cdot t^{-1} + 3xe^{3xy} \cdot (-ts^{-2})$$

$$= 3(ts^{-1})e^3 \cdot t^{-1} - 3(st^{-1})e^3 \cdot (ts^{-2})$$

$$= 0$$

A similar calculation will show that $\partial z/\partial t = 0$ also.

15. Partial derivative calculations using the Chain Rule

$$\frac{\partial z}{\partial s} = \frac{\partial z}{\partial x} \cdot \frac{\partial x}{\partial s} + \frac{\partial z}{\partial y} \cdot \frac{\partial y}{\partial s}$$

$$= \frac{x}{\sqrt{x^2 - y^2}} \cdot \left(-\frac{2s}{(s^2 - t^2)^2} \right) - \frac{y}{\sqrt{x^2 - y^2}} \cdot \left(-\frac{2s}{(s^2 + t^2)^2} \right)$$

$$= \left(\frac{2s}{(s^2 + t^2)^3} - \frac{2s}{(s^2 - t^2)^3} \right) \Big/ \sqrt{x^2 - y^2}$$

$$= -\frac{4st^2(3s^4 + t^4)}{(s^2 + t^2)^3(s^2 - t^2)^3} \Big/ \sqrt{\frac{1}{(s^2 - t^2)^2} - \frac{1}{(s^2 + t^2)^2}}$$

$$= -\frac{2st^2(3s^4 + t^4)}{(s^2 + t^2)^2(s^2 - t^2)^3} \sqrt{\frac{(s^2 - t^2)^2}{s^2 t^2}}$$

$$= -\frac{2st^2(3s^4 + t^4)}{(s^2 + t^2)^2(s^2 - t^2)^3} \left| \frac{s^2 - t^2}{st} \right|$$

If s and t are the same sign and $s^2 > t^2$, then $\frac{\partial z}{\partial s} = -\frac{2t(3s^4 + t^4)}{(s^4 - t^4)^2}$. A similar calculation will show that

$$\frac{\partial z}{\partial t} = \frac{2s^2 t(s^4 + 3t^4)}{(s^2 + t^2)^2(s^2 - t^2)^3} \left| \frac{s^2 - t^2}{st} \right|,$$

and if s and t are the same sign and $s^2 > t^2$, then $\frac{\partial z}{\partial t} = \frac{2s(s^4 + 3t^4)}{(s^4 - t^4)^2}$.

17. Partial derivative calculations using the Chain Rule

$$\frac{\partial z}{\partial s} = \frac{\partial z}{\partial x} \cdot \frac{\partial x}{\partial s} + \frac{\partial z}{\partial y} \cdot \frac{\partial y}{\partial s}$$

$$= 2e^{2x-y} \cdot \frac{1}{s - t} - e^{2x-y} \cdot \left(-\frac{2}{2s + 3t} \right)$$

$$= 2e^{2x-y} \cdot \left(\frac{1}{s - t} - \frac{1}{2s + 3t} \right)$$

$$= 2\frac{(s - t)^2}{2s + 3t} \cdot \frac{s + 4t}{(s - t)(2s + 3t)}$$

$$= 2\frac{(s - t)(s + 4t)}{(2s + 3t)^2}$$

A similar calculation will show that $\partial z / \partial t = -(s - t)(7s + 3t)/(2s + 3t)^2$.

19. Derivative calculation using the Chain Rule

$$\frac{dw}{ds} = \frac{\partial w}{\partial x} \cdot \frac{\partial x}{\partial s} + \frac{\partial w}{\partial y} \cdot \frac{\partial y}{\partial s} + \frac{\partial w}{\partial z} \cdot \frac{\partial z}{\partial s}$$

$$= yz^2 \cdot e^s + xz^2 \cdot 3s^2 + 2xyz \cdot 3(s-2)^2$$

$$= s^3(s-2)^6 \cdot e^s + e^s(s-2)^6 \cdot 3s^2 + 2e^s s^3(s-2)^3 \cdot 3(s-2)^2$$

$$= s^2(s-2)^5(s^2 + 7s - 6)e^s$$

21. Derivative calculation using the Chain Rule

$$\frac{dw}{ds} = \frac{\partial w}{\partial x} \cdot \frac{\partial x}{\partial s} + \frac{\partial w}{\partial y} \cdot \frac{\partial y}{\partial s} + \frac{\partial w}{\partial z} \cdot \frac{\partial z}{\partial s}$$

$$= \frac{x}{\sqrt{x^2 - y^3 + z}} \cdot 9s^2 - \frac{3y^2}{2\sqrt{x^2 - y^3 + z}} \cdot 4s + \frac{1}{2\sqrt{x^2 - y^3 + z}} \cdot s^{-1}$$

$$= \frac{54s^5 - 48s^5 + s^{-1}}{2\sqrt{9s^6 - 8s^6 + \ln(s)}} = \frac{6s^5 + s^{-1}}{2\sqrt{s^6 + \ln(s)}}$$

23. Derivative calculation using the Chain Rule

$$\frac{dw}{ds} = \frac{\partial w}{\partial x} \cdot \frac{\partial x}{\partial s} + \frac{\partial w}{\partial y} \cdot \frac{\partial y}{\partial s} + \frac{\partial w}{\partial z} \cdot \frac{\partial z}{\partial s}$$

$$= yz\cos(xyz) \cdot 3e^{3s} + xz\cos(xyz) \cdot (-4s^{-5}) + xy\cos(xyz) \cdot 7s^6$$

$$= \cos(s^3 e^{3s})(yz \cdot (3e^{3s}) + xz \cdot (-4s^{-5}) + xy \cdot 7s^6)$$

$$= e^{3s}\cos(s^3 e^{3s})(3s^3 + 3s^2)$$

$$= 3s^2(s+1)e^{3s}\cos(s^3 e^{3s})$$

25. Derivative calculation using the Chain Rule

$$\frac{dw}{ds} = \frac{\partial w}{\partial x} \cdot \frac{\partial x}{\partial s} + \frac{\partial w}{\partial y} \cdot \frac{\partial y}{\partial s} + \frac{\partial w}{\partial z} \cdot \frac{\partial z}{\partial s}$$

$$= \frac{2x}{y^2 + z^2} \cdot 8^s \ln(8) - \frac{2x^2 y}{(y^2 + z^2)^2} \cdot 2^s \ln(2) - \frac{2x^2 z}{(y^2 + z^2)^2} \cdot 4^s \ln(4)$$

$$= \frac{2x(y^2 + z^2) \cdot x \ln(8) - 2x^2 y \cdot y \ln(2) - 2x^2 z \cdot z \ln(4)}{(y^2 + z^2)^2}$$

$$= \frac{6x^2(y^2 + z^2)\ln(2) - 2x^2(y^2 + 2z^2)\ln(2)}{(y^2 + z^2)^2}$$

$$= \frac{4x^2 y^2 + 2x^2 z^2}{(y^2 + z^2)^2}\ln(2) = \frac{4 \cdot 8^{2s} \cdot 2^{2s} + 2 \cdot 8^{2s} \cdot 4^{2s}}{(2^{2s} + 4^{2s})^2}\ln(2)$$

$$= \frac{2^2 \cdot 2^{6s} \cdot 2^{2s} + 2 \cdot 2^{6s} \cdot 2^{4s}}{(2^{2s} + 4^{2s})^2}\ln(2) = \frac{2^{8s+1}(2 + 2^{2s})}{2^{4s}(1 + 2^{2s})^2}\ln(2)$$

$$= \frac{2^{4s+1}(2 + 2^{2s})}{(1 + 2^{2s})^2}\ln(2)$$

27. Observe that $f(1, \pi) = 0$. Moreover,

$$f_x(x, y) = y/x - y\cos(xy) \qquad \Longrightarrow \qquad f_x(1, \pi) = 2\pi$$
$$f_y(x, y) = \ln(x) - x\cos(xy) \qquad \Longrightarrow \qquad f_y(1, \pi) = 1$$
$$f_{xx}(x, y) = -y/x^2 + y^2\sin(xy) \qquad \Longrightarrow \qquad f_{xx}(1, \pi) = -\pi$$
$$f_{xy}(x, y) = 1/x + xy\sin(xy) - \cos(xy) \qquad \Longrightarrow \qquad f_{xy}(1, \pi) = 2$$
$$f_{yy}(x, y) = x^2\sin(xy) \qquad \Longrightarrow \qquad f_{yy}(1, \pi) = 0$$

Therefore, the second order Taylor polynomial of f at $(1, \pi)$ is

$$2\pi(x - 1) + y - \pi + \frac{1}{2}(\pi(x - 1)^2 + 2(x - 1)(y - \pi))$$

29. Observe that $f(0, 0) = 1$. Moreover,

$$f_x(x, y) = \frac{-y}{(1+xy)^2} \qquad \Longrightarrow \qquad f_x(0, 0) = 0$$
$$f_y(x, y) = \frac{-x}{(1+xy)^2} \qquad \Longrightarrow \qquad f_y(0, 0) = 0$$
$$f_{xx}(x, y) = \frac{2y^2}{(1+xy)^3} \qquad \Longrightarrow \qquad f_{xx}(0, 0) = 0$$
$$f_{xy}(x, y) = \frac{xy-1}{(1+xy)^3} \qquad \Longrightarrow \qquad f_{xy}(0, 0) = -1$$
$$f_{yy}(x, y) = \frac{2x^2}{(1+xy)^3} \qquad \Longrightarrow \qquad f_{yy}(0, 0) = 0$$

All third partials evaluate to 0 at $(0, 0)$ (verify). Therefore, the third order Taylor polynomial of f at $(0, 0)$ is

$$1 - xy$$

31. Observe that $f(\pi/4, \pi/4) = 1$. Moreover,

$$f_x(x, y) = 2\sec^2(2x - y) \qquad \Longrightarrow \qquad f_x(\pi/4, \pi/4) = 4$$
$$f_y(x, y) = -\sec^2(2x - y) \qquad \Longrightarrow \qquad f_y(\pi/4, \pi/4) = -2$$
$$f_{xx}(x, y) = 4\sec^2(2x - y)\tan(2x - y) \qquad \Longrightarrow \qquad f_{xx}(\pi/4, \pi/4) = 16$$
$$f_{xy}(x, y) = -2\sec^2(2x - y)\tan(2x - y) \qquad \Longrightarrow \qquad f_{xy}\pi/4, \pi/4) = -8$$
$$f_{yy}(x, y) = \sec^2(2x - y)\tan(2x - y) \qquad \Longrightarrow \qquad f_{yy}(\pi/4, \pi/4) = 4$$

Therefore, the second order Taylor polynomial of f at $(\pi/4, \pi/4)$ is

$$1 + 4(x - \pi/4) - 2(y - \pi/4) + \frac{1}{2}(16(x - \pi/4)^2 - 16(x - \pi/4)(y - \pi/4) + 4(y - \pi/4)^2)$$

Further Theory and Practice

33. The calculation of $\partial z/\partial u$ can go like this.

$$\frac{\partial z}{\partial u} = \frac{\partial z}{\partial x}\frac{\partial x}{\partial u} + \frac{\partial z}{\partial y}\frac{\partial y}{\partial u}$$

$$= \frac{\partial z}{\partial x}\left(\frac{\partial x}{\partial s}\frac{\partial s}{\partial u} + \frac{\partial x}{\partial t}\frac{\partial t}{\partial u}\right) + \frac{\partial z}{\partial y}\left(\frac{\partial y}{\partial s}\frac{\partial s}{\partial u} + \frac{\partial y}{\partial t}\frac{\partial t}{\partial u}\right)$$

$$= \frac{\partial z}{\partial x}\frac{\partial x}{\partial s}\frac{\partial s}{\partial u} + \frac{\partial z}{\partial x}\frac{\partial x}{\partial t}\frac{\partial t}{\partial u} + \frac{\partial z}{\partial y}\frac{\partial y}{\partial s}\frac{\partial s}{\partial u} + \frac{\partial z}{\partial y}\frac{\partial y}{\partial t}\frac{\partial t}{\partial u}$$

A schematic diagram like the one in Figure 5 is helpful. A similar calculation will show that

$$\frac{\partial z}{\partial v} = \frac{\partial z}{\partial x}\frac{\partial x}{\partial s}\frac{\partial s}{\partial v} + \frac{\partial z}{\partial x}\frac{\partial x}{\partial t}\frac{\partial t}{\partial v} + \frac{\partial z}{\partial y}\frac{\partial y}{\partial s}\frac{\partial s}{\partial v} + \frac{\partial z}{\partial y}\frac{\partial y}{\partial t}\frac{\partial t}{\partial v}.$$

35. Let S denote the area of the surface of the box, $S = 2(lw + lh + wh)$. Using the Chain Rule,

$$\frac{dS}{dt} = \frac{\partial S}{\partial l}\frac{dl}{dt} + \frac{\partial S}{\partial w}\frac{dw}{dt} + \frac{\partial S}{\partial h}\frac{dh}{dt}$$

$$= 2(w + h)\frac{dl}{dt} + 2(l + h)\frac{dw}{dt} + 2(w + l)\frac{dh}{dt}.$$

Therefore,

$$\left.\frac{dS}{dt}\right|_{\substack{l=15 \\ w=8 \\ h=5}} = 26\frac{dl}{dt} + 40\frac{dw}{dt} + 46\frac{dh}{dt}$$

$$= 26 \cdot (0.01) + 40 \cdot (-0.02) + 46 \cdot (0.005)$$

$$= -0.31.$$

The surface area is decreasing at the rate of 0.31 square centimeters per second.

37. A formula for the partial derivative $g_x(x, y)$ can be obtained by holding y constant and differentiating the function $x \to g(x, y) = \phi(f(x, y))$. Apply the Chain Rule for functions of a single variable to obtain

$$g_x(x, y) = \phi'(f(x, y))\, f_x(x, y).$$

Similarly,

$$g_y(x, y) = \phi'(f(x, y))\, f_y(x, y).$$

Let $w = \phi(z)$ and $z = f(x, y)$ to obtain the following schematic diagram.

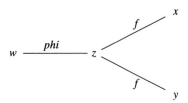

For example, if $g(x, y) = \sin(x^2 \ln(y))$, then

$$g_x(x, y) = \cos(x^2 \ln(y)) \cdot 2x \ln(y),$$

and

$$g_y(x, y) = \cos(x^2 \ln(y)) \cdot x^2/y.$$

39. Apply the differentiation formula in Exercise 37 to obtain

$$\frac{\partial u}{\partial x}(x,y) = a\phi(ax+by) \qquad \text{and} \qquad \frac{\partial u}{\partial y}(x,y) = b\phi(ax+by).$$

It follows that $bu_x(x,y) = ab\phi(ax+by) = au_y(x,y)$.

41. Apply the differentiation formula in Exercise 37 to obtain

$$\frac{\partial v}{\partial x}(x,y) = 2x\phi'(x^2+y^2) \qquad \text{and} \qquad \frac{\partial v}{\partial y}(x,y) = 2y\phi'(x^2+y^2).$$

It follows that $yv_x(x,y) = 2xy\phi'(x^2+y^2) = xv_y(x,y)$.

43. Apply the Chain Rule to obtain

$$v_\xi(\xi,\eta) = \frac{1}{2}y_x\left(\frac{\xi+\eta}{2},\frac{\xi-\eta}{2c}\right) + \frac{1}{2c}y_t\left(\frac{\xi+\eta}{2},\frac{\xi-\eta}{2c}\right).$$

Consequently, another application of the Chain Rule yields

$$v_{\xi\eta}(\xi,\eta) = \frac{1}{4}y_{xx}\left(\frac{\xi+\eta}{2},\frac{\xi-\eta}{2c}\right) - \frac{1}{4c}y_{xt}\left(\frac{\xi+\eta}{2},\frac{\xi-\eta}{2c}\right)$$
$$+ \frac{1}{4c}y_{tx}\left(\frac{\xi+\eta}{2},\frac{\xi-\eta}{2c}\right) - \frac{1}{4c^2}y_{tt}\left(\frac{\xi+\eta}{2},\frac{\xi-\eta}{2c}\right)$$
$$= 0.$$

The fact that $v_{\xi\eta}(\xi,\eta) = 0$ implies that $v_\xi(\xi,\eta) = f(\xi)$. From this we may infer that $v(\xi,\eta) = \phi(\xi) + \psi(\eta)$, where $\phi' = f$.

Since $(\xi+\eta)/2 = x$ and $(\xi-\eta)/2c = t$ it follows that $y(x,t) = v(\xi,\eta) = \phi(\xi) + \psi(\eta) = \phi(x+ct) + \psi(x-ct)$.

45. The function $\phi : x \mapsto f(x,y(x))$ is constant. Therefore, $\phi'(x) = 0$. Apply the Chain Rule to obtain $f_x(x,y) + f_y(x,y)y'(x) = 0$. Consequently, $y'(x) = -f_x(x,y)/f_y(x,y)$.

47. This takes some care. Let $z = f(x,y)$.

$$\frac{\partial}{\partial x}f(z,f(x,z)) = f_x(z,f(x,z))\frac{\partial z}{\partial x} + f_y(z,f(x,z))\frac{\partial}{\partial x}f(x,z)$$

$$= f_x(z,f(x,z))f_x(x,y) + f_y(z,f(x,z))\left(f_x(x,z) + f_y(x,z)\frac{\partial z}{\partial x}\right)$$

$$= f_x(f(x,y),f(x,z))f_x(x,y) + f_y(f(x,y),f(x,f(x,y)))f_x(x,f(x,y))$$
$$+ f_y(f(x,y),f(x,f(x,y)))f_y(x,f(x,y)))f_x(x,y)$$

Calculator/Computer Exercises

49. The first entry defines the functions and the constants. A step size of 0.001 is used for the derivative approximations. Only the function formulas are displayed.

```
> F,rho,sigma := unapply((x+y)/(x+1),x,y),
                 unapply(sqrt(s*t+t),s,t),
                 unapply(2^(s-t),s,t):
  f := (s,t) -> F(rho(s,t),sigma(s,t)):
  s0,t0,h,k := 3.,1.,0.001,0.001:
  x0,y0 := rho(s0,t0),sigma(s0,t0):
  'F(x,y)'=F(x,y), 'rho(s,t)'=rho(s,t), 'sigma(s,t)'=sigma(s,t);
  'f(s,t)' = f(s,t);
```

$$F(x,y) = \frac{x+y}{x-y}, \ \rho(s,t) = \sqrt{st+t}, \ \sigma(s,t) = 2^{s-t}$$

$$f(s,t) = \frac{\sqrt{st+t} + 2^{s-t}}{\sqrt{st+t}+1}$$

The partial derivatives of ρ, σ, and F are calculated using central difference quotients.

```
> rh_s := (rho(s0+h,t0) - rho(s0-h,t0))/2/h;
  rh_t := (rho(s0,t0+k) - rho(s0,t0-k))/2/k;
  si_s := (sigma(s0+h,t0) - sigma(s0-h,t0))/2/h;
  si_t := (sigma(s0,t0+k) - sigma(s0,t0-k))/2/k;
  F_x := (F(x0+h,y0) - F(x0-h,y0))/2/h;
  F_y := (F(x0,y0+k) - F(x0,y0-k))/2/k;
```

$$rh\_s := 0.2500000000$$

$$rh\_t := 1.000000000$$

$$si\_s := 2.772589000$$

$$si\_t := -2.772589000$$

$$F\_x := -.3333330000$$

$$F\_y := 0.3333330000$$

$f_s(s_0, t_0)$ and $f_t(s_0, t_0)$ are calculated using central difference quotients.

```
> f_s := (f(s0+h,t0) - f(s0-h,t0))/2/h;
  f_t := (f(s0,t0+k) - f(s0,t0-k))/2/k;
```

$$f\_s := 0.8408630000$$

$$f\_t := -1.257530000$$

$f_s(s_0, t_0)$ and $f_t(s_0, t_0)$ are found using the Chain Rule and the approximate derivatives.

```
> CRf_s = F_x*rh_s + F_y*si_s;
  CRf_t = F_x*rh_t + F_y*si_t;
```

$$CRf\_s := 0.8408621591$$

$$CRf\_t := -1.257528409$$

Here are the values of $f_s(s_0, t_0)$ and $f_t(s_0, t_0)$ using the exact derivative formulas.

```
> D[1](f)(s0,t0) = evalf(D[1](f)(s0,t0));
  D[2](f)(s0,t0) = evalf(D[2](f)(s0,t0));
```

$$D_1(f)(3., 1.) = 0.8408629074$$

$$D_2(f)(3., 1.) = -1.257529574$$

51. See 49 and change the function definitions to fit this problem. The values of the partial derivatives using the exact derivative formulas are shown below.

```
> D[1](f)(s0,t0) = evalf(D[1](f)(s0,t0));
  D[2](f)(s0,t0) = evalf(D[2](f)(s0,t0));
```

$$D_1(f)(1., -1.) = 0.7220161984$$

$$D_2(f)(1., -1.) = 0.$$

11.6   Gradients and Directional Derivatives

Problems for Practice

1. $\nabla f(x, y) = \langle f_x(x, y), f_y(x, y) \rangle = \langle \sin(y), x \cos(y) \rangle$.

3. $\nabla f(x, y) = \langle f_x(x, y), f_y(x, y) \rangle = \langle y^2 \sec^2(xy^2), 2xy \sec^2(xy^2) \rangle$.

5. $\nabla f(x, y, z) = \langle f_x(x, y, z), f_y(x, y, z), f_z(x, y, z) \rangle$
$= \langle -y \sin(xy) \sin(yz), z \cos(xy) \sin(yz) - x \sin(xy) \sin(yz), y \cos(xy) \cos(yz) \rangle$.

7. $\nabla f(x, y) = \langle f_x(x, y), f_y(x, y) \rangle = \langle y^2 / \cos(y), xy(2 \cos(y) + y \sin(y)) / \cos^2(y) \rangle$.

9. $\nabla f(x, y, z) = \langle f_x(x, y, z), f_y(x, y, z), f_z(x, y, z) \rangle$
$= \langle yz \cos(xy), xz \cos(xy), \sin(xy) \rangle$.

11. $\nabla f(x, y) = \langle 2xy - y^3, x^2 - 3xy^2 \rangle$. Therefore,

$$D_{\mathbf{u}} f(2, -3) = \langle 15, -50 \rangle \cdot \langle -3/5, 4/5 \rangle = -49 .$$

13. $\nabla f(x, y) = \langle -2x \sin(x^2 - y^2), 2y \sin(x^2 - y^2) \rangle$. Therefore,

$$D_{\mathbf{u}} f(\sqrt{\pi}, \sqrt{\pi}/2) = \langle -\sqrt{2\pi}, \sqrt{2\pi}/2 \rangle \cdot \langle 1/\sqrt{2}, -1/\sqrt{2} \rangle = -3\sqrt{\pi}/2.$$

15. $\nabla f(x, y) = e^{4x - 7y} \langle 4, -7 \rangle$. Therefore,

$$D_{\mathbf{u}} f(0, \ln(2)) = \langle 1/32, -7/128 \rangle \cdot \langle 1, 0 \rangle = 1/32.$$

17. $\nabla f(x, y) = \langle -2x/(x^2 + y^2)^2, -2y/(x^2 + y^2)^2 \rangle$. Therefore,

$$D_{\mathbf{u}} f(2, 2) = \langle -1/16, -1/16 \rangle \cdot \langle 0, 1 \rangle = -1/16.$$

19. $\nabla f(x, y) = (x+y)^{-2} \langle \cos(xy)(xy+y^2) - \sin(xy), \cos(xy)(xy+x^2) - \sin(xy) \rangle$. Therefore,

$$D_{\mathbf{u}} f(0, 1) = \langle 1, 0 \rangle \cdot \langle 1/\sqrt{2}, -1/\sqrt{2} \rangle = 1/\sqrt{2}.$$

21. $\nabla f(x, y) = 3 \sin^2(x + y) \cos(x + y) \langle 1, 1 \rangle$. Therefore,

$$D_{\mathbf{u}} f(\pi/3, -\pi/6) = 3\sqrt{3}/8 \langle 1, 1 \rangle \cdot \langle -1, 0 \rangle = -3\sqrt{3}/8.$$

23. (a) $\mathbf{u}$ is the unit vector having the same direction as

$$\nabla f(P_0) = \langle 3x^2 y^2 - y, 2x^3 y - x \rangle \big|_{(2, \frac{1}{2})} = \langle 5/2, 6 \rangle.$$

Therefore, $\mathbf{u} = \langle 5/13, 12/13 \rangle$.

(b) $D_{\mathbf{u}}(f)(P_0) = \|\nabla f(P_0)\| = 13/2$.

(c) $\mathbf{v} = -\mathbf{u} = -\langle 5/13, 12/13 \rangle$.

(d) $D_{\mathbf{v}}(f)(P_0) = -13/2$.

25. (a) $\mathbf{u}$ is the unit vector having the same direction as

$$\nabla f(P_0) = \frac{\ln(x + y) - 1}{(x + y)^2} \langle 1, 1 \rangle \big|_{(1,1)} = \frac{\ln(2) - 1}{4} \langle 1, 1 \rangle.$$

Therefore, $\mathbf{u} = \langle 1/\sqrt{2}, 1/\sqrt{2} \rangle$.

(b) $D_{\mathbf{u}}(f)(P_0) = \|\nabla f(P_0)\| = \sqrt{2}\,(1 - \ln(2))/4$.

(c) $\mathbf{v} = -\mathbf{u} = -\langle 1/\sqrt{2}, 1/\sqrt{2} \rangle$.

(d) $D_{\mathbf{v}}(f)(P_0) = -\sqrt{2}\,(1 - \ln(2))/4$.

27. (a) $\mathbf{u}$ is the unit vector having the same direction as

$$\nabla f(P_0) = \frac{1}{\cos^2(x - y)} \langle \cos(2y), \cos(2x) \rangle \big|_{(\pi/2, \pi/3)} = -\frac{2}{3} \langle 1, 2 \rangle.$$

Therefore, $\mathbf{u} = -\frac{1}{\sqrt{5}} \langle 1, 2 \rangle$.

(b) $D_{\mathbf{u}}(f)(P_0) = \|\nabla f(P_0)\| = 2\sqrt{5}/3$.

(c) $\mathbf{v} = -\mathbf{u} = \frac{1}{\sqrt{5}}\langle 1, 2\rangle$.

(d) $D_{\mathbf{v}}(f)(P_0) = -2\sqrt{5}/3$.

29. (a) $\mathbf{u}$ is the unit vector having the same direction as

$$\nabla f(P_0) = -e^{-x}\langle \cos(y), \sin(y)\rangle\Big|_{(1,\pi)} = e^{-1}\langle 1, 0\rangle.$$

Therefore, $\mathbf{u} = \langle 1, 0\rangle$.

(b) $D_{\mathbf{u}}(f)(P_0) = \|\nabla f(P_0)\| = e^{-1}$.

(c) $\mathbf{v} = -\mathbf{u} = -\langle 1, 0\rangle$.

(d) $D_{\mathbf{v}}(f)(P_0) = -e^{-1}$.

31. (a) $\mathbf{u}$ is the unit vector having the same direction as

$$\nabla f(P_0) = 5(x^2 + y)^4 \langle 2x, 1\rangle\Big|_{(1,1)} = 80\langle 2, 1\rangle.$$

Therefore, $\mathbf{u} = \frac{1}{\sqrt{5}}\langle 2, 1\rangle$.

(b) $D_{\mathbf{u}}(f)(P_0) = \|\nabla f(P_0)\| = 80\sqrt{5}$.

(c) $\mathbf{v} = -\frac{1}{\sqrt{5}}\langle 2, 1\rangle$.

(d) $D_{\mathbf{v}}(f)(P_0) = -80\sqrt{5}$.

33. Since $\nabla F(x, y, z) = \langle yz^3, xz^3, 3xyz^2\rangle$,

$$D_{\mathbf{u}}(F)(2, 1, -3) = \langle -27, -54, 54\rangle \cdot \langle 2/3, -2/3, 1/3\rangle = 36.$$

35. Since $\nabla F(x, y, z) = \langle yz\cos(xz), \sin(xz), xy\cos(xz)\rangle$,

$$D_{\mathbf{u}}(F)(1/2, 3, \pi) = \langle 0, 1, 0\rangle \cdot \langle \sqrt{3}/8, -6/8, 5/8\rangle = -3/4.$$

37. Since $\nabla F(x, y, z) = (xy + z^3)^{-1}\langle y, x, 3z^2\rangle$,

$$D_{\mathbf{u}}(F)(1, 1, 1) = (1/2)\langle 1, 1, 3\rangle \cdot \langle -1/2, 3/4, \sqrt{3}/4\rangle = (1 + 3\sqrt{3})/8.$$

39. Since $\nabla F(x, y, z) = \langle 2x(y^3 - yz), x^2(3y^2 - z), -x^2\rangle$,

(a) $\mathbf{u}$ is the unit vector having the same direction as

$$\nabla F(2, 1, 4) = \langle 12, -4, 4\rangle = 4\langle 3, -1, 1\rangle.$$

Therefore, $\mathbf{u} = \frac{1}{\sqrt{11}}\langle 3, -1, 1\rangle$.

(b) $D_{\mathbf{u}}(F)(P_0) = \|\nabla f(P_0)\| = 4\sqrt{11}$.

(c) $\mathbf{v} = -\frac{1}{\sqrt{11}}\langle 3, -1, 1\rangle$.

(d) $D_{\mathbf{v}}(F)(P_0) = -4\sqrt{11}$.

41. Since $\nabla F(x, y, z) = \cos(x^2 - y + z)\langle 2x, -1, 1\rangle$,

(a) $\mathbf{u}$ is the unit vector having the same direction as

$$\nabla F(0, \pi, 2\pi) = \langle 0, 1, -1\rangle.$$

Therefore, $\mathbf{u} = \frac{1}{\sqrt{2}}\langle 0, 1, -1\rangle$.

(b) $D_{\mathbf{u}}(F)(P_0) = \|\nabla f(P_0)\| = \sqrt{2}$.

(c) $\mathbf{v} = -\frac{1}{\sqrt{2}}\langle 0, 1, -1\rangle$.

(d) $D_{\mathbf{v}}(F)(P_0) = -\sqrt{2}$.

Further Theory and Practice

43. It says that there is no direction of greatest increase or decrease of f at P. The graph is "flat" at the point $(P, f(P))$. Examples: $f(x, y) = x^2 + y^2$, $f(x, y) = xy$, $P = (0, 0)$.

45. Since $\mathbf{v} = \overrightarrow{PQ} = \langle 3, 4\rangle$, the direction vector is $\mathbf{u} = \langle 3/5, 4/5\rangle$. The gradient vector at (x, y) is

$$\nabla f(x, y) = \frac{1}{2\sqrt{1 + xy}}\langle 2 + 3xy, x^2\rangle.$$

Therefore, the directional derivative from P to Q is

$$\begin{aligned} D_{\mathbf{u}}f(P) &= \nabla f(2, 4) \cdot \mathbf{u} \\ &= \frac{1}{6}\langle 26, 4\rangle \cdot \langle 3/5, 4/5\rangle \\ &= 58/15. \end{aligned}$$

47. Since $\mathbf{v} = \overrightarrow{PQ} = \langle -4\pi, 3\pi\rangle$, the direction vector is $\mathbf{u} = \langle -4/5, 3/5\rangle$. The gradient vector at (x, y) is

$$\nabla f(x, y) = \sec^2(xy^2)\langle y^2, 2xy\rangle.$$

Therefore, the directional derivative from P to Q is

$$\begin{aligned} D_{\mathbf{u}}f(P) &= \nabla f(\pi, 1/2) \cdot \mathbf{u} \\ &= 2\langle 1/4, \pi\rangle \cdot \langle -4/5, 3/5\rangle \\ &= -2/5 + 6\pi/5. \end{aligned}$$

49, 51. If $\mathbf{u} = \langle u_1, u_2 \rangle$ and $\mathbf{v} = \langle v_1, v_2 \rangle$, then the equations $D_{\mathbf{u}} f(P) = a$ and $D_{\mathbf{v}} f(P) = b$ can be expressed in the following form:

$$u_1 \frac{\partial f}{\partial x}(P) + u_2 \frac{\partial f}{\partial y}(P) = a$$

$$v_1 \frac{\partial f}{\partial x}(P) + v_2 \frac{\partial f}{\partial y}(P) = b .$$

The values of the two partial derivatives can be found by solving these two linear equations simultaneously.

49. In this case, the linear system can be simplified to

$$3 \frac{\partial f}{\partial x}(P) + 4 \frac{\partial f}{\partial y}(P) = 5$$

$$\frac{\partial f}{\partial x}(P) + \frac{\partial f}{\partial y}(P) = 2 .$$

Solve to obtain $(\partial f / \partial x)(P) = 3$, $(\partial f / \partial y)(P) = -1$.

51. In this case, the linear system can be simplified to

$$\frac{\partial f}{\partial x}(P) - \sqrt{3} \frac{\partial f}{\partial y}(P) = 6$$

$$\frac{\partial f}{\partial x}(P) + \frac{\partial f}{\partial y}(P) = 0 .$$

Solve to obtain $(\partial f / \partial x)(P) = 3\sqrt{3} - 3$, $(\partial f / \partial y)(P) = 3 - 3\sqrt{3}$.

53. Since $\nabla f(x, y) = \langle y, x - 4y^3 \rangle$, $\nabla f(8, -2) = \langle -2, 40 \rangle$. Therefore, the maximum value of the directional derivative at P is $\|\nabla f(8, -2)\| = \sqrt{1604}$ which is approximately 40.05.

55. The missile will move in the direction of the gradient, $\nabla T(100, 15, 8)$. Since $\nabla T(x, y, z) = \langle -20(x - 90), -8(y - 12)^3, -2z \rangle$,

$$\nabla T(100, 15, 8) = -\langle 200, 216, 16 \rangle = -8 \langle 25, 27, 2 \rangle .$$

Therefore, the direction is

$$\mathbf{u} = -\frac{1}{\sqrt{1358}} \langle 25, 27, 2 \rangle .$$

57. Since $\text{grad}(\phi)(x, y) = \langle \phi_x(x, y), \phi_y(x, y) \rangle$,

$$\text{div}(\text{grad}(\phi))(x, y) = \phi_{xx}(x, y) + \phi_{yy}(x, y) = \Delta \phi(x, y) .$$

59. Since $D_{\mathbf{u}}f(P_0) = f_x(P_0)u_1 + f_y(P_0)u_2$,

$$D_{\mathbf{u}}(D_{\mathbf{u}}f)(P_0) = \frac{\partial}{\partial x}(f_x(P_0)u_1 + f_y(P_0)u_2)u_1 + \frac{\partial}{\partial y}(f_x(P_0)u_1 + f_y(P_0)u_2)u_2$$
$$= f_{xx}(P_0)u_1^2 + 2f_{xy}(P_0)u_1u_2 + f_{yy}(P_0)u_2^2.$$

Similarly,

$$D_{\mathbf{v}}(D_{\mathbf{v}}f)(P_0) = f_{xx}(P_0)v_1^2 + 2f_{xy}(P_0)v_1v_2 + f_{yy}(P_0)v_2^2,$$

while

$$D_{\mathbf{u}}(D_{\mathbf{v}}f)(P_0) = \frac{\partial}{\partial x}(f_x(P_0)v_1 + f_y(P_0)v_2)u_1 + \frac{\partial}{\partial y}(f_x(P_0)v_1 + f_y(P_0)v_2)u_2$$
$$= f_{xx}(P_0)u_1v_1 + f_{xy}(P_0)(u_1v_2 + u_2v_1) + f_{yy}(P_0)u_2v_2.$$

If $D_{\mathbf{u}}(D_{\mathbf{u}}f)(P_0)$ is positive, then the slice of the graph of the surface $z = f(x, y)$ directly above the line through P_0 in the direction of $\mathbf{u}$ is concave up at $(x_0, y_0, f(x_0, y_0))$. If $D_{\mathbf{u}}(D_{\mathbf{u}}f)(P_0)$ is negative, then the slice of the surface is concave down at that point.

Calculator/Computer Exercises

61. The first entry loads the plots and VectorCalculus packages, defines the function f, and creates the plot of the level curves with the name LC.

```
> with(plots): with(VectorCalculus):
  f := (x,y) -> 1 - x^2*y/2 - x^2 - y^2 - 2*y:
  LC := contourplot( f(x,y), x=-1.25..4.5, y=-5..0.75,
                  contours=[(2*k-1)/2$k=0..4], color=black):
```

What follows defines the function UGrad : $\mathbf{x} \mapsto (\|\nabla f(\mathbf{x})\|^{-1})\nabla f(\mathbf{x})$, defines a list containing the position vectors of the points, then displays the level curve plot and the arrows.

```
> UGrad := unapply(
         Normalize(<D[1](f)(x[1],x[2]),D[2](f)(x[1],x[2])>),x
         ):
  P := [<3,-3/2>,<2,-2-sqrt(2)/2>,<2,-2+sqrt(2)/2>,
        <1,-1>,<1,-3/2>,<3,-3>,<3,-7/2>,<4,-5+sqrt(26)/2>]:
  display( LC,
         'arrow(P[k],UGrad(P[k]),head_length=0.15)'$k=1..8,
         scaling=constrained, tickmarks=[5,5]);
```

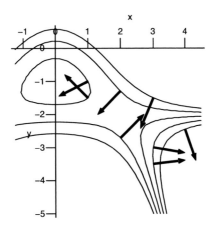

11.7 Tangent Planes

Problems for Practice

1. Since $\nabla f(x,y) = \langle y + 2x, x - 3y^2 \rangle$, $\nabla f(1,4) = \langle 6, -47 \rangle$. Therefore, the vector $\mathbf{N} = \langle 6, -47, -1 \rangle$ is normal to the plane and, since $f(1,4) = -59$, its equation is $6(x-1) - 47(y-4) - (z+59) = 0$ or $z = 6x - 47y + 123$.

3. Since $\nabla f(x,y) = (2 + x^2 + y^2)^{-1} \langle 2x, 2y \rangle$, $\nabla f(1,2) = \langle 2/7, 4/7 \rangle$. Therefore, the vector $\mathbf{N} = \langle 2/7, 4/7, -1 \rangle$ is normal to the plane and, since $f(1,2) = \ln(7)$, its equation is $\frac{2}{7}(x-1) + \frac{4}{7}(y-2) - (z - \ln(7)) = 0$ or $z = \frac{2}{7}x + \frac{4}{7}y - \frac{10}{7} + \ln(7)$.

5. Since $\nabla f(x,y) = (x+y)^{-2} \langle y, -x \rangle$, $\nabla f(2,1) = \langle 1/9, -2/9 \rangle$. Therefore, the vector $\mathbf{N} = \langle 1/9, -2/9, -1 \rangle$ is normal to the plane and, since $f(2,1) = 2/3$, its equation is $\frac{1}{9}(x-2) - \frac{2}{9}(y-1) - (z - 2/3) = 0$ or $z = \frac{1}{9}x - \frac{2}{9}y + \frac{2}{3}$.

7. Since $\nabla f(x,y) = e^{x+y} \langle y(1+x), x(1+y) \rangle$, $\nabla f(1,2) = e^3 \langle 4, 3 \rangle$. Therefore, the vector $\mathbf{N} = \langle 4e^3, 3e^3, -1 \rangle$ is normal to the plane and, since $f(1,2) = 2e^3$, its equation is $4e^3(x-1) + 3e^3(y-2) - (z - 2e^3) = 0$ or $z = e^3(4x + 3y - 8)$.

9. Since $\nabla f(x,y) = (x+y)^{-2} \langle x^2 + 2xy - y^2, y^2 - 2xy - x^2 \rangle$, $\nabla f(1,1) = \langle 1/2, 1/2 \rangle$. Therefore, the vector $\mathbf{N} = \langle 1/2, 1/2, -1 \rangle$ is normal to the plane and, since $f(1,1) = 1$, its equation is $\frac{1}{2}(x-1) + \frac{1}{2}(y-1) - (z-1) = 0$ or $z = \frac{1}{2}x + \frac{1}{2}y$.

11. Since $\nabla f(x,y) = (1/3)(x-y)^{-4/3} \langle -1, 1 \rangle$, $\nabla f(9,1) = \langle -1/48, 1/48 \rangle$. Therefore, the vector $\mathbf{N} = \langle -1/48, 1/48, -1 \rangle$ is normal to the plane and, since $f(9,1) = 1/2$, its equation is $-\frac{1}{48}(x-9) + \frac{1}{48}(y-1) - (z - 1/2) = 0$ or $z = \frac{1}{48}(-x + y + 32)$.

13. Since $\nabla F(x, y, z) = \langle 2x, -8y, 2z \rangle$ is normal to the surface at the point (x, y, z), $\nabla F(1, 2, 1) = \langle 2, -16, 2 \rangle$ is a normal at Q_0 (as is $\langle 1, -8, 1 \rangle$). The tangent plane has the equation $(x - 1) - 8(y - 2) + (z - 1) = 0$ or $x - 8y + z = -14$.

15. Since $\nabla F(x, y, z) = (x+z)^{-2} \langle -y, x+z, -y \rangle$ is normal to the surface at the point (x, y, z), $\nabla F(1, 8, 3) = \langle -1/2, 1/4, -1/2 \rangle$ is a normal at Q_0 (as is $\langle 2, -1, 2 \rangle$). The tangent plane has the equation $2(x-1)-(y-8)+2(z-3) = 0$ or $2x - y + 2z = 0$.

17. Let $F(x, y, z) = \ln(1 + x + y + z)$. Observe that the surface $F(x, y, z) = 3$ is the same as the surface $1 + x + y + z = e^3$ or $x + y + z = e^3 - 1$. Thus $\langle 1, 1, 1 \rangle$ is a normal vector at all points on the surface and the surface is its own tangent plane.

19. Let $F(x, y, z) = (x + y + z^2)^6$. The surface $F(x, y, z) = 1$ can be split into two pieces: $x+y+z^2 = 1$ and $x+y+z^2 = -1$. The point Q_0 is on the first piece. The normal at $Q_0 = (x_0, y_0, z_0)$ is the vector $\langle 1, 1, 2z_0 \rangle = \langle 1, 1, 2 \rangle$. Therefore, the equation of the tangent plane is $(x-3)+(y+3)+2(z-1) = 0$ or $x + y + 2z = 2$.

21. The surface is the level set $F(x, y, z) = 0$ for $F(x, y, z) = x^3 - y^5 - z$. Since $\nabla F(x, y, z) = \langle 3x^2, -5y^4, -1 \rangle$, $\nabla F(2, 1, 7) = \langle 12, -5, -1 \rangle$ is a normal vector at $(2, 1, 7)$. The normal line has the symmetric equations

$$\frac{x - 2}{12} = \frac{y - 1}{-5} = \frac{z - 7}{-1}.$$

23. The portion of the surface containing the point $(\pi/4, 0, \pi/4)$ is the level set $x + y + z = \pi/2$. All normal lines are parallel to $\mathbf{N} = \langle 1, 1, 1 \rangle$. The symmetric equations for this normal line are

$$x - \pi/4 = y = z - \pi/4.$$

25. The surface is the level set $F(x, y, z) = 0$ for $F(x, y, z) = \cos(x) - \sin(y) - z$. Since $\nabla F(x, y, z) = \langle -\sin(x), -\cos(y), -1 \rangle$, $\nabla F(0, \pi, 1) = \langle 0, 1, -1 \rangle$ is a normal vector at $(0, \pi, 1)$. The normal line has the symmetric equations

$$x = 0, \; y - \pi = 1 - z.$$

27. Since $f_x(x, y) = 2x^3/\sqrt{x^4 + y^2}$ and $f_y(x, y) = y/\sqrt{x^4 + y^2}$, the partial derivatives are $f_x(2, 3) = 16/5$ and $f_y(2, 3) = 3/5$. Moreover, $f(2, 3) = 5$, so the tangent plane approximation is

$$L(2.1, 2.9) = 5 + \frac{16}{5}(2.2 - 2) + \frac{3}{5}(2.9 - 3)$$

$$= 5.26$$

Observe that $f(2.1, 2.9) = 5.278\ldots$.

29. Since $f_x(x, y) = \cos(2x-y) - 2x\sin(2x-y)$ and $f_y(x, y) = x\sin(2x-y)$, the partial derivatives are $f_x(1, 2) = 1$ and $f_y(1, 2) = 0$. Moreover, $f(1, 2) = 1$, so the tangent plane approximation is

$$L(0.9, 2.2) = 1 + (0.9 - 1)$$
$$= 0.9$$

Observe that $f(0.9, 2.2) = 0.828\ldots$.

31. The area is $A = bh/2$. The error ΔA in the area induced by an error Δb in the base and an error Δh in the height is $\Delta A \approx (b\Delta h + h\Delta b)/2$. The percent error is obtained by dividing by A:

$$\frac{\Delta A}{A} \approx \frac{b\Delta h/2}{bh/2} + \frac{h\Delta b/2}{bh/2}$$
$$= \frac{\Delta h}{h} + \frac{\Delta b}{b}.$$

Therefore, the error in the area measurement might be as great as approximately $2 \times 1 = 2\%$. Since the exact area would be 250,000 square feet, the error might be as much as 5,000 square feet.

33. The volume is $V = lwh$. The error ΔV in the volume induced by an error Δl in the length, an error Δw in the width, and an error Δh in the height is $\Delta V \approx lw\Delta h + lh\Delta w + wh\Delta l$. The percent error is obtained by dividing by V:

$$\frac{\Delta V}{V} \approx \frac{lw\Delta h}{lwh} + \frac{lh\Delta w}{lwh} + \frac{wh\Delta l}{lwh}$$
$$= \frac{\Delta h}{h} + \frac{\Delta w}{w} + \frac{\Delta l}{l}.$$

Therefore, the error in the measurement of the volume might be as great as approxmately $3 \times 0.5 = 1.5\%$. Since the exact volume would be 72 cubic centimeters, the error might be as much as $72 \times 0.015 = 1.08$ cubic centimeters.

Further Theory and Practice

35. The value of the function at the point $(1, -1)$ has been omitted on the right side of the equation. It should be $z = 2 + 6(x - 1) + 6(y + 1)$.

37. The coefficients of the terms x and $y + 3$ are the partial derivatives, but they are should be evaluated at the point $(0, -3)$. The correct equation is $z = -3x + (y + 3)$.

39. If $x = t\cos(\theta)$, $y = t\sin(\theta)$, and $z = t$, then $x^2 + y^2 = t^2\cos^2(\theta) + t^2\sin^2(\theta) = t^2 = z^2$. Consequently, every point on the curve $t \mapsto \mathbf{r}_\theta(t)$ is on the surface $z^2 = x^2 + y^2$. Observe that the surface is circular cone with two nappes. Each curve is a straight line on the cone passing through the origin.

Since $\mathbf{r}'_\theta(t) = \cos(\theta)\,\mathbf{i} + \sin(\theta)\,\mathbf{j} + \mathbf{k}$, $\mathbf{r}'_\theta(0) = \cos(\theta)\,\mathbf{i} + \sin(\theta)\,\mathbf{j} + \mathbf{k}$. Let $\mathbf{v} = x\mathbf{i} + y\mathbf{j} + z\mathbf{k}$ be a vector that is perpendicular to $\mathbf{r}'_\theta(0)$ for all θ. Then $\mathbf{v} \cdot \mathbf{r}'_\theta(0) = 0$ so

$$x\cos(\theta) + y\sin(\theta) = -z$$

for all θ. Consequently, $x = -z$ and $-x = -z$ (let $\theta = 0$ then let $\theta = \pi$), and both x and z must be 0. But then $y = 0$ also (let $\theta = \pi/2$). Thus $\mathbf{v}$ is the zero vector.

The trajectories are all tangent lines to the surface at the origin. If the surface had a tangent plane at the origin, then these lines would be perpendicular to its normal. Since only the zero vector is perpendicular to all the lines, we conclude that this surface does not have a tangent plane at $(0,0,0)$.

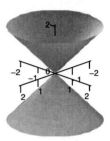

A circular cone with two nappes.

41. At a point (x, y, z), the first surface has normal $\langle 2x - 3z, 4, -3x \rangle$ and the second one has normal $\langle 3yz - 8, 2y + 3xz, 3xy \rangle$. At the intersection point $(0,0,0)$ the normals are $\langle 0, 4, 0 \rangle$ and $\langle -8, 0, 0 \rangle$. These vectors are perpendicular, so the angle between the tangent planes is $\pi/2$.

43. Substitute the equations defining the line into the equation for the sphere to obtain $50t^2 + 150t + 100 = 0$ or $t^2 + 3t + 2 = 0$. The intersection points (when $t = -1$ and $t = -2$) are $P = (3, 4, 0)$ and $Q = (0, 0 - 5)$. The angle between the line and the sphere is the angle between the line and the tangent plane to the sphere. This will be the complement of the angle between the line and a normal vector to the sphere at the point of intersection.

For the point P an outward normal to the sphere is $\mathbf{n}_P = \langle 3, 4, 0 \rangle$. The direction vector for the line as it exits the sphere is $\mathbf{v} = \langle 3, 4, 5 \rangle$, so the angle between the line and the normal vector is

$$\arccos\left(\frac{\mathbf{n}_P \cdot \mathbf{v}}{\|\mathbf{n}_P\|\,\|\mathbf{v}\|}\right) = \arccos\left(\frac{25}{\sqrt{25}\,\sqrt{50}}\right) = \arccos\left(\frac{1}{\sqrt{2}}\right) = \frac{\pi}{4}.$$

Consequently, this is also the angle between the line and the sphere at the point P.

For the point Q, an outward normal is $\mathbf{n}_Q = \langle 0, 0, -5 \rangle$. The direction vector for the line as it exits the sphere is $\mathbf{v} = -\langle 3, 4, 5 \rangle$, so the angle between the line and the normal vector is

$$\arccos\left(\frac{\mathbf{n}_Q \cdot \mathbf{v}}{\|\mathbf{n}_Q\|\,\|\mathbf{v}\|}\right) = \arccos\left(\frac{25}{\sqrt{25}\,\sqrt{50}}\right) = \arccos\left(\frac{1}{\sqrt{2}}\right) = \frac{\pi}{4}.$$

Consequently, this is also the angle between the line and the sphere at the point Q.

45. The volume of gold in the bead is $V = (4/3)\pi r^3 p$ where r is its radius and $100p$ is the percent of gold by volume. Let ΔV denote the error in V corresponding to an error of Δr in radius and Δp in gold content. Using the differential approximation, $\Delta V = 4\pi r^2 p \Delta r + (4/3)\pi r^3 \Delta p$. Divide both sides by V to obtain $\Delta V/V = 3(\Delta r/r) + \Delta p/p$. Consequently, to guarantee that $\Delta V/V$ is less than 0.01, the relative errors in r and p must satisfy $3 \cdot \frac{\Delta r}{r} + \frac{\Delta p}{p} < 0.01$. For example, a 0.3% error in the radius and a 0.1% error in the percentage of gold would be acceptable.

47. The surface is the graph of $z = \frac{x+y+2xy^2}{x-y}$ or $F(x,y,z) = 0$ where $F(x,y,z) = xz - yz - x - y - 2xy^2$. The vector $\nabla F(3,1,5) = \langle 2, -18, 2\rangle$ is normal to the surface at the point $(3,1,5)$, as is the vector $\langle 1, -9, 1\rangle$. Therefore, the tangent plane has the equation $(x-3) - 9(y-1) + (z-5) = 0$.

49. The surface is the graph of $z = (2x^3 + x^2 y - xy^3)^{1/3}$ or $F(x,y,z) = 0$ where $F(x,y,z) = z^3 - 2x^3 - x^2 y + xy^3$. The vector $\nabla F(-1,2,2) = \langle 6, -13, 12\rangle$ is normal to the surface at the point $(-1,2,2)$. Therefore, the tangent plane has the equation $6(x+1) - 13(y-2) + 12(z-2) = 0$.

51. The surface is the graph of $z = \sqrt{\frac{x+2y}{x-y}}$ or $F(x,y,z) = 0$ where $F(x,y,z) = z^2(x-y) - x - 2y$. The vector $\nabla F(2,1,2) = \langle 3, -6, 4\rangle$ is normal to the surface at the point $(2,1,2)$. Therefore, the tangent plane has the equation $3(x-2) - 6(y-1) + 4(z-2) = 0$.

53. The surface is the graph of $z = \frac{2+\ln(y+2)}{1+\ln(x)}$ or $F(x,y,z) = 0$ where $F(x,y,z) = z(1+\ln(x)) - 2 - \ln(y+2)$. The vector $\nabla F(1,-1,2) = \langle 2, -1, 1\rangle$ is normal to the surface at the point $(1,-1,2)$. Therefore, the tangent plane has the equation $2(x-1) - (y+1) + (z-2) = 0$.

55. The surface is the graph of $z = \arctan\left(\frac{y-x}{3x-y}\right)$ or $F(x,y,z) = 0$ where $F(x,y,z) = \tan(z)(3x-y) - y + x$. The vector $\nabla F(1,2,\pi/4) = \langle 4, -2, 2\rangle$ is normal to the surface at the point $(1,2,\pi/4)$ as is $\langle 2, -1, 1\rangle$. Therefore, the tangent plane has the equation $2(x-1) - (y-2) + (z-\pi/4) = 0$.

Calculator/Computer Exercises

57. Define f and the tangent plane function L. Then display the plot of the surface and the plot of the tangent plane. The ranges for x and y were determined by experimentation.

```
> f := (x,y) -> 3*x*y^2/(1+x^2+y^2):
  L := (x,y) -> f(2,1) + D[1](f)(2,1)*(x-2) + D[2](f)(2,1)*(y-1):
  plots[display](
  plot3d( f(x,y), x=1..3, y=0..2, grid=[13,13], lightmodel=light3),
  plot3d( L(x,y), x=1..3, y=0..2, style=patchnogrid, grid=[3,3],
          color = green), axes=boxed, orientation=[25,60],
          tickmarks=[5,5,5]   );
```

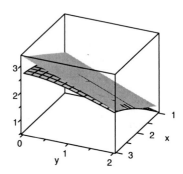

59. Change the definitions of f and L in Exercise 47. The following picture was obtained using the viewing cube $[1, 3] \times [2, 4] \times [-3, 9]$.

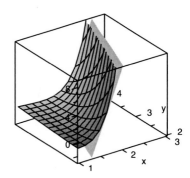

11.8   Maximum-Minimum Problems

Problems for Practice

1. Since $f_x(x, y) = -4x + 4$ and $f_y(x, y) = -4y + 8$, there is only one critical point, $(1, 2)$. It is a local maximum of f because

$$\mathcal{D}(f, (1, 2)) = (-4)(-4) - (0)(0) = 16 > 0,$$

and $f_{xx}(1, 2) = -4 < 0$.

3. Since $f_x(x, y) = 3x^2 - 6y$ and $f_y(x, y) = -3y^2 - 6x$, the critical points are the simultaneous solutions for the equations

$$3x^2 - 6y = 0$$
$$-3y^2 - 6x = 0.$$

Solve the first equation for y: $y = x^2/2$, and substitute into the second equation to see that $x^4/4 + 2x = 0$. Consequently, $x^4 = -8x$ so either $x = 0, y = 0$ or $x = -2, y = 2$.

Test $(0,0)$. The discriminant evaluates as

$$\mathcal{D}(f,(0,0)) = (0)(0) - (-6)(-6) = -36\,,$$

this is a saddle point.

Test $(2,-2)$. Since

$$\mathcal{D}(f,(2,-2)) = (-12)(-12) - (-6)(-6) > 0\,,$$

and $f_{xx}(2,-2) = -12$, this is a local maximum of f.

5. Since $f_x(x,y) = 6x - 6y$ and $f_y(x,y) = 6y^2 - 6x$, the critical points are the simultaneous solutions for the equations

$$x - y = 0$$
$$y^2 - x = 0\,.$$

Solve the first equation for y: $y = x$, and substitute into the second equation to see that $x^2 - x = 0$. Consequently, $x^2 = x$ so either $x = 0, y = 0$ or $x = 1, y = 1$.

Test $(0,0)$. The discriminant evaluates as

$$\mathcal{D}(f,(0,0)) = (6)(0) - (-6)(-6) = -36\,,$$

this is a saddle point.

Test $(1,1)$. Since

$$\mathcal{D}(f,(1,1)) = (6)(12) - (-6)(-6) > 0\,,$$

and $f_{xx}(2,-2) = 6$, this is a local minimum of f.

7. Since $f_x(x,y) = 2xy + 3y + y^2$ and $f_y(x,y) = 2xy + 3x + x^2$, the critical points are the simultaneous solutions for the equations

$$2xy + 3y + y^2 = 0$$
$$2xy + 3x + x^2 = 0\,.$$

Clearly one possibility is $x = y = 0$. Another is $y = 0$ and $3x + x^2 = 0$ which implies that $y = 0, x = -3$. A third possibility is $x = 0$ and $3y + y^2 = 0$ which implies that $x = 0, y = -3$. Finally, if x and y are both not zero, then the system simplifies to

$$2x + 3 + y = 0$$
$$2y + 3 + x = 0\,.$$

Double the first equation and subtract it from the second to eliminate y: $-3x - 3 = 0$. Therefore, $x = -1, y = -1$. There are four points to test, $(0,0)$, $(-3,0)$, $(0,-3)$, $(-1,-1)$.

(x,y)	$\mathcal{D}(f,(x,y))$	Classification
$(0,0)$	$(0)(0) - (3)(3) < 0$	Saddle Point
$(-3,0)$	$(0)(-6) - (-3)(-3) < 0$	Saddle Point
$(0,-3)$	$(-6)(0) - (-3)(-3) < 0$	Saddle Point
$(-1,-1)$	$(-2)(-2) - (-1)(-1) > 0$	Local Maximum

9. Since $f_x(x,y) = 4x - 12$ and $f_y(x,y) = 8y - 24$, there is one critical point: $x = 3, y = 3$. Note that $f_{xx}(x,y) = 4$, $f_{yy}(x,y) = 8$, and $f_{xy}(x,y) = 0$.

Test $(3,3)$. The discriminant evaluates as

$$\mathcal{D}(f,(3,3)) = (4)(8) - (0)^2 = 32,$$

this is a local minimum.

11. Since $f_x(x,y) = 2x + 2xy - 2 - 2y$ and $f_y(x,y) = x(x-2)$, the critical points are the simultaneous solutions for the equations

$$x + xy - 1 - y = 0$$
$$x(x-2) = 0.$$

Solve the first equation for y: $y = x$, and substitute into the second equation to see that $x^2 - x = 0$. Consequently, $x^2 = x$ so either $x = 0, y = 0$ or $x = 1, y = 1$. Note that $f_{xx}(x,y) = 2 + 2y$, $f_{yy}(x,y) = 0$, and $f_{xy}(x,y) = 2x - 2$.

Test $(0,-1)$. The discriminant evaluates as

$$\mathcal{D}(f,(0,0)) = (0)(0) - (-2)(-2) < 0,$$

this is a saddle point.

Test $(2,-1)$. Since

$$\mathcal{D}(f,(2,-1)) = (0)(0) - (2)(2) < 0,$$

and this is also a saddle point.

13. Since $f_x(x,y) = y^2 + 2xy + 8y$ and $f_y(x,y) = 2xy + x^2 + 8x$, the critical points are the simultaneous solutions for the equations

$$y^2 + 2xy + 8y = 0$$
$$2xy + x^2 + 8x = 0.$$

Clearly one possibility is $x = y = 0$. Another is $y = 0$ and $x^2 + 8x = 0$ which implies that $y = 0, x = -8$. A third possibility is $x = 0$ and $y^2 + 8y = 0$ which implies that $x = 0, y = -8$. Finally, if x and y are both not zero, then the system simplifies to

$$y + 2x + 8 = 0$$
$$2y + x + 8 = 0.$$

Double the first equation and subtract it from the second to eliminate y: $-3x - 8 = 0$. Therefore, $x = -3/8, y = -3/8$ (verify). There are four points to test, $(0,0)$, $(-8,0)$, $(0,-8)$, $(-8/3,-8/3)$. Note that $f_{xx}(x,y) = 2y$, $f_{yy}(x,y) = 2x$, and $f_{xy}(x,y) = 2y + 2x + 8$.

(x,y)	$\mathcal{D}(f,(x,y))$	Classification
$(0,0)$	$(0)(0) - (8)^2 < 0$	Saddle Point
$(-8,0)$	$(0)(-16) - (-8)^2 < 0$	Saddle Point
$(0,-8)$	$(-16)(0) - (-8)^2 < 0$	Saddle Point
$(-8/3,-8/3)$	$(-16/3)(-16/3) - (-8/3)^2 > 0$	Local Maximum

15. Since $f_x(x,y) = 6xy + 6x$ and $f_y(x,y) = 3y^2 + 3x^2 - 15$, the critical points are the simultaneous solutions for the equations

$$6xy + 6x = 0$$
$$3y^2 + 3x^2 - 15 = 0.$$

Clearly one possibility is $x = 0$. In this case, $y^2 = 5$, so there are two solutions: $x = 0$, $y = \sqrt{5}$ and $x = 0$, $y = -\sqrt{5}$. If x is not 0, then the first equation implies that $y = -1$. Substitute this into the second equation and $x^2 = 4$. Consequently, there are two more solutions: $y = -1$, $x = 2$ and $y = -1$, $x = -2$. There are four points to test, $(0,\sqrt{5})$, $(0,-\sqrt{5})$, $(2,-1)$, $(-2,-1)$. Note that $f_{xx}(x,y) = 6y$, $f_{yy}(x,y) = 6y$, and $f_{xy}(x,y) = 6x$.

(x,y)	$\mathcal{D}(f,(x,y))$	Classification
$(0,\sqrt{5})$	$(6\sqrt{5})(6\sqrt{5}) - (0)^2 > 0$	Local Minimum
$(0,-\sqrt{5})$	$(-6\sqrt{5})(-6\sqrt{5}) - (0)^2 > 0$	Local Maximum
$(2,-1)$	$(-6)(-6) - (12)^2 < 0$	Saddle Point
$(-2,-1)$	$(-6)(-6) - (-12)^2 < 0$	Saddle Point

17. Since $f_x(x,y) = 4x^3 - 8y$ and $f_y(x,y) = 4y^3 - 8x$, the critical points are the simultaneous solutions for the equations

$$4x^3 - 8y = 0$$
$$4y^3 - 8x = 0.$$

Clearly one possibility is $x = 0$ and $y = 0$. If this not the case, then neither x nor y can be zero. Solve the first equation for y: $y = x^3/2$, and substitute into the second equation: $4(x^3/2)^3 = 8x$, so $x^8 = 32 = 2^4$ and $x = \pm 2^{1/2} = \pm\sqrt{2}$, $y = \pm 2^{3/2}/2 = \pm\sqrt{2}$ (plus with plus and minus with minus). There are three points to test, $(0,0)$, $(\sqrt{2},\sqrt{2})$, $(-\sqrt{2},-\sqrt{2})$. Note that $f_{xx}(x,y) = 12x^2$, $f_{yy}(x,y) = 12y^2$, and $f_{xy}(x,y) = -8$.

(x, y)	$\mathcal{D}(f, (x, y))$	Classification
$(0, 0)$	$(0)(0) - (-8)^2 < 0$	Saddle Point
$(\sqrt{2}, \sqrt{2})$	$(24)(24) - (-8)^2 > 0$	Local Minimum
$(-\sqrt{2}, -\sqrt{2})$	$(24)(24) - (-8)^2 > 0$	Local Minimum

19. Since $\phi(u) = \ln(u)$ is an increasing function the local extreme values and saddle points for f are the same as those for the function $g(x, y) = 1 + x^2 + y^2$. Because $g_x(x, y) = 2x$ and $g_y(x, y) = 2y$, the only critical point is $(0, 0)$. It is obviously a local minimum.

21. We wish to maximize $f(x, y, z) = xyz$ where $x + y + z = 100$ and x, y, z are positive. Since $z = 100 - x - y$ this can be accomplished by finding numbers x and y that maximize $g(x, y) = xy(100 - x - y)$ for $x > 0$, $y > 0$, and $x + y < 100$.

Since $g_x(x, y) = 100y - 2xy - y^2$ and $g_y(x, y) = 100x - x^2 - 2xy$, the critical points for g are the simultaneous solutions for the equations

$$100y - 2xy - y^2 = 0$$
$$100x - x^2 - 2xy = 0.$$

Since x and y are positive they can be cancelled to yield the system

$$100 - 2x - y = 0$$
$$100 - x - 2y = 0.$$

There is one solution, $x = y = 100/3$ and it does satisfy the conditions on x and y that are given above. It must be an absolute maximum for g because $g(x, y)$ clearly approach 0 as (x, y) approaches each axis in the first quadrant and as (x, y) approaches the line $x + y = 100$. The three positive numbers are $x = y = z = \frac{100}{3}$.

23. Let the dimensions of the box be x, y, and z (length, width, height). The cost to make the box is $C(x, y, z) = 2xy + 0.5(2xz + 2yz) + xy$ cents. Moreover, $xyz = 64$. Since $z = 64/xy$ we look for positive numbers x and y that minimize $g(x, y) = 3xy + 64/y + 64/x$.

Since $g_x(x, y) = 3y - 64x^{-2}$ and $g_y(x, y) = 3x - 64y^{-2}$, the critical points for g are the simultaneous solutions for the equations

$$3y - 64x^{-2} = 0$$
$$3x - 64y^{-2} = 0.$$

Substitute $y = 64x^{-2}/3$ into the second equation to obtain

$$x = \frac{64}{3}(64x^{-2}/3)^{-2} = \frac{3}{64}x^4$$

so $x^3 = 64/3$ and $x = 4/3^{1/3}$. Moreover, $y = 64(4/3^{1/3})^{-2}/3 = 4/3^{1/3}$ also. Therefore, the box should have a square base with side length $4/3^{1/3}$ and height $z = 64/(xy) = 64/(4/3^{1/3})^2 = 4 \cdot 3^{2/3}$.

It will cost about 69 cents to make the box and, since $4 \cdot 3^{2/3}/(4/3^{1/3}) = 3$, it will be 3 times higher than it is wide. It's those cheap sides.

25. Let x, y, and z be the lengths of the sides of the triangle and θ the angle between the sides of length x and y. The area of the triangle is $A = \frac{1}{2}xy\sin(\theta)$ so $\sin(\theta) = \frac{2A}{xy}$ and $\cos(\theta) = \sqrt{1 - \frac{4A^2}{x^2y^2}} = \frac{\sqrt{x^2y^2 - 4A^2}}{xy}$. Using the Law of Cosines,

$$z = \sqrt{x^2 + y^2 - 2xy\cos(\theta)} = \sqrt{x^2 + y^2 - 2\sqrt{x^2y^2 - 4A^2}},$$

so we minimize $g(x, y) = x + y + \sqrt{x^2 + y^2 - 2\sqrt{x^2y^2 - 4A^2}}$ for $0 < x$, $0 < y$, and $xy > 2A$. The symmetry of g (note that $g(x, y) = g(y, x)$) suggests that its extreme values must be attained when $x = y$. This is confirmed by examining the critical point equations $g_x = 0$ and $g_y = 0$ displayed below.

$$1 + \frac{2x - \frac{2xy^2}{\sqrt{x^2y^2 - 4A^2}}}{\sqrt{x^2 + y^2 - 2\sqrt{x^2y^2 - 4A^2}}} = 0, \quad 1 + \frac{2y - \frac{2x^2y}{\sqrt{x^2y^2 - 4A^2}}}{\sqrt{x^2 + y^2 - 2\sqrt{x^2y^2 - 4A^2}}} = 0.$$

Equate the left sides, cancel the 1s, then the common denominator, to obtain the equation $2x - \frac{2xy^2}{\sqrt{x^2y^2 - 4A^2}} = 2y - \frac{2x^2y}{\sqrt{x^2y^2 - 4A^2}}$ which is equivalent to $x - y = \frac{xy^2 - x^2y}{\sqrt{x^2y^2 - 4A^2}}$. This implies that $x = y$, for if not, then the term $x - y$ can be cancelled from both sides yielding $1 = \frac{-xy}{\sqrt{x^2y^2 - 4A^2}}$, which is not possible because the right hand side of this equation is negative.

So, the triangle with minimum perimeter is isosceles with $x = y$. The same argument will show that $y = z$, and the triangle is actually equilateral. Since $A = \frac{1}{2}x^2\sin(\pi/3)$, the side length is $x = \sqrt{2A \cdot \frac{2}{\sqrt{3}}} = \frac{20}{3^{1/4}}$.

27. Calculate

$$\mathbf{x} \cdot \mathbf{y} = (5.013)(0.270) + (10.124)(0.277) + \cdots + (44.862)(0.345) = 31.60$$

$$\mathbf{x} \cdot \mathbf{x} = 5.013^2 + 10.124^2 + \cdots + 44.862^2 = 2986.99$$

$$\bar{x} = \frac{1}{5}(5.013 + 10.123 + \cdots + 44.862) = 19.99$$

$$\bar{y} = \frac{1}{5}(0.270 + 0.277 + \cdots + 0.345) = 0.297$$

The least squares line is $y = mx + b$ where

$$m = \frac{\mathbf{x} \cdot \mathbf{y} - 5\bar{x}\bar{y}}{\mathbf{x} \cdot \mathbf{x} - 5\bar{x}\bar{x}} = 0.001955$$

and

$$b = \bar{y} - m\bar{x} = 0.2577\,.$$

The data points and regression line are plotted below, $y(100) = 0.4533$.

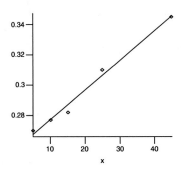

29. Calculate

$$\mathbf{x} \cdot \mathbf{y} = (32.2)(238.1) + (17.0)(118.1) + \cdots + (18.9)(150.3) = 42488.5$$
$$\mathbf{x} \cdot \mathbf{x} = 32.2^2 + 17.0^2 + \cdots + 18.9^2 = 6201.1$$
$$\bar{x} = \frac{1}{8}(32.2 + 17.0 + \cdots + 18.9) = 26.8$$
$$\bar{y} = \frac{1}{8}(238.1 + 118.1 + \cdots + 150.3) = 185.1$$

The least squares line is $y = mx + b$ where

$$m = \frac{\mathbf{x} \cdot \mathbf{y} - 8\bar{x}\bar{y}}{\mathbf{x} \cdot \mathbf{x} - 8\bar{x}\bar{x}} = 6.12$$

and

$$b = \bar{y} - m\bar{x} = 21.1\,.$$

The data points and the regression line are plotted below.

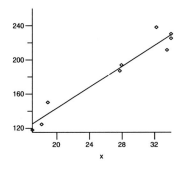

Further Theory and Practice

31. Calculate

$$\mathbf{t} \cdot \ln(\mathbf{y}) = (0.0)(\ln(1.0)) + (1.0)(\ln(0.91)) + \cdots + (8.0)(\ln(0.34) = -28.22$$
$$\mathbf{t} \cdot \mathbf{t} = 0^2 + 1^2 + \cdots + 8^2 = 204$$
$$\bar{t} = \frac{1}{9}(0 + 1 + \cdots + 8) = 4$$
$$\overline{\ln(y)} = \frac{1}{9}(\ln(1.0) + \ln(0.91) + \cdots + \ln(0.34)) = -0.55$$

The least squares line is $\ln(y) = mt + b$ where

$$m = \frac{\mathbf{t} \cdot \ln(\mathbf{y}) - 9\bar{t}\,\overline{\ln(y)}}{\mathbf{t} \cdot \mathbf{t} - 9\bar{t}\bar{t}} = -0.14$$

and

$$b = \overline{\ln(y)} - m\bar{t} = 0.00934\,.$$

The data points and the regression line are plotted below. The approximation formula is exponential: $y = g(t) = e^b e^{mt} = 0.009 e^{-0.14t}$; $g(10) = 0.25$.

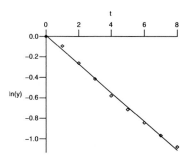

33. Completing the square can go like this.

$$\begin{aligned}
f(x,y) &= Ax^2 + Cy^2 + Dx + Ey + F \\
&= A(x^2 + Dx/A) + C(y^2 + Ey/C) + F \\
&= A(x + D/2A)^2 + C(y + E/2C)^2 + F - D^2/4A - E^2/4C
\end{aligned}$$

Observe that if A and C are both positive, then f has a global minimum at $x = -D/2A$, $y = -E/2C$. If A and C are both negative, then f attains a global maximum value at this point. If A and C have opposite signs, then $(-D/2A, -E/2D)$ is a saddle point and f has no extreme values.

35. Let $P = (a, b, c)$, $Q = (x_0, y_0, z_0)$, and $\mathbf{r}(t) = \langle x(t), y(t), z(t) \rangle$ be a smooth curve on the surface that passes through Q at $t = 0$. Observe that the function $\phi(t) = (x(t) - a)^2 + (y(t) - b)^2 + (z(t) - c)^2$ attains a minimum value at $t = 0$ so $\phi'(0) = 0$. Since

$$\phi'(t) = 2(x(t) - a)x'(t) + 2(y(t) - b)y'(t) + 2(z(t) - c)z'(t)\,,$$

let $t = 0$ to obtain

$$0 = 2(x_0 - a)x'(0) + 2(y_0 - b)y'(0) + 2(z_0 - c)z'(0) \,.$$

Consequently $\overrightarrow{QP} \cdot \mathbf{r}'(0) = 0$ and the line from Q to P is perpendicular to the tangent vector to the curve at the point Q. The fact that this is true for all smooth curves on the surface implies that the line from P to Q is perpendicular to the tangent plane for the surface at Q.

37. (a) Using formulas in the proof of Theorem 3, $\frac{\partial^2}{\partial^2 m} SSE(m, b) = 2\sum_{j=1}^{N} x_j^2$, $\frac{\partial^2}{\partial^2 b} SSE(m, b) = 2N$, and $\frac{\partial^2}{\partial m \partial b} SSE(m, b) = 2\sum_{j=1}^{N} x_j$. Therefore,

$$\mathcal{D}(SSE, P) = SSE_{mm}(P)SSE_{bb}(P) - SSE_{mb}(P)^2$$

$$= 4N \sum_{j=1}^{N} x_j^2 - 4\left(\sum_{j=1}^{N} x_j\right)^2$$

$$= 4\left(N\sum_{j=1}^{N} x_j^2 - \sum_{j=1}^{N} x_j^2 - 2\sum_{1 \le i < j \le N} x_i x_j\right)$$

(b) We show that if $a_1, a_2, \ldots, a_N$ are real numbers ($N \ge 2$), then

$$\sum_{i=1}^{N-1} \sum_{j=i+1}^{N} (a_i + a_j) = (N-1)\sum_{j=1}^{N} a_j \,.$$

By induction on N. The statement is clearly true if $N = 2$. Assume $N > 2$ and that the statement is true for $N - 1$ terms. We then have

$$\sum_{i=1}^{N-1} \sum_{j=i+1}^{N} (a_i + a_j) = \sum_{i=1}^{N-2} \sum_{j=i+1}^{N} (a_i + a_j) + a_{N-1} + a_N$$

$$= \sum_{i=1}^{N-2} \left(\sum_{j=i+1}^{N-1} (a_i + a_j) + a_i + a_N\right) + a_{N-1} + a_N$$

$$= \sum_{i=1}^{N-2} \sum_{j=i+1}^{N-1} (a_i + a_j) + \sum_{i=1}^{N-2}(a_i + a_N) + a_{N-1} + a_N$$

$$= (N-2)\sum_{i=1}^{N-1} a_i + \sum_{i=1}^{N-1} a_i + (N-1)a_N$$

$$= (N-1)\sum_{i=1}^{N-1} a_i + (N-1)a_N$$

$$= (N-1)\sum_{i=1}^{N} a_i = (N-1)\sum_{j=1}^{N} a_j$$

(c) Combining parts a and b we have

$$\mathcal{D}(SSE, P) = 4\left(N\sum_{j=1}^{N} x_j^2 - \sum_{j=1}^{N} x_j^2 - 2\sum_{1\leq i<j\leq N} x_i x_j\right)$$

$$= 4\left(\sum_{i=1}^{N-1}\sum_{j=i+1}^{N}(x_i^2 + x_j^2) - 2\sum_{1\leq i<j\leq N} x_i x_j\right)$$

$$= 4\sum_{1\leq i<j\leq N}(x_i^2 - 2x_i x_j + x_j^2)$$

$$= 4\sum_{1\leq i<j\leq N}(x_i - x_j)^2 > 0.$$

Therefore, both $\mathcal{D}(SSE, P)$ and $\frac{\partial^2}{\partial^2 b}SSE(P)$ are positive implying that P is a point where SSE has a local minimum.

Calculator/Computer Exercises

39. Define and plot the function.

```
> f := (x,y) -> x^4 - y^5 + x^2*y + x;
  plot3d( f(x,y), x=-1..0, y=0..1, axes=boxed, grid=[13,13],
        orientation=[-60,70], tickmarks=[5,5,5] );
```

$$f := (x, y) \to x^4 - y^5 + x^2 y + x$$

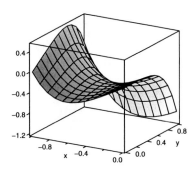

Locate the critical point and verify that it is a saddle point.

```
> CP := fsolve( {D[1](f)(x,y)=0, D[2](f)(x,y)=0}, {x=-0.5,y=0.5});
  Disc = eval( D[1,1](f)(x,y)*D[2,2](f)(x,y) - D[1,2](f)(x,y)^2, CP);
```

$$CP := \{x = -0.5060833700, y = 0.4757387555\}$$

$$Disc = -9.691965832$$

41. Define and plot the function.

```
> f := (x,y) -> ln(1+x^4) - sin(y);
  plot3d( f(x,y), x=-1..1, y=1..2, axes=boxed, grid=[20,15],
     orientation=[-60,70], tickmarks=[5,5,5] );
```

$$f := (x, y) \to \ln(1 + x^4) - \sin(y)$$

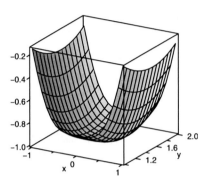

Locate the critical point and verify that the second derivative test is inconclusive.

```
> CP := solve( {D[1](f)(x,y)=0, D[2](f)(x,y)=0}, {x,y});
  Disc = eval( D[1,1](f)(x,y)*D[2,2](f)(x,y) - D[1,2](f)(x,y)^2, CP);
```

$$CP := \left\{ x = 0, y = \frac{1}{2}\pi \right\}$$

$$Disc = 0$$

The plot indicates that the critical point is a local minimum for f.

11.9   Lagrange Multipliers

Problems for Practice

1. We seek x, y, and λ such that $\nabla f(x, y) = \lambda \nabla g(x, y)$ and $g(x, y) = 4$. That is, $\langle 2, -3 \rangle = \lambda \langle 2x, 4y \rangle$ and $x^2 + 2y^2 = 4$, yielding three scalar equations:

$$2 = 2\lambda x, \quad -3 = 4\lambda y, \quad \text{and} \quad x^2 + 2y^2 = 4.$$

Since λ cannot be 0 it can be safely eliminated by dividing the second equation by the first to obtain $y = -3x/4$. Use this to eliminate y from the constraint equation: $x^2 + 2 \cdot (-3x/4)^2 = 4$. Consequently, $17x^2/8 = 4$ and $x^2 = 32/17$. It follows that $y^2 = 9x^2/16 = 18/17$.

Since x and y have opposite signs, there are only two critical points: $P = (4\sqrt{34}/17, -3\sqrt{34}/17)$ and $Q = (-4\sqrt{34}/17, 3\sqrt{34}/17)$. Substitute into f to obtain the maximum value $f(P) = 6 + \sqrt{34}$ and the minimum value $f(Q) = 6 - \sqrt{34}$.

3. We seek x, y, and λ such that $\nabla f(x, y) = \lambda \nabla g(x, y)$ and $g(x, y) = 16$. That is, $\langle 2x + 2, 2y \rangle = \lambda \langle 2x, 8y \rangle$ and $x^2 + 4y^2 = 16$. There are three scalar equations:

$$2x + 2 = 2\lambda x, \quad 2y = 8\lambda y, \quad \text{and} \quad x^2 + 4y^2 = 16.$$

One possibility is $y = 0$ implying that $x = \pm 4$. This yields two critical points: $P = (4, 0)$ and $Q = (-4, 0)$. If $y \neq 0$, then $\lambda = 1/4$ and $x = -4/3$. This implies that $y = \pm 4\sqrt{2}/3$ (verify) and yields two more critical points: $R = (-4/3, 4\sqrt{2}/3)$ and $S = (-4/3, -4\sqrt{2}/3)$. Substitute into f to obtain the maximum value $f(P) = 25$ and the minimum value $f(R) = f(S) = 11/3$.

5. We seek x, y, and λ such that $\nabla f(x, y) = \lambda \nabla g(x, y)$ and $g(x, y) = 16$. That is, $\langle 3y^2, 6xy \rangle = \lambda \langle 2x, 2y \rangle$ and $x^2 + y^2 = 16$. There are three scalar equations:

$$3y^2 = 2\lambda x, \quad 6xy = 2\lambda y, \quad \text{and} \quad x^2 + y^2 = 16.$$

One possibility is $y = 0$ implying that $\lambda = 0$ and $x = \pm 4$. This yields two critical points: $P = (4, 0)$ and $Q = (-4, 0)$. If $y \neq 0$, then $\lambda = 3x$ and $3y^2 = 6x^2$. This implies that $y^2 = 2x^2$ and $x^2 + 2x^2 = 16$. Consequently, $x^2 = 16/3$, $x = \pm 4/\sqrt{3}$, and $y = \pm 4\sqrt{6}/3$ (verify). This yields four more critical points: $R_{1,2} = (4/\sqrt{3}, \pm 4\sqrt{6}/3)$ and $S_{1,2} = (-4/\sqrt{3}, \pm 4\sqrt{6}/3)$. Substitute into f to obtain the maximum value $f(R_{1,2}) = 128/\sqrt{3} - 24$ and the minimum value $f(S_{1,2}) = -128/\sqrt{3} - 24$.

7. We seek x, y, and λ such that $\nabla f(x, y) = \lambda \nabla g(x, y)$ and $g(x, y) = 16$. That is, $\langle 1, 2y \rangle = \lambda \langle 2x, 2y \rangle$ and $x^2 + y^2 = 9$. There are three scalar equations:

$$1 = 2\lambda x, \quad 2y = 2\lambda y, \quad \text{and} \quad x^2 + y^2 = 9.$$

One possibility is $y = 0$ implying that $x = \pm 3$. This yields two critical points: $P = (3, 0)$ and $Q = (-3, 0)$. If $y \neq 0$, then $\lambda = 1$ and $x = 1/2$. This implies that $y^2 = 9 - 1/4 = 35/4$ and $y = \pm\sqrt{35}/2$. This yields two more critical points: $R_{1,2} = (1/2, \pm\sqrt{35}/2)$. Substitute into f to obtain the maximum value $f(R_{1,2}) = 37/4$ and the minimum value $f(Q) = -3$.

9. We seek x, y, and λ such that $\nabla f(x, y) = \lambda \nabla g(x, y)$ and $g(x, y) = 16$. That is, $\langle 8x, 8y \rangle = \lambda \langle 4x^3, 4y^3 \rangle$ and $x^4 + y^4 = 16$. There are three scalar equations:

$$8x = 4\lambda x^3, \quad 8y = 4\lambda y^3, \quad \text{and} \quad x^4 + y^4 = 16.$$

One possibility is $y = 0$ implying that $x = \pm 2$. This yields two critical points: $P_{1,2} = (\pm 2, 0)$. Similarly, perhaps $x = 0$, yielding two more points:

$Q_{1,2} = (0, \pm 2)$. If x and y are not zero, then $\lambda = 2x^{-2}$ and $\lambda = 2y^{-2}$, implying that $y^2 = x^2$. Consequently, $x^4 = 8$ so $x = y = \pm 2^{3/4}$. This yields four more critical points: $R_{1,2,3,4} = (\pm 2^{3/4}, \pm 2^{3/4})$. Substitute into f to obtain the maximum value $f(R_{1,2,3,4}) = 16\sqrt{2}$ and the minimum value $f(P_{1,2}) = f(Q_{1,2}) = 16$.

11. We seek x, y, z, and λ such that $\nabla f(x, y, z) = \lambda \nabla g(x, y, z)$ and $g(x, y, z) = 4$. That is, $\langle yz, xz, xy \rangle = \lambda \langle 2x, 2y, 2z \rangle$ and $x^2 + y^2 + z^2 = 4$. There are four scalar equations:

$$ yz = 2\lambda x, \quad xz = 2\lambda y, \quad xy = 2\lambda z, \quad \text{and} \quad x^2 + y^2 + z^2 = 4. $$

It is clear that the maximum and minimum values of f will occur at points (x, y, z) where x, y, and z are not 0. Therefore, λ is not 0 and we may safely divide the first equation by the second to see that $y/x = x/y$ and $x^2 = y^2$. Similarly $y^2 = z^2$, so the constraint equation can be reduced to $3x^2 = 4$. Thus $x^2 = y^2 = z^2 = 4/3$. This yields 8 points on the sphere: $(\pm 2/\sqrt{3}, \pm 2/\sqrt{3}, \pm 2/\sqrt{3})$. Substitute into f to see that the maximum value is $8/(3\sqrt{3})$ and the minimum value is $-8/(3\sqrt{3})$.

13. We seek x, y, and λ such that $\nabla f(x, y) = \lambda \nabla g(x, y)$ and $g(x, y) = 2x - 6y = 5$. That is, $\langle 2x, 4y \rangle = \lambda \langle 2, -6 \rangle$ and $2x - 6y = 5$. There are three scalar equations:

$$ 2x = 2\lambda, \quad 4y = -6\lambda, \quad \text{and} \quad 2x - 6y = 5. $$

Clearly λ is not 0 so it can be eliminated between the first two equations (divide one by the other) to obtain $y = -3x/2$. Use this to eliminate y from the constraint equation: $2x - 6 \cdot (-3x/2) = 5$. Consequently, $11x = 5$ and $x = 5/11$. It follows that $y = -15/22$.

There is one critical point $P = (5/11, -15/22)$. Substitute into f to obtain the value $f(P) = 223/22$. Examination of the graph of f (an ellliptic paraboloid opening upward) shows that this is the minimum value of f when (x, y) is constrained to lie on the line $2x - 6y = 5$.

15. Let $f(x, y) = x^2 + y^2 + 4y$ and $g(x, y) = x^2 + 2y^2$. We seek x, y, and λ such that $\nabla f(x, y) = \lambda \nabla g(x, y)$ and $g(x, y) = 8$. That is, $\langle 2x, 2y + 4 \rangle = \lambda \langle 2x, 4y \rangle$ and $x^2 + 2y^2 = 8$. There are three scalar equations:

$$ 2x = 2\lambda x, \quad 2y + 4 = 4\lambda y, \quad \text{and} \quad x^2 + 2y^2 = 8. $$

One possibility is $x = 0$. Then $y = \pm 2$, yielding two critical points: $P_{1,2} = (0, \pm 2)$. If $x \neq 0$, then $\lambda = 1$ and $y = 2$ once more. But this forces x to be 0. We conclude that x must be 0 and there are only two critical points.

Substitute into f to obtain the minimum value $f(0, -2) = -4$.

17. Let $f(x,y) = x^3 + 2y$ and $g(x,y) = x^2 + y^2$. We seek x, y, and λ such that $\nabla f(x,y) = \lambda \nabla g(x,y)$ and $g(x,y) = 4/3$. That is, $\langle 3x^2, 2 \rangle = \lambda \langle 2x, 2y \rangle$ and $x^2 + y^2 = 4/3$. There are three scalar equations:

$$3x^2 = 2\lambda x, \ 2 = 2\lambda y, \quad \text{and} \quad x^2 + y^2 = 4/3.$$

One possibility is $x = 0$. Then $y = \pm 2/\sqrt{3}$, yielding two critical points: $P = (0, 2/\sqrt{3})$ and $Q = (0, -2/\sqrt{3})$. If $x \neq 0$, then $2\lambda = 3x = 2y^{-1}$ and $x = 2(3y)^{-1}$. Substitute this into the constraint equation to obtain

$$\frac{4}{3y^2} + y^2 = \frac{4}{3} \implies 3y^4 - 4y^2 + 4 = 0.$$

The quadratic in y^2 factors nicely to yield $(3y^2 - 2)(y^2 + 2) = 0$, so $y = \pm\sqrt{2/3}$ and $x = \pm(2/3)\sqrt{3/2} = \pm\sqrt{2/3}$ (plus with plus, minus with minus). Thus there are two more critical points, $R_{1,2} = \pm(\sqrt{2/3}, \sqrt{2/3})$.

Substitute into f to obtain the maximum value $f(P) = 4/\sqrt{3}$.

19. Let S be the surface area and V the volume of the can. We seek r, h, and λ such that $\nabla S = \lambda \nabla V$ and $V = V_0$. That is, $\langle 4\pi r + 2\pi h, 2\pi r \rangle = \lambda \langle 2\pi rh, \pi r^2 \rangle$ and $\pi r^2 h = V_0$. There are three scalar equations:

$$4\pi r + 2\pi h = 2\lambda\pi rh, \ 2\pi r = \lambda\pi r^2, \quad \text{and} \quad \pi r^2 h = V_0.$$

The second equation implies that $\lambda = 2/r$. Substitute this into the first equation to obtain $2r + h = 2h$. Consequently, $h = 2r$, and the can with the minimum surface area has its height equal to the diameter of its base.

The minimum surface area is $S = 2\pi r^2 + 2\pi r \cdot 2r = 6\pi r^2$. This can be expressed in terms of the volume V_0 by observing that $V_0 = \pi r^2 h = 2\pi r^3$ so $r = (V_0/2\pi)^{1/3}$ and $S = 6\pi(V_0/2\pi)^{2/3} \left(= 3\sqrt[3]{2\pi V_0^2} \right)$.

21. Let $f(x, y, z) = x^4 + y^4 + z^4$. The maximum of f occurs at a point (x, y, z) where $\langle 4x^3, 4y^3, 4z^3 \rangle = \lambda \langle 2x, 2y, 2z \rangle$ for some λ. Therefore,

$$2x^3 = \lambda x, \ 2y^3 = \lambda y, \ 2z^3 = \lambda z, \quad \text{and} \quad x^2 + y^2 + z^2 = 12.$$

Clearly $\lambda \neq 0$.

If $x = 0$ and $y = 0$, then $z = \pm\sqrt{12}$. This yields two points: $(0, 0, \pm\sqrt{12})$. By symmetry there are four more: $(0, \pm\sqrt{12}, 0)$ and $(\pm\sqrt{12}, 0, 0)$. The value of f at all four points is 144.

If $x = 0$ and y and z are not zero, then their squares must be equal (this is because $\lambda = 2y^2 = 2z^2$). Therefore, $2z^2 = 12$ and $z = \pm\sqrt{6}$. This yields four points: $(0, \pm\sqrt{6}, \pm\sqrt{6})$. By symmetry there are eight more. The value of f at all twelve points is 72.

If x, y, and z are all not zero, then (as above) their squares are equal so $3z^2 = 12$ and $z = \pm2$. This yields 8 points: $(\pm2, \pm2, \pm2)$. The value of f at all eight points is 64.

The maximum value of f is 144.

23. The square of the distance from a point (x, y, z) on the ellipsoid to the point $(1, 0, 0$ is $f(x, y, z) = (x-1)^2 + y^2 + z^2$. We will minimize this function subject to the constraint $g(x, y, z) = 4$ where $g(x, y, z) = x^2 + 2y^2 + 4z^2$. The scalar equations derived from the Lagrange condition $\nabla f = \lambda \nabla g$ and the constraint are

$$2(x - 1) = 2\lambda x, \quad 2y = 4\lambda y, \quad 2z = 8\lambda z, \quad \text{and} \quad x^2 + 2y^2 + 4z^2 = 4.$$

Observe that $x \neq 0$ and λ cannot be zero either for if it were, then $x = 1$, $y = 0$, $z = 0$, and the constraint equation is not satisfied.

One possibility is $y = z = 0$. If this is the case, then $x = \pm 2$ and the values of f are 1 and 9.

Continuing our analysis of y and z, note that one of them *must* be zero. Indeed, if $y \neq 0$ and $z \neq 0$, then $\lambda = 1/2$ and $\lambda = 1/4$, so this is not possible. Consider the two remaining possibilities.

$y = 0$, $z \neq 0$

In this case, $\lambda = 1/4$ implying that $2(x - 1) = x/2$ so $x = 4/3$ and, using the constraint equation, $z = \pm\sqrt{5}/3$. The value of f at both points is $2/3$.

$y \neq 0$, $z = 0$

In this case, $\lambda = 1/2$ implying that $2(x - 1) = x$ so $x = 2$ and, using the constraint equation, $y = 0$, which is not possible.

The minimum distance is $\sqrt{2/3}$ attained at the points $(4/3, 0, \pm\sqrt{5}/3)$ on the ellipsoid.

25. Let $f(x, y, z) = xy^2z$ and $g(x, y, z) = x^2 + 3y^2 + 2z^2$. The scalar equations derived from the Lagrange condition $\nabla f = \lambda \nabla g$ and the constraint are

$$y^2z = 2\lambda x, \quad 2xyz = 6\lambda y, \quad xy^2 = 4\lambda z, \quad \text{and} \quad x^2 + 3y^2 + 2z^2 = 64.$$

Clearly the maximum value will be attained at a point where x, y, and z are not zero. This implies that $\lambda \neq 0$ also.

Divide the first equation by the second to see that $3y^2 = 2x^2$. Divide the first equation by the third to see that $2z^2 = x^2$. Substitute into the constraint equation to obtain $4x^2 = 64$ so $x = \pm 4$, $y = \pm 4\sqrt{2/3}$, and $z = \pm 4/\sqrt{2}$.

The maximum value of xy^2z is $256\sqrt{2}/3$.

Further Theory and Practice

27. We wish to maximize $f(x, y) = y$ subject to the constraint $g(x, y) = 3x^2 + 2xy + 3y^2 = 24$. The Lagrange condition $\nabla f = \lambda \nabla g$ yields the first two scalar equations below. The constraint equation is also listed.

$$0 = \lambda(6x + 2y), \ 1 = \lambda(2x + 6y), \ \text{and} \ 3x^2 + 2xy + 3y^2 = 24.$$

Since λ cannot be zero the first equation implies that $y = -3x$. Substitute this into the constraint equation to obtain $3x^2 - 6x^2 + 27x^2 = 24$ so $x^2 = 1$ and $x = \pm 1$. Consequently, $y = \pm 3$ and the maximum y value is 3.

29. Since the graph of the altitude function is a plane its highest and lowest values will be attained on the boundary of the elliptical county. We wish to find the extreme values of $a(x, y) = 80x - 70y + 150$ subject to the constraint $g(x, y) = 4x^2 + 2y^2 = 16$. The Lagrange condition $\nabla a = \lambda \nabla g$ and the constraint equation yield the scalar equations

$$80 = 8\lambda x, \ -70 = 4\lambda y, \ 4x^2 + 2y^2 = 16.$$

Since λ, x, and y cannot be 0, divide the first equation by the second to obtain $-\frac{8}{7} = 2\frac{x}{y}$ so $y = -\frac{7}{4}x$. Substitute this into the constraint equation and $4x^2 + 2 \cdot \frac{49}{16}x^2 = 16$ implying that $x = \pm\frac{8}{9}\sqrt{2}$ and $y = \mp\frac{14}{9}\sqrt{2}$. There are two critical points $P = \left(\frac{8}{9}\sqrt{2}, -\frac{14}{9}\sqrt{2}\right)$ and $Q = -P$. The maximum altitude is $a(P) = 150 + 180\sqrt{2}$ and the minimum altitude is $a(Q) = 150 - 180\sqrt{2}$.

31. Let $C(r, h)$ be the cost to make the can when its radius is r and its height is h. We assume that the side material costs \$1 per unit area and the top and bottom material costs \$k per unit area so $C(r, h) = 2\pi rh + k \cdot 2\pi r^2$. This is to be minimized subject to the constraint $V(r, h) = \pi r^2 h = V_0$.

We seek r, h, and λ such that $\nabla C = \lambda \nabla V$ and $V = V_0$. That is, $\langle 2\pi h + 4\pi kr, 2\pi r\rangle = \lambda\langle 2\pi rh, \pi r^2\rangle$ and $\pi r^2 h = V_0$. There are three scalar equations:

$$2\pi h + 4\pi kr = 2\lambda\pi rh, \ 2\pi r = \lambda\pi r^2, \quad \text{and} \quad \pi r^2 h = V_0.$$

The second equation implies that $\lambda = 2/r$. Substitute this into the first equation to obtain $h + 2kr = 2h$. Consequently, $h = 2kr$, and the can that costs the least has its height equal to k times the diameter of its base.

The minimum cost is $2\pi r \cdot 2kr + 2k\pi r^2 = 6k\pi r^2$. This can be expressed in terms of the volume V_0 by observing that $V_0 = \pi r^2 h = 2k\pi r^3$ so $r = (V_0/2k\pi)^{1/3}$ and $S = 6k\pi(V_0/2k\pi)^{2/3} \left(= 3\sqrt[3]{2k\pi V_0^2}\right)$.

33. Let $g(x, y, z) = x + y + z$ and $h(x, y, z) = x + 2y - 3z$. Extreme values of f are attained at points where $\nabla f = \lambda \nabla g + \mu \nabla h$. That is, $\langle 2x, 2y, 2z\rangle = \lambda\langle 1, 1, 1\rangle + \mu\langle 1, 2, -3\rangle$. This yields 3 scalar equations and the two constraints.

$$2x = \lambda + \mu, \ 2y = \lambda + 2\mu, \ 2z = \lambda - 3\mu, \ x + y + z = 6, \ x + 2y - 3z = 12.$$

Substitute $\lambda = 2x - \mu$ into the second and third equations to obtain

$$2y = 2x + \mu \quad \text{and} \quad 2z = 2x - 4\mu.$$

Multiply the left equation by 4 and add it to the right equation to see that $8y + 2z = 10x$ or $z = 5x - 4y$. This equation can be used to eliminate z from the two constraint equations. Doing so yields.

$$6x - 3y = 6 \quad \text{and} \quad -14x + 14y = 12.$$

We leave it to you to verify that $x = 20/7$ and $y = 26/7$. The value of z is obtained from the equation above: $z = -4/7$, and $x^2 + y^2 + z^2$ evaluates to $156/7$.

This is a minimum because the intersection of the planes determined by the constraint equations is a line and $156/7$ is the square of the distance from this line to the origin.

35. Look for (x, y, z) satisfying $\nabla F = \lambda \nabla g + \mu \nabla h$ where $g(x, y, z) = 2 + 2y + 3z$ and $h(x, y, z) = 5(x^2 + y^2 + z^2)$. This yields 3 scalar equations and the two constraints:

$$0 = \lambda + 10\mu x, \ 0 = 2\lambda + 10\mu y, \ 1 = 3\lambda + 10\mu z$$

$$x + 2y + 3z = 6, \ 5(x^2 + y^2 + z^2) = 14.$$

Substitute $\lambda = -10\mu x$ into the second and third equations to obtain

$$\mu(y - 2x) = 0 \quad \text{and} \quad \mu(z - 3x) = 1/10.$$

Since μ cannot be zero, this implies that $y = 2x$. Use this to reduce the constraint equations to two variables:

$$5x + 3z = 6 \quad \text{and} \quad 5(5x^2 + z^2) = 14.$$

Solve simultaneously (substitute $z = 2 - 5x/3$ into the right equation) to obtain $x = 3/5$ and $x = 9/35$. When $x = 3/5$, $y = 6/5$, and $z = 1$. When $x = 9/35$, $y = 18/35$, and $z = 11/7$. This is the maximum value for F.

37. Look for (x, y, z) satisfying $\nabla F = \lambda \nabla g + \mu \nabla h$ where $g(x, y, z) = x + 2y + z$ and $h(x, y, z) = x^2 + y^2 - z$. This yields 3 scalar equations and the two constraints:

$$2x = \lambda + 2\mu x, \ 2y = 2\lambda + 2\mu y, \ 2z = \lambda - \mu$$

$$x + 2y + z = 10, \ x^2 + y^2 - z = 0.$$

Divide the second equation by 2 and then subtract it from the first equation to eliminate λ and obtain $2x - y = \mu(2x - y)$. If $y \neq 2x$, then $\mu = 1$ implying that $\lambda = 0$ and $z = -\frac{1}{2}$. However, this is not possible because the second constraint equation could not be satisfied. Therefore, $y = 2x$.

Substitute for y in the constraint equations to obtain $5x + z = 10$ and $5x^2 - z = 0$ implying that $5x^2 + 5x - 10 = 0$ and either $x = 1$ or $x = -2$. If $x = 1$, then $y = 2$ and $z = 5$. If $x = -2$, then $y = -4$ and $z = 20$. The function F is maximum at the point $P = (-2, -4, 20)$, $F(P) = 420$.

39. The consumer wants to maximize $f(x, y) = x^p y^q$ subject to the constraint $Ax + By = T$. Using Lagrange multipliers he looks for x and y positive such that $\nabla f = \lambda \nabla g$ where $g(x, y) = Ax + By$. The scalar equations are

$$px^{p-1}y^q = \lambda A, \quad qx^p y^{q-1} = \lambda B, \quad \text{and} \quad Ax + By = T.$$

Divide the first equation by the second to see that $Ax = pBy/q$. Substitute into the constraint equation and $(pB/q + B)y = T$ so

$$y = \frac{T}{pB/q + B} = \frac{qT}{B(p + q)} \quad \text{and} \quad x = \frac{pT}{A(p + q)}.$$

41. We are to maximize $f(x, y)$ under the constraint that $ax + by = T$. This occurs when $\nabla f = \lambda \nabla g$ where $g(x, y) = ax + by$. That is, $\langle f_x, f_y \rangle = \lambda \langle a, b \rangle$. The scalar equations are $f_x = \lambda a$ and $f_y = \lambda b$. Divide the first by the second to see that $f_x/f_y = a/b$.

43. We are to maximize $f(x, y) = xy^{3/4}$ under the constraint that $15x + 12y = c$. This occurs when $\nabla f = \lambda \nabla g$ where $g(x, y) = 15x + 12y$. That is, $\langle y^{3/4}, (3/4)xy^{-1/4} \rangle = \lambda \langle 15, 12 \rangle$. The three scalar equations are

$$y^{3/4} = 15\lambda, \quad (3/4)xy^{-1/4} = 12\lambda, \quad \text{and} \quad 15x + 12y = c.$$

Divide the first equation by the second to see that $y/x = 15/16$ so $15x = 16y$. Consequently, $16y + 12y = c$, and $y(c) = c/28$, $x(c) = 16y(c)/15 = 4c/105$.

The maximum value is $M(c) = (\sqrt[4]{28}/735)c^{7/4}$ and $\lambda(c) = (\sqrt[4]{28}/420)c^{3/4}$. When $c = 10000$,

$$M(c + 1) - M(c) \approx 5.47717$$

and

$$\lambda(c) \approx 5.47696.$$

45. *The Milkmaid Problem*

Given a point R on the river, the ellipse $\mathcal{E}_R$ passing through R with foci at H and B contains all of the points P for which the distance $|\overline{HP}| + |\overline{PB}|$ is the same as $|\overline{HR}| + |\overline{RB}|$. If the river is not tangent to $\mathcal{E}_R$ at the point R, then it passes inside this ellipse, and the milkmaid's walk will be shorter if she walks to a point on the river that lies inside $\mathcal{E}_R$.

47. Parts a, b, and c of the argument require no more comments. The tran-

sition from c to d is clarified by appealing to the Chain Rule.

$$\frac{d}{dt}f(\mathbf{r}(t)) = \frac{d}{dt}f(x(t), y(t))$$

$$= f_x(x(t), y(t))x'(t) + f_y(x(t), y(t))y'(t)$$

$$= \langle f_x(x(t), y(t)), f_y(x(t), y(t)) \rangle \cdot \langle x'(t), y'(t) \rangle$$

$$= \nabla f(\mathbf{r}(t)) \cdot \mathbf{r}'(t)$$

Part e is clarified by combining part d: $\nabla f(P_0) \cdot \mathbf{r}'(t_0) = 0$, with the observation that $\nabla g(P_0) \cdot \mathbf{r}'(t_0) = 0$ as well. This is a consequence of the fact that the function $t \mapsto g(\mathbf{r}(t))$ is constant and also differentiates to 0 at $t = t_0$. Because $\nabla f(P_0)$ and $\nabla g(P_0)$ are both perpendicular to $\mathbf{r}'(t_0)$, they must be parallel to one another. That is, there is a scalar λ such that $\nabla f(P_0) = \lambda \nabla g(P_0)$.

49. We wish to maximize $f(x_1, x_2, \ldots, x_N) = x_1 + x_2 + \cdots + x_N$ subject to the constraint $g(x_1, x_2, \ldots, x_N) = x_1^2 + x_2^2 + \cdots + x_N^2 = 1$. Using Lagrange multipliers we seek a solution to the vector equation

$$\langle 1, 1, \ldots, 1 \rangle = \lambda \langle 2x_1, 2x_2, \ldots, 2x_N \rangle.$$

The scalar equations are

$$1 = 2\lambda x_1,\ 1 = 2\lambda x_2, \ldots, 1 = 2\lambda x_N,\ \text{and}\ x_1^2 + x_2^2 + \cdots + x_N^2 = 1.$$

Clearly it must be the case that $x_1 = x_2 = \cdots = x_{N-1} = x_N$. Since their squares sum to 1, $x_j^2 = 1/N$ for each j, and $x_j = 1/\sqrt{N}$ for $j = 1, 2, \ldots N$.

The maximum value of f is $1/\sqrt{N} + 1/\sqrt{N} + \cdots + 1/\sqrt{N} = N \cdot (1/\sqrt{N}) = \sqrt{N}$. This can be expressed in the form

$$x_1^2 + x_2^2 + \cdots + x_N^2 = 1 \implies x_1 + x_2 + \cdots + x_N \leq \sqrt{N}.$$

The inequality in the problem can be obtained by observing that, given any numbers $x_1, x_2, \ldots, x_N$ (not all zero), if we let $x_1^2 + x_2^2 + \cdots + x_N^2 = a^2$, then $(x_1/a)^2 + (x_2/a)^2 + \cdots + (x_N/a)^2 = 1$ so

$$\frac{x_1}{a} + \frac{x_2}{a} + \cdots + \frac{x_N}{a} \leq \sqrt{N}.$$

Consequently,

$$x_1 + x_2 + \cdots + x_N \leq a\sqrt{N}$$

and, dividing both sides by N and replacing a with $x_1^2 + x_2^2 + \cdots + x_N^2$, we have

$$\frac{x_1 + x_2 + \cdots + x_N}{N} \leq \frac{\sqrt{x_1^2 + x_2^2 + \cdots + x_N^2}}{\sqrt{N}}$$

$$= \sqrt{\frac{x_1^2 + x_2^2 + \cdots + x_N^2}{N}}.$$

Calculator/Computer Exercises

51. Maximize $f(x,y) = x \exp(x^2 - xy)$ subject to $g(x,y) = x^2 + y^2 - 1 = 0$.

The first entry defines the functions and the set of Lagrange equations.

```
> f := (x,y) -> x*exp(x^2 - x*y):
  g := (x,y) -> x^2 + y^2 - 1:
  eqns := {D[1](f)(x,y) = lambda*D[1](g)(x,y),
           D[2](f)(x,y) = lambda*D[2](g)(x,y),
           g(x,y) = 0};
```

$$eqns := \left\{ e^{x^2 - yx} + x(2x - y)x^{x^2 - yx} = 2\lambda x, \; -x^2 e^{x^2 - yx} + 2x^2 e^{x^2 - y} = 2\lambda y, \; x^2 + y^2 - 1 = 0 \right\}$$

Application of the **solve** procedure yields very complicated solution formulas that evaluate numerically to complex-valued approximations. We will use **fsolve** instead. In order to obtain all critical points we need accurate approximate values. The plot below, showing the constraint circle and some level curves for f, shows us where to look.

```
> with(plots):
  display( contourplot( f(x,y), x=-1.5..1.5, y=-1.5..1.5,
           color=black, contours=[$-4..4] ),
               implicitplot( g(x,y), x=-1.5..1.5, y=-1.5..1.5,
               color=black ), scaling=constrained );
```

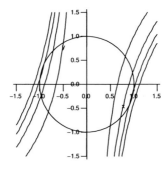

Using this picture the following entries generate the critical points, then evaluate the function f to see which one yields a maximum value.

```
> CP1 := fsolve( eqns, {x=1,y=-0.5,lambda=1} );
  CP2 := fsolve( eqns, {x=-1,y=0.5,lambda=-1} );
```

$$CP1 := \{\lambda = 5.339925875, x = 0.9589803844, y = -.2834724367\}$$

$$CP2 := \{\lambda = -5.339925875, x = -.9589803844, y = 0.2834724367\}$$

```
> eval(f(x,y),CP1);
  eval(f(x,y),CP2);
```

$$3.156940068$$
$$-3.156940068$$

53. Minimize $f(x,y) = (1 + x^2 + xy)/(1 + x^2 + y^4)$ subject to $g(x,y) = x^2 + y^2/4 - 1 = 0$.

The first entry defines the functions and the set of Lagrange equations.

```
> f := (x,y) -> (1+x^2+x*y)/(1+x^2+y^4):
  g := (x,y) -> x^2 + y^2/4 - 1:
  eqns := {D[1](f)(x,y) = lambda*D[1](g)(x,y),
           D[2](f)(x,y) = lambda*D[2](g)(x,y),
           g(x,y) = 0};
```

$$eqns := \left\{ \frac{x}{1 + x^2 + y^4} - \frac{-4(1 + x^2 + yx)y^3}{(1 + x^2 + y^4)^2} = \frac{1}{2}\lambda y, \; x^2 + \frac{1}{4}y^2 - 1 = 0, \right.$$
$$\left. \frac{2x + y}{1 + x^2 + y^4} - \frac{-2(1 + x^2 + yx)x}{(1 + x^2 + y^4)^2} = 2\lambda x \right\}$$

As in Exercise 49 **fsolve** is used to solve the equations. A plot is needed to show us where to look.

```
> with(plots):
  display( contourplot( f(x,y), x=-1.5..1.5, y=-2.5..2.5,
           color=black, contours=[k/4$k=-5..5] ),
               implicitplot( g(x,y), x=-1.5..1.5, y=-2.5..2.5,
               color=black ), scaling=constrained );
```

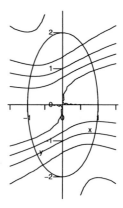

Using this picture the following entries generate the critical points, then evaluate the function f to see which one yields a minimum value.

```
> CP1 := fsolve( eqns, {x=1,y=-2,lambda=1} );
  CP2 := fsolve( eqns, {x=-1,y=2,lambda=-1} );
```

$$CP1 := \{x = 0.3955913280, \lambda = -.1081236129, y = -1.836853289\}$$

$$CP2 := \{\lambda = -.1081236129, x = -.3955913280, y = 1.836853289\}$$

```
> eval(f(x,y),CP1);
  eval(f(x,y),CP2);
```

$$0.03427669025$$

$$0.03427669025$$

55. We wish to obtain the extreme values of $F(x, y, z) = x + y + z$ subject to the constraints $G(x, y, z) = e^x - z^3 - z = 0$ and $H(x, y, z) = x^2 + y^2 - 1 = 0$. The constraint surfaces are a cylinder and a flat surface resembling a plane.

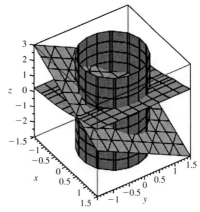

See the picture on the right. The sharply tilted plane is the set of points where $F(x, y, z) = 0$.

Applying *fsolve* to the set of equations

$$\{1 = \lambda e^x + 2\mu x, 1 = 2\mu y, 1 = \lambda \cdot (-3z^2 - 1), e^x - z^3 - z = 0, x^2 + y^2 - 1 = 0\}$$

yields solutions $x = 0.8350985386$, $y = 0.5501003825$ and $z = 1.072250862$ where $F(x, y, z) = 2.457449783$. Applying *fsolve* to the same set of equations with the hint that $x = -0.8, y = -0.5, z = -1, \lambda = 1, \mu = -1$ yields solutions $x = -0.7947305799$, $y = -0.6069623591$, $z = 0.3916348177$ where $F(x, y, z) = -1.010058121$.

Chapter 12

Multiple Integrals

12.1 Double Integrals over Rectangular Regions

Problems for Practice

1. The midpoints of the four rectangles are $(1/2, 2)$, $(3/2, 2)$, $(1/2, 4)$, and $(3/2, 4)$. Each rectangle has area 2 so the approximation is

$$V \approx (1/2 + 2 \cdot 2) \cdot 2 + (3/2 + 2 \cdot 2) \cdot 2$$
$$+ (1/2 + 2 \cdot 4) \cdot 2 + (3/2 + 2 \cdot 4) \cdot 2$$
$$= 56.$$

3. The midpoints of the four rectangles are $(0, 1/2)$, $(2, 1/2)$, $(0, 3/2)$, and $(2, 3/2)$. Each rectangle has area 2 so the approximation is

$$V \approx (3 \cdot 0^2 + 2 \cdot 1/2) \cdot 2 + (3 \cdot 2^2 + 2 \cdot 1/2) \cdot 2$$
$$+ (3 \cdot 0^2 + 2 \cdot 3/2) \cdot 2 + (3 \cdot 2^2 + 2 \cdot 3/2) \cdot 2$$
$$= 64.$$

5. $\mathcal{I}(x) = \int_1^5 (x + 2y)\, dy = \left(xy + y^2\right)\big|_{y=1}^{y=5} = 4x + 24$

$\mathcal{J}(y) = \int_0^2 (x + 2y)\, dx = \left(x^2/2 + 2yx\right)\big|_{x=0}^{x=2} = 2 + 4y$

7. $\mathcal{I}(x) = \int_{-2}^2 (3x^2 + 2y)\, dy = \left(3x^2 y + y^2\right)\big|_{y=-2}^{y=2} = 12x^2$

$\mathcal{J}(y) = \int_{-1}^3 (3x^2 + 2y)\, dx = \left(x^3 + 2yx\right)\big|_{x=-1}^{x=3} = 28 + 8y$

9. $\mathcal{I}(x) = \int_0^1 (ye^{xy^2})\, dy = \left(e^{xy^2}/(2x)\right)\big|_{y=0}^{y=1} = (e^x - 1)/(2x)$

   $\mathcal{J}(y) = \int_0^1 (ye^{xy^2})\, dx = \left(e^{xy^2}/y\right)\big|_{x=0}^{x=1} = (e^{y^2} - 1)/y$

11. $\int_1^4 \left(\int_{-2}^5 (x^2 - y)\, dx\right) dy = \int_1^4 \left((x^3/3 - yx)\right)\big|_{x=-2}^{x=5}\, dy$

   $= \int_1^4 \left(133/3 - 7y\right) dy = \left(133y/3 - 7y^2/2\right)\big|_1^4 = 161/2$

13. $\int_0^\pi \left(\int_0^\pi x\cos(y)\, dx\right) dy = \int_0^\pi \left((x^2/2)\cos(y)\right)\big|_{x=0}^{x=\pi}\, dy$

   $= \int_0^\pi \left((\pi^2/2)\cos(y)\right) dy = \left(\pi^2/2\right)\sin(y)\big|_0^\pi = 0$

15. $\int_{-2}^0 \left(\int_0^1 \exp(x-y)\, dx\right) dy = \int_{-2}^0 \left(\exp(x-y)\right)\big|_{x=0}^{x=1}\, dy$

   $= \int_{-2}^0 \left(e^{1-y} - e^{-y}\right) dy = \left(-e^{1-y} + e^{-y}\right)\big|_{-2}^0 = e^3 - e^2 - e + 1$

17. $\int_\pi^{2\pi} \left(\int_{\pi/2}^\pi (\cos(x)/y - \cos(y)/x)\, dx\right) dy = \int_\pi^{2\pi} \left(\sin(x)/y - \cos(y)\ln(x)\right)\big|_{x=\pi/2}^{x=\pi}\, dy$

   $= \int_\pi^{2\pi} \left(-\cos(y)\ln(\pi) - 1/y\right) dy = \left(\sin(y)\ln(\pi) - \ln(y)\right)\big|_\pi^{2\pi} = -\ln(2)$

19. $\int_1^4 \left(\int_1^4 (y\sqrt{x} - x\sqrt{y})\, dx\right) dy = \int_1^4 \left(2yx^{3/2}/3 - x^2\sqrt{y}/2\right)\big|_{x=1}^{x=4}\, dy$

   $= \int_1^4 \left(14y/3 - 15\sqrt{y}/2\right) dy = \left(7y^2/3 - 5y^{3/2}\right)\big|_1^4 = 0$

21. $\int_{-4}^{-2} \left(\int_{-3}^{-1} (x/y^2 + y/x^3)\, dx\right) dy = \int_{-4}^{-2} \left(x^2/(2y^2) - y/(2x^2)\right)\big|_{x=-3}^{x=-1}\, dy$

   $= \int_{-4}^{-2} (-4y/9 - 4/y^2)\, dy = \left(-2y^2/9 + 4/y\right)\big|_{-4}^{-2} = 5/3$

23. $\int_0^\pi \left(\int_{\pi/2}^\pi \cos^2(x)\sin^2(y)\, dx\right) dy = \int_0^\pi \sin^2(y)(x/2 + \sin(2x)/4)\big|_{x=\pi/2}^{x=\pi}\, dy$

   $= \int_0^\pi (\pi\sin^2(y)/4)\, dy = \left(\pi(y/2 - \sin(2y)/4)/4\right)\big|_0^\pi = \pi^2/8$

25. $\int_0^{\pi/2} \left(\int_0^\pi \cos^3(x)\sin(y)\, dy\right) dx = \int_0^{\pi/2} \left(-\cos^3(x)\cos(y)\right)\big|_{y=0}^{y=\pi}\, dx$

   $= \int_0^{\pi/2} 2\cos^3(x)\, dx = 2\int_0^{\pi/2} (1 - \sin^2(x))\cos(x)\, dx$

   $= 2(\sin(x) - \sin^3(x)/3)\big|_0^{\pi/2} = 4/3$

27. $\int_0^1 \left(\int_0^1 (x^3y^2\, dx\right) dy = \int_0^1 \left(x^4y^2/4\big|_{x=0}^{x=1}\right) dy = \int_0^1 y^2/4\, dy = y^3/12\big|_0^1 = 1/12$

   $\int_0^1 \left(\int_0^1 x^3y^2\, dy\right) dx = \int_0^1 \left(x^3y^3/3\big|_{y=0}^{y=1}\right) dx = \int_0^1 x^3/3\, dx = x^4/12\big|_0^1 = 1/12$

29. $\int_1^e \left(\int_0^2 1/(x+y)\, dx\right) dy = \int_1^e \left(\ln(x+y)\big|_{x=0}^{x=2}\right) dy$

   $= \int_1^e (\ln(2+y) - \ln(y))\, dy = ((y+2)\ln(y+2) - y\ln(y))\big|_1^e$

   $= (e+2)\ln(e+2) - e - 3\ln(3)$

   $\int_0^2 \left(\int_1^e 1/(x+y)\, dy\right) dx = \int_0^2 \left(\ln(x+y)\big|_{y=1}^{y=e}\right) dx$

   $= \int_0^2 (\ln(x+e) - \ln(x+1))\, dx = ((x+e)\ln(x+e) - (x+1)\ln(x+1))\big|_0^2$

   $= (e+2)\ln(e+2) - e - 3\ln(3)$

Further Theory and Practice

31. $\int_1^e \left(\int_1^e \ln(xy)\, dx \right) dy = \int_1^e \left((x\ln(x) - x + x\ln(y))\big|_{x=1}^{x=e} \right) dy$
$$= \int_1^e ((e-1)\ln(y) + 1)\, dy = ((e-1)(y\ln y - y) + y)\big|_1^e = 2e - 2$$

33. $\int_1^e \left(\int_1^e xy\ln(y)\, dx \right) dy = \int_1^e \left((x^2/2)y\ln(y)\big|_{x=1}^{x=e} \right) dy$
$$= \frac{e^2-1}{2}\int_1^e y\ln(y)\, dy = \frac{e^2-1}{2}\left(y^2\left(\frac{\ln(y)}{2} - \frac{1}{4}\right) \right)\Big|_1^e = \frac{e^2-1}{2}\cdot\frac{e^2+1}{4} = \frac{e^4-1}{8}$$

35. $\int_0^1 \left(\int_0^1 xe^{xy}\, dy \right) dx = \int_0^1 \left(e^{xy}\big|_{y=0}^{y=1} \right) dx = \int_0^1 (e^x - 1)\, dx = (e^x - x)\big|_0^1 = e - 2$

37. Substitute $u = x^2y^2$, $du = 2x^2y\, dy$ to obtain
$$\int_1^4 \left(\int_{x^2}^{4x^2} \ln(u)\, du \right) dx = \int_1^4 \left((u\ln(u) - u)\big|_{u=x^2}^{u=4x^2} \right) dx$$
$$= \int_1^4 ((8\ln(2) - 3)x^2 + 6x^2\ln(x))\, dx = 21(8\ln(2) - 3) + 6\int_1^4 x^2\ln(x)\, dx\,.$$

The integral that remains can be evaluated using integration by parts $(u = \ln(x))$ to yield $424\ln(2) - 105$.

39. $\int_{-1}^1 \left(\int_{-1}^1 x^2\cos(xy)\, dy \right) dx = \int_{-1}^1 \left(x\sin(xy)\big|_{y=-1}^{y=1} \right) dx$
$$= \int_{-1}^1 (2x\sin(x))\, dx = (2\sin(x) - 2x\cos(x))\big|_{-1}^1 = 4(\sin(1) - \cos(1))$$

41. Substitute $u = xy$, $du = y\, dx$ in the dx integral to obtain
$$\int_1^e \left(\int_1^e xy\ln(xy)\, dx \right) dy = \int_1^e \left(\int_y^{ey} u\ln(u)/y\, du \right) dy$$
$$= 1/4 \int_1^e \left((2u^2\ln(u) - u^2)/y)\big|_{u=y}^{u=ey} \right) dy$$
$$= (e^2 - 1)/2 \int_1^e y\ln(y)\, dy + (e^2 + 1)/4 \int_1^e y\, dy$$
$$= (e^2 - 1)/2 \cdot (e^2 + 1)/4 + (e^2 + 1)/4 \cdot (e^2 - 1)/2$$
$$= (e^4 - 1)/4$$

43. Observe that
$$\int_a^b \left(\int_c^d \phi(x)\psi(y)\, dy \right) dx = \int_a^b \phi(x)\left(\int_c^d \psi(y)\, dy \right) dx$$
$$= \left(\int_a^b \phi(x)\, dx \right)\left(\int_c^d \psi(y)\, dy \right)$$

Applying this to the given integral,
$$\iint_{\mathcal{R}} (xy)^2\sqrt{(1-x^2)^3(4-y^2)}\, dA = \int_0^1 \left(\int_0^2 x^2(1-x^2)^{3/2}y^2(4-y^2)^{1/2}\, dy \right) dx$$
$$= \left(\int_0^1 x^2(1-x^2)^{3/2}\, dx \right)\left(\int_0^2 y^2(4-y^2)^{1/2}\, dy \right)$$
$$= \pi/32 \cdot \pi = \pi^2/32\,.$$

45. Since $0 \le g(x,y) \le f(x,y)$ over the rectangle $\mathcal{R}$, the volume can be obtained by subtracting the volume under the graph of g from the volume

under the graph of f. This yields the following volume intergal.

$$\text{Vol} = \iint_{\mathcal{R}} (f(x,y) - g(x,y))\, dA = \int_0^1 \left(\int_1^2 (1 + x - y)\, dx \right) dy$$

$$= \int_0^1 \left(x + \frac{1}{2}x^2 - yx \right) \Big|_{x=1}^{x=2} dy = \int_0^1 \left(\frac{5}{2} - y \right) dy$$

$$= \left(\frac{5}{2}y - \frac{1}{2}y^2 \right) \Big|_0^1 = 2$$

47. Since $0 \le g(x,y) \le f(x,y)$ over the rectangle $\mathcal{R}$, the volume can be obtained by subtracting the volume under the graph of g from the volume under the graph of f. This yields the following volume intergal.

$$\text{Vol} = \iint_{\mathcal{R}} (f(x,y) - g(x,y))\, dA = \int_0^1 \left(\int_1^2 (2x + y - (x + 2y))\, dx \right) dy$$

$$= \int_0^1 \left(\frac{1}{2}x^2 - yx \right) \Big|_{x=1}^{x=2} dy = \int_0^1 \left(\frac{3}{2} - y \right) dy$$

$$= \left(\frac{3}{2}y - \frac{1}{2}y^2 \right) \Big|_0^1 = 1$$

Calculator/Computer Exercises

49. Adapting the code in the text we have

```
> f := (x,y) -> cos(sqrt(1+x) + y):
  a,b,c,d,N := -1,1,0,1,50:
  Delta := [(b-a)/N,(d-c)/N];
```

$$\Delta := \left[\frac{1}{25}, \frac{1}{50} \right]$$

```
> sum(sum(evalf(
      f(a+(i-1/2)*Delta[1],c+(j-1/2)*Delta[2])),
          j=1..N),i=1..N)*Delta[1]*Delta[2];
```

$$0.2248113852$$

The following entries show how to obtain Maple's approximation for the integral. It is accurate to 10 digits.

```
> Int(Int(f(x,y),x=a..b),y=c..d);
  evalf(%);
```

$$\int_0^1 \int_{-1}^1 \cos(\sqrt{x+1} + y)\, dx\, dy$$

$$0.2250379252$$

51. Use the code displayed in Exercise 49 with the appropriate changes in the data. Maple's approximation for the integral is shown below.

```
> Int(Int(exp(-x^2-y^2),x=-1..1),y=-1..1);
  evalf(%);
```

$$\int_{-1}^{1} \int_{-1}^{1} e^{-x^2-y^2} \, dx \, dy$$

2.230985141

12.2   Integration over More General Regions

Problems for Practice

1. This region is x-simple. The functions that form its boundary are

$$y \mapsto -2y - 1$$

and

$$y \mapsto -y^2 + 2$$

for $-1 \leq y \leq 3$.

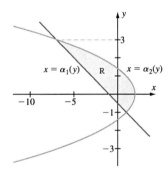

3. This region is x-simple. The functions that form its boundary are

$$y \mapsto y^2$$

and

$$y \mapsto -2y^2 - 3y + 18$$

for $-3 \leq y \leq 2$.

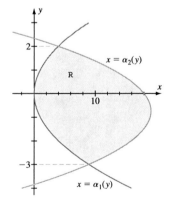

5. This region is y-simple. The functions that form its boundary are

$$x \mapsto x^{3/2}$$

and

$$x \mapsto 6x - 1$$

for $1 \le x \le 3$.

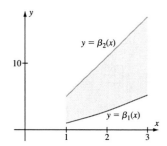

7. $\int_{-1}^{1} \left(\int_{(x+1)^2}^{8-(x+1)^2} 3x \, dy \right) dx = \int_{-1}^{1} \left((3xy)\big|_{y=(x+1)^2}^{y=8-(x+1)^2} \right) dy$

$= \int_{-1}^{1} \left(3x \big(8 - 2(x+1)^2 \big) \right) dx = \int_{-1}^{1} (18x - 6x^3 - 12x^2) \, dx$

$= \left(9x^2 - 3x^4/2 - 4x^3 \right)\big|_{-1}^{1} = -8$

9. $\int_{0}^{2} \left(\int_{x/4}^{4x} (2x - 1)y^{1/2} \, dy \right) dx = 2/3 \int_{0}^{2} \left((2x - 1)y^{3/2} \big|_{y=x/4}^{y=4x} \right) dx$

$= 21/4 \int_{0}^{2} (2x^{5/2} - x^{3/2}) \, dx = 21/4((4/7)x^{7/2} - (2/5)x^{5/2})\big|_{0}^{2}$

$= 3 \cdot 2^3 \sqrt{2} - (21/10) \cdot 2^2 \sqrt{2} = 78\sqrt{2}/5$

11. $\int_{\pi/2}^{\pi} \left(\int_{0}^{\sin(x)} 4x \, dy \right) dx = \int_{\pi/2}^{\pi} \left((4xy)\big|_{y=0}^{y=\sin(x)} \right) dx$

$= 4 \int_{\pi/2}^{\pi} x \sin(x) \, dx = 4(\sin(x) - x \cos(x))\big|_{\pi/2}^{\pi} = 4\pi - 4$

13. $\int_{\pi/4}^{\pi/2} \left(\int_{0}^{\sin(y)} 5 \, dx \right) dy = \int_{\pi/4}^{\pi/2} \left((5x)\big|_{x=0}^{x=\sin(y)} \right) dy$

$= 5 \int_{\pi/4}^{\pi/2} \sin(y) \, dy = 5(-\cos(y))\big|_{\pi/4}^{\pi/2} = 5(0 + 1/\sqrt{2}) = 5/\sqrt{2}$

15. The boundary curves intersect at $(0,0)$ and $(2,4)$. The region is x- and y-simple. As a y-simple region: $\mathcal{R} = \{(x,y) : 0 \le x \le 2, x^2 \le y \le 2x\}$, and the integral evaluates as follows.

$$\int_{0}^{2} \left(\int_{x^2}^{2x} (6y - 4x) \, dy \right) dx = \int_{0}^{2} \left((3y^2 - 4xy)\big|_{y=x^2}^{y=2x} \right) dx$$

$$= \int_{0}^{2} \left(12x^2 - 3x^4 - 4x(2x - x^2) \right) dx$$

$$= \left(4x^3/3 - 3x^5/5 + x^4 \right)\bigg|_{0}^{2} = 112/15$$

17. The region is y-simple: $\mathcal{R} = \{(x,y) : 4 \le x \le 7, 0 \le y \le 2x - 6\}$. The

integral evaluates as follows.

$$\int_4^7 \left(\int_0^{2x-6} xe^y \, dy \right) dx = \int_4^7 \left((xe^y)\big|_{y=0}^{y=2x-6} \right) dx$$

$$= \int_4^7 (xe^{2x-6} - x) \, dx$$

$$= \left. \left((1/4)(2x-1)e^{2x-6} - x^2/2 \right) \right|_4^7$$

$$= e^2(13e^6 - 7)/4 - 33/2$$

19 . The region is y-simple: $\mathcal{R} = \{(x, y) : 3 \le x \le 4, x^2 - 3x \le y \le x^2 + 8\}$. The integral evaluates as follows.

$$\int_3^4 \left(\int_{x^2-3x}^{x^2+8} y\sqrt{x} \, dy \right) dx = \int_3^4 \left((\sqrt{x}\, y^2/2)\big|_{y=x^2-3x}^{y=x^2+8} \right) dx$$

$$= \int_3^4 (3x^{7/2} + 7x^{5/2}/2 + 32x^{1/2}) \, dx$$

$$= \left. \left(2x^{9/2}/3 + x^{7/2} + 64x^{3/2}/3 \right) \right|_3^4$$

$$= 640 - 145\sqrt{3}$$

21. The region is y-simple, $\mathcal{R} = \{(x, y) : 0 \le x \le 1, x/6 \le y \le x\}$. The integral evaluates as follows.

$$\int_0^1 \left(\int_{x/6}^x e^x \, dy \right) dy = \int_0^1 \left((ye^x)\big|_{y=x/6}^{y=x} \right) dx = \int_0^1 (5xe^x/6) \, dx$$

$$= 5/6(xe^x - e^x)\big|_0^1 = 5/6$$

23. The region is x-simple and y-simple. As an x-simple region, $\mathcal{R} = \{(x, y) : 0 \le y \le \pi/2, 0 \le x \le y\}$. The integral evaluates as follows.

$$\int_0^{\pi/2} \left(\int_0^y \sin(y)/y \, dx \right) dy = \int_0^{\pi/2} \left((x\sin(y)/y)\big|_{x=0}^{x=y} \right) dy$$

$$= \int_0^{\pi/2} \sin(y) \, dy = 1$$

25. The region is x-simple and y-simple. As an x-simple region, $\mathcal{R} = \{(x, y) : 0 \le y \le 4, 0 \le x \le \sqrt{y}\}$. The integral evaluates as follows.

$$\int_0^4 \left(\int_0^{\sqrt{y}} e^y/\sqrt{y} \, dx \right) dy = \int_0^4 \left((xe^y/\sqrt{y})\big|_{x=0}^{x=\sqrt{y}} \right) dy$$

$$= \int_0^4 e^y \, dy = e^4 - 1$$

Further Theory and Practice

27. The region is x-simple and y-simple. As an x-simple region, $\mathcal{R} = \{(x, y) : 0 \le y \le \sqrt{\pi}, 0 \le x \le y\}$. The integral evaluates as follows.

$$\int_0^{\sqrt{\pi}} \left(\int_0^y \sin(y^2)\, dx \right) dy = \int_0^{\sqrt{\pi}} \left((x\sin(y^2))\big|_{x=0}^{x=y} \right) dy$$

$$= \int_0^{\sqrt{\pi}} y \sin(y^2)\, dy = (-\cos(y^2)/2)\Big|_0^{\sqrt{\pi}} = 1$$

29. The region is x-simple and y-simple. As a y-simple region, $\mathcal{R} = \{(x, y) : 0 \le x \le 1, 0 \le y \le x^2\}$. The integral evaluates as follows.

$$\int_0^1 \left(\int_0^{x^2} y(1 - x^2)^2/x^3\, dy \right) dx = \int_0^1 \left(((y^2/2)(1 - x^2)^2/x^3)\big|_{y=0}^{y=x^2} \right) dx$$

$$= 1/2 \int_0^1 x(1 - x^2)^2\, dx$$

$$= 1/2(-(1 - x^2)^3/6)\Big|_0^1 = 1/12$$

31. The region is y-simple, $\mathcal{R} = \{(x, y) : 0 \le x \le \pi/4, \sin(x) \le y \le \cos(x)\}$. The integral evaluates as follows.

$$\int_0^{\pi/4} \left(\int_{\sin(x)}^{\cos(x)} (x - y)\, dy \right) dx = \int_0^{\pi/4} \left((xy - y^2/2)\big|_{y=\sin(x)}^{y=\cos(x)} \right) dx$$

$$= \int_0^{\pi/4} (x\cos(x) - x\sin(x) - (\cos^2(x) - \sin^2(x))/2)\, dx$$

$$= \int_0^{\pi/4} x\cos(x)\, dx - \int_0^{\pi/4} x\sin(x)\, dx - 1/2 \int_0^{\pi/4} \cos(2x)\, dx$$

$$= (\cos(x) + x\sin(x))\Big|_0^{\pi/4} - (\sin(x) - x\cos(x))\Big|_0^{\pi/4} - (1/4)\sin(2x)\Big|_0^{\pi/4}$$

$$= \frac{\pi}{2\sqrt{2}} - \frac{5}{4} = \frac{\sqrt{2}\,\pi - 5}{4}$$

33. The region is y-simple, $\mathcal{R} = \{(x, y) : 0 \le x \le 1, 0 \le y \le x\}$. The integral evaluates as follows.

$$\int_0^1 \left(\int_0^x x^2 \sin(xy)\, dy \right) dx = \int_0^1 \left((-x\cos(xy))\big|_{y=0}^{y=x} \right) dx$$

$$= \int_0^1 (-x\cos(x^2) + x)\, dx$$

$$= (x^2/2 - \sin(x^2)/2)\Big|_0^1 = (1 - \sin(1))/2$$

35. This is an x-simple region, $\mathcal{R} = \{(x, y) : 0 \leq y \leq 1, -y \leq x \leq \sqrt{y}\}$. The integral evaluates as follows.

$$\int_0^1 \left(\int_{-y}^{\sqrt{y}} y \, dx \right) dy = \int_0^1 \left((yx)\big|_{x=-y}^{x=\sqrt{y}} \right) dy$$

$$= \int_0^1 (y^{3/2} + y^2) \, dy$$

$$= (2y^{5/2}/5 + y^3/3)\big|_0^1 = 11/15$$

37. The region is y-simple, $\mathcal{R} = \{(x, y) : 0 \leq x \leq 1, -x \leq y \leq x^2\}$. The integral evaluates as follows.

$$\int_0^1 \left(\int_{-x}^{x^2} xy \, dy \right) dx = \int_0^1 \left((xy^2/2)\big|_{y=-x}^{y=x^2} \right) dx$$

$$= 1/2 \int_0^1 (x^5 - x^3) \, dx$$

$$= 1/2(x^6/6 - x^4/4)\big|_0^1 = -1/24$$

39. The region is y-simple, $\mathcal{R} = \{(x, y) : -1 \leq x \leq 2, x^2 \leq y \leq x + 2\}$. The integral evaluates as follows.

$$\int_{-1}^2 \left(\int_{x^2}^{x+2} x \, dy \right) dx = \int_{-1}^2 \left((xy)\big|_{y=x^2}^{y=x+2} \right) dx$$

$$= \int_{-1}^2 (x^2 + 2x - x^3) \, dx$$

$$= (x^3/3 + x^2 - x^4/4)\big|_{-1}^2 = 9/4$$

41. The region is y-simple, $\mathcal{R} = \{(x, y) : -1 \leq x \leq 1, -\sqrt{2 - x^2} \leq y \leq x\}$. The integral evaluates as follows.

$$\int_{-1}^1 \left(\int_{-\sqrt{2-x^2}}^{x} x \, dy \right) dx = \int_{-1}^1 \left((xy)\big|_{y=-\sqrt{2-x^2}}^{y=x} \right) dx$$

$$= \int_{-1}^1 (x^2 + x\sqrt{2 - x^2}) \, dx$$

$$= (x^3/3 - (2 - x^2)^{3/2}/3)\big|_{-1}^1 = 2/3$$

43. The region is y-simple, $\mathcal{R} = \{(x, y) : 0 \leq x \leq \pi/4, \sin(x) \leq y \leq \cos(x)\}$. The integral evaluates as follows.

$$\int_0^{\pi/4} \left(\int_{\sin(x)}^{\cos(x)} y \, dy \right) dx = \int_0^{\pi/4} \left((y^2/2)\big|_{y=\sin(x)}^{y=\cos(x)} \right) dx$$

$$= 1/2 \int_0^{\pi/4} (\cos^2(x) - \sin^2(x)) \, dx$$

$$= 1/2 \int_0^{\pi/4} \cos(2x) \, dx$$

$$= (1/4) \sin(2x)\Big|_0^{\pi/4} = 1/4$$

45. The region $\mathcal{R}$ is neither x-simple nor y-simple. Draw a picture to see that it is the union of the following two disjoint y-simple regions:

$$\mathcal{R}_1 = \{(x, y) : 0 < x < 1, 2x/5 \leq y \leq 3x\}$$

and

$$\mathcal{R}_2 = \{(x, y) : 1 < x < 10, 2x/5 \leq y \leq (x + 26)/9\} \, .$$

Consequently, the integral can be evaluated as $\iint_{\mathcal{R}_1} f(x, y) \, dA + \iint_{\mathcal{R}_2} f(x, y) \, dA$. Since

$$\iint_{\mathcal{R}_1} f(x, y) \, dA = \int_0^1 \left(\int_{2x/5}^{3x} (ax + by) \, dy \right) dx$$

$$= \int_0^1 \left((axy + by^2/2)\big|_{y=2x/5}^{y=3x} \right) dx$$

$$= (13a/5 + 221b/50) \int_0^1 x^2 \, dx = 13a/15 + 221b/150$$

and

$$\iint_{\mathcal{R}_2} f(x, y) \, dA = \int_1^{10} \left(\int_{2x/5}^{(x+26)/9} (ax + by) \, dy \right) dx$$

$$= \int_1^{10} \left((axy + by^2/2)\big|_{y=2x/5}^{y=(x+26)/9} \right) dx$$

$$= \int_1^{10} \left(338b/81 + (26a/9 + 26b/81)x - (13a/45 + 299b/4050)x^2 \right) dx$$

$$= 234a/5 + 1443b/50 \, ,$$

the integral over the entire triangle is

$$\iint_{\mathcal{R}} (ax + by) \, dA = \left(\frac{13}{15} + \frac{243}{5} \right) a + \left(\frac{234}{5} + \frac{1443}{50} \right) b$$

$$= \frac{143}{3} a + \frac{91}{3} b \, .$$

47. The region $\mathcal{R}$ is neither x-simple nor y-simple. Draw a picture to see that it is the union of the following two disjoint y-simple regions:

$$\mathcal{R}_1 = \{(x,y) : 0 < x < 1, 0 \le y \le x^2 + 1\}$$

and

$$\mathcal{R}_2 = \{(x,y) : 1 < x < 2, 5(x-1) \le y \le x^2 + 1\}.$$

Consequently, the integral can be evaluated as $\iint_{\mathcal{R}_1} f(x,y)\, dA + \iint_{\mathcal{R}_2} f(x,y)\, dA$. Since

$$\iint_{\mathcal{R}_1} f(x,y)\, dA = \int_0^1 \left(\int_0^{x^2+1} (x+y)\, dy \right) dx$$

$$= \int_0^1 \left((xy + y^2/2)\big|_{y=0}^{y=x^2+1} \right) dx$$

$$= \int_0^1 (x(x^2+1) + (x^2+2)^2/2)\, dx = 101/60$$

and

$$\iint_{\mathcal{R}_2} f(x,y)\, dA = \int_1^2 \left(\int_{5(x-1)}^{x^2+1} (x+y)\, dy \right) dx$$

$$= \int_1^2 (x^4/4 + x^3 - 33x^2/2 + 31x - 12)\, dx = 57/20,$$

the integral over the entire region is

$$\iint_{\mathcal{R}} (x+y)\, dA = \frac{101}{60} + \frac{57}{20} = \frac{68}{15}.$$

49. The region is both x-simple and y-simple. As a y-simple region, $\mathcal{R} = \{(x,y) : 0 \le x \le 1, 0 \le y \le \sqrt{1-x^2}\}$. Therefore, the integral can be evaluated as follows.

$$\iint_{\mathcal{R}} f(x,y)\, dA = \int_0^1 \left(\int_0^{\sqrt{1-x^2}} (x+y)\, dy \right) dx$$

$$= \int_0^1 \left((xy + y^2/2)\big|_{y=0}^{y=\sqrt{1-x^2}} \right) dx$$

$$= \int_0^1 \left(x\sqrt{1-x^2} + (1-x^2)/2 \right) dx$$

$$= \left(-(1-x^2)^{3/2}/3 + x/2 - x^3/6 \right)\Big|_0^1 = 2/3$$

51. The region is both x-simple and y-simple. As a y-simple region, $\mathcal{R} = \{(x, y) : 0 \leq x \leq \pi/2, 0 \leq y \leq 1 - \cos(x)\}$. Therefore, the integral can be evaluated as follows.

$$\iint_{\mathcal{R}} f(x, y) \, dA = \int_0^{\pi/2} \left(\int_0^{1-\cos(x)} y \sin(x) \, dy \right) dx$$

$$= \int_0^{\pi/2} \left((\sin(x) y^2 / 2 \big|_{y=0}^{y=1-\cos(x)}) \right) dx$$

$$= 1/2 \int_0^{\pi/2} \sin(x)(1 - \cos(x))^2 \, dx$$

$$= 1/2 \big((1 - \cos(x))^3 / 3\big) \Big|_0^{\pi/2} = 1/6$$

53. The region is both x-simple and y-simple. As a y-simple region, $\mathcal{R} = \{(x, y) : 0 < x < 1, 0 \leq y \leq 2x\}$. Therefore, the integral can be set up as follows.

$$\iint_{\mathcal{R}} f(x, y) \, dA = \int_0^1 \left(\int_0^{2x} |y - x| \, dy \right) dx$$

To evaluate the dy integral, observe that for a fixed value of x between 0 and 1, $|y - x| = x - y$ when $0 \leq y \leq x$, and $|y - x| = y - x$ when $x \leq y \leq 2x$. Therefore, for these values of x,

$$\int_0^{2x} |y - x| \, dy = \int_0^x (x - y) \, dy + \int_x^{2x} (y - x) \, dy$$

$$= (xy - y^2/2) \Big|_0^x + (xy - y^2/2) \Big|_x^{2x} = x^2 \, .$$

Consequently,

$$\int_0^1 \left(\int_0^{2x} |y - x| \, dy \right) dx = \int_0^1 x^2 \, dx = 1/3 \, .$$

55. Draw a picture to see that, for positive z,

$$\int_0^z \text{erf}(x) \, dx = \int_0^z \left(\frac{2}{\sqrt{\pi}} \int_0^x e^{-y^2} \, dy \right) dx = \frac{2}{\sqrt{\pi}} \int_0^z \left(\int_y^z e^{-y^2} \, dx \right) dy$$

$$= \frac{2}{\sqrt{\pi}} \int_0^z \left((xe^{-y^2}) \big|_{x=y}^{x=z} \right) dy = \frac{2}{\sqrt{\pi}} \int_0^z (ze^{-y^2} - ye^{-y^2}) \, dy$$

$$= z \, \text{erf}(z) - \frac{2}{\sqrt{\pi}} \int_0^z ye^{-y^2} \, dy$$

$$= z \, \text{erf}(z) - \frac{1}{\sqrt{\pi}} (-e^{-y^2}) \Big|_0^z$$

$$= z \, \text{erf}(z) - \frac{1}{\sqrt{\pi}} (1 - e^{-z^2}) \, .$$

Consequently,

$$\int_0^z (1 - \operatorname{erf}(x))\, dx = z - \int_0^z \operatorname{erf}(x)\, dx$$

$$= z - \left(z \operatorname{erf}(z) - \frac{1}{\sqrt{\pi}}(1 - e^{-z^2}) \right)$$

$$= z(1 - \operatorname{erf}(z)) + \frac{1}{\sqrt{\pi}}(1 - e^{-z^2}).$$

Since $\lim_{z \to \infty} e^{-z^2} = 0$,

$$\int_0^\infty (1 - \operatorname{erf}(x))\, dx = \lim_{z \to \infty} \left(z(1 - \operatorname{erf}(z)) + \frac{1}{\sqrt{\pi}}(1 - e^{-z^2}) \right)$$

$$= \frac{1}{\sqrt{\pi}} + \lim_{z \to \infty} \frac{1 - \operatorname{erf}(z)}{z^{-1}} = \frac{1}{\sqrt{\pi}}.$$

To confirm that the limit in the last line is 0, apply L'Hôpital's Rule (it is indeterminate of the form $0/0$).

$$\lim_{z \to \infty} \frac{1 - \operatorname{erf}(z)}{z^{-1}} = \lim_{z \to \infty} \frac{-\frac{d}{dz}\operatorname{erf}(z)}{-z^{-2}} = \lim_{z \to \infty} \frac{-2e^{-z^2}/\sqrt{\pi}}{-z^{-2}}$$

$$= \frac{2}{\sqrt{\pi}} \lim_{z \to \infty} \frac{z^2}{e^{z^2}} = \frac{2}{\sqrt{\pi}} \lim_{z \to \infty} \frac{2z}{2ze^{z^2}} = \frac{2}{\sqrt{\pi}} \lim_{z \to \infty} e^{-z^2} = 0.$$

Calculator/Computer Exercises

57. A plot of the graphs of $1 + 3x$ and e^x reveals that they intersect at $x = 0$ and near to $x = 2$. In between, the exponential function lies below the line. The entries below locate the intersection point near $x = 2$, then evaluate the iterated integral.

```
> b := fsolve(1+3*x = exp(x), x=2):
  Int(Int( x*y, y=exp(x)..1+3*x), x=0..b) =
  int(int( x*y, y=exp(x)..1+3*x), x=0..b);
```

$$\int_0^{1.903813694} \int_{e^x}^{1+3x} xy\, dy\, dx = 6.652540906$$

59. A plot of the graphs of $1/(1 + x^2)$ and x^{10} reveals they intersect near to $x = 1$. The entries below locate the intersection point near $x = 1$, then evaluate the iterated integral.

```
> b := fsolve(1/(1+x^2) = x^10, x=1):
  Int(Int( x, y=x^10..1/(1+x^2)), x=0..b) =
  int(int( x, y=x^10..1/(1+x^2)), x=0..b);
```

$$\int_0^{0.9387605987} \int_{x^{10}}^{\frac{1}{x^2+1}} x\, dy\, dx = 0.2769368786$$

61–63. Simpson's Rule will be applied using the procedure that was defined in Exercise 59 of Chapter 10, Section 3. Given a function g, an interval $[a, b]$, and n (= half the number of subintervals), the output is the Simpson Rule Approximation

$$\int_a^b g(x)\, dx \approx \frac{h}{3}\left(g(x_0) + 4g(x_1) + 2g(x_2) + 4g(x_3) + \cdots + 4g(x_{2n-1}) + g(x_{2n})\right)$$

where $h = (b-a)/2n$ and $x_k = a + k \cdot h$.

```
> Simp := proc(g,a,b,n)
    local m,h;
      m := evalf(2*n);
      h := evalf((b-a)/m);
      h/3*(g(a)+g(b)+4*add(g(a+(2*k-1)*h),k=1..n)
                    +2*add(g(a+2*k*h),k=1..n-1));
      evalf(%);
    end proc:
```

61. The iterated integral is

$$I = \int_0^1 \left(\int_0^{\sqrt{1-x^2}} \sqrt{x^2 + y}\, dy \right) dx$$

$$= 2/3 \int_0^1 \left((x^2 + y)^{3/2}\big|_{y=0}^{y=\sqrt{1-x^2}} \right) dx$$

$$= 2/3 \int_0^1 \left(\left(x^2 + \sqrt{1-x^2}\right)^{3/2} - x^3 \right) dx \, .$$

The following entry defines the integrand as the function $x \mapsto g(x)$ and generates three Simpson Rule approximations for the value of I.

```
> g := x -> ((x^2 + sqrt(1-x^2))^(3/2) - x^3):
  '2/3*Simp(g,0,1,n)' $ n=10..12;
```

$$0.6227339973, 0.6229081265, 0.6230465427$$

63. As a x-simple region, $\mathcal{R} = \{(x, y) : 0 \le y \le 1, \arctan(y) \le x \le \arccos(y/\sqrt{2})\}$. The iterated integral is

$$I = \int_0^1 \left(\int_{\arctan(y)}^{\arccos(y/\sqrt{2})} e^{-y^2}\, dx \right) dy$$

$$= \int_0^1 e^{-y^2}\left(\arccos(y/\sqrt{2}) - \arctan(y)\right) dy \, .$$

The following entry defines the integrand as the function $y \mapsto g(y)$ and generates three Simpson Rule approximations for the value of I.

```
> g := y -> exp(-y^2)*(arccos(y/sqrt(2)) - arctan(y)):
  'Simp(g,0,1,n)' $ n=10..12;
```

$$0.6575488237, 0.6575489981, 0.6575491086$$

65–67. The following entry defines the values for the coefficients $c_{i,j}$. It is set up to work for any value of N.

```
> N := 4:
  for i from 0 to N
    do
      for j from 0 to N
        do
          if {i,j} subset {0,N} then c[i,j]:=1
            elif {i,j} subset {$1..N-1} then c[i,j]:=4
            else c[i,j]:=2
          end if
        end do
    end do;
```

Print the coefficients to verify that the correct values are assigned for $N = 4$.

```
> Matrix(5,5,(i,j)->c[i-1,j-1]);
```

$$\begin{bmatrix} 1 & 2 & 2 & 2 & 1 \\ 2 & 4 & 4 & 4 & 2 \\ 2 & 4 & 4 & 4 & 2 \\ 2 & 4 & 4 & 4 & 2 \\ 1 & 2 & 2 & 2 & 1 \end{bmatrix}$$

65. First define a, b, β_1, β_2, and f. (N is already defined.)

```
> a,b,beta[1],beta[2],f :=
  -1,1,unapply(0,x),unapply(1-x^2,x),unapply(y,x,y);
```

$$a, b, \beta_1, \beta_2, f := -1, 1, x \to 0, x \to 1 - x^2, (x, y) \to y$$

Define the increment dx, the increments dy_i, and the rest of the data needed to calculate the approximation.

```
> dx := (b-a)/N:
  for i from 0 to N
    do
      dy[i] := (beta[2](a+i*dx) - beta[1](a+i*dx))/N:
      X[i] := a + i*dx:
      for j from 0 to N
        do
          Y[i,j] := beta[1](X[i]) + j*dy[i]:
        end do
    end do:
```

Now make the approximation, calculate the exact value, and then the error.

```
> Approx := dx/2*add( dy[i]/2*
                 add( c[i,j]*f(X[i],Y[i,j]), j=0..N),
                 i=0..N);
  Exact := int(int(y, y=0..1-x^2), x=-1..1);
  Error = abs(Approx-Exact); evalf(%);
```

$$Approx := \frac{17}{32}$$

$$Exact := \frac{8}{15}$$

$$Error = \frac{1}{480}$$

$$Error = 0.002083333333$$

67. Adapt the code for Exercise 65.

12.3 Calculation of Volumes of Solids

Problems for Practice

1. The curves intersect at $x = -4$ and $x = 3$, the region is y-simple: $\mathcal{R} = \{(x, y) : -4 \le x \le 3, 2x^2 + 2x - 12 \le y \le x^2 + x\}$. The integral evaluates as follows.

$$\int_{-4}^{3} \left(\int_{2x^2+2x-12}^{x^2+x} (5+x) \, dy \right) dx = \int_{-4}^{3} \left((5y + xy) \big|_{y=2x^2+2x-12}^{y=x^2+x} \right) dx$$

$$= \int_{-4}^{3} (5+x)(12 - x^2 - x) \, dx$$

$$= \left(60x + 7x^2/2 - 2x^3 - x^4/4 \right) \Big|_{-4}^{3} = 1029/4$$

3. The curves intersect at $x = -1$ and $x = 2$, the region is y-simple: $\mathcal{R} = \{(x, y) : -1 \le x \le 2, x^2 - 2 \le y \le x\}$. The integral evaluates as follows.

$$\int_{-1}^{2} \left(\int_{x^2-2}^{x} (5+x) \, dy \right) dx = \int_{-1}^{2} \left((5y + xy) \big|_{y=x^2-2}^{y=x} \right) dx$$

$$= \int_{-1}^{2} (5+x)(x - x^2 + 2) \, dx$$

$$= \left(10x + 7x^2/2 - 4x^3/3 - x^4/4 \right) \Big|_{-1}^{2} = 99/4$$

5. The curves intersect at $x = 1$ and $x = 3$, the region is y-simple: $\mathcal{R} = \{(x, y) : 1 \leq x \leq 3, x^3 \leq y \leq 5x^2 - 7x + 3\}$. The integral evaluates as follows.

$$\int_1^3 \left(\int_{x^3}^{5x^2-7x+3} (5 + x)\, dy \right) dx = \int_1^3 \left((5y + xy)\big|_{y=x^3}^{y=5x^2-7x+3} \right) dx$$

$$= \int_1^3 (5 + x)(5x^2 - 7x + 3 - x^3)\, dx$$

$$= \left(6x^3 - 16x^2 + 15x - x^5/5 \right)\Big|_1^3 = 48/5$$

7. The curves intersect at $x = -3$ and $x = 2$, the region is y-simple: $\mathcal{R} = \{(x, y) : -3 \leq x \leq 2, x^2 - 4 \leq y \leq 2 - x\}$. The integral evaluates as follows.

$$\int_{-3}^2 \left(\int_{x^2-4}^{2-x} (5 + x)\, dy \right) dx = \int_{-3}^2 \left((5y + xy)\big|_{y=x^2-4}^{y=2-x} \right) dx$$

$$= \int_{-3}^2 (5 + x)(6 - x - x^2)\, dx$$

$$= \left(30x + x^2/2 - 2x^3 - x^4/4 \right)\Big|_{-3}^2 = 375/4$$

9. Treating the region as y-simple we have

$$\int_8^{10} \left(\int_2^4 (x - 2y)\, dy \right) dx = \int_8^{10} \left((xy - y^2)\big|_{y=2}^{y=4} \right) dx$$

$$= \int_8^{10} (2x - 12)\, dy$$

$$= \left(x^2 - 12x \right)\Big|_8^{10} = 12 \,.$$

11. As a y-simple region the volume integral evaluates as follows.

$$\int_{-4}^{-2} \left(\int_{-5}^{-2} x^2 y^2\, dy \right) dx = \int_{-4}^{-2} \left((x^2 y^3/3)\big|_{y=-5}^{y=-2} \right) dx$$

$$= \int_{-4}^{-2} 39x^2\, dy = 728$$

13. The region is y-simple.

$$\int_3^6 \left(\int_{3x}^{2x+8} (10x - 3y + 40)\, dy \right) dx = \int_3^6 \left((10xy - 3y^2/2 + 40y)\big|_{y=3x}^{y=2x+8} \right) dx$$

$$= \int_3^6 (224 - 5x^2/2 - 8x)\, dx$$

$$= \left(224x - 5x^3/6 - 4x^2 \right)\Big|_3^6 = 813/2 \,.$$

15. The region in the xy-plane is y-simple.

$$\int_3^5 \left(\int_{-6x}^{-2x} (x - y)\, dy \right) dx = \int_3^5 \left((xy - y^2/2)\big|_{y=-6x}^{y=-2x} \right) dx$$

$$= \int_3^5 20x^2\, dx = 1960/3$$

17. The boundary curves in the xy-plane intersect at $x = -2$ and $x = 4$. The parabola is below the line, $\mathcal{R} = \{(x, y) : -2 \le x \le 4, x^2 + x \le y \le 3x + 8\}$. The volume integral evaluates as follows.

$$\int_{-2}^4 \left(\int_{x^2+x}^{3x+8} (x + 2)\, dy \right) dx = \int_{-2}^4 \left((x + 2)y\big|_{y=x^2+x}^{y=3x+8} \right) dx$$

$$= \int_{-2}^4 (x + 2)(2x + 8 - x^2)\, dx = 108$$

19. The two slanted lines in the xy-plane intersect at $x = -3$. Therefore, $\mathcal{R} = \{(x, y) : -3 \le x \le 2, x \le y \le 2x + 3\}$. The volume integral evaluates as follows.

$$\int_{-3}^2 \left(\int_x^{2x+3} e^x\, dy \right) dx = \int_{-3}^2 \left(e^x y\big|_{y=x}^{y=2x+3} \right) dx$$

$$= \int_{-3}^2 e^x (x + 3)\, dx$$

$$= (xe^x - e^x + 3e^x)\Big|_{-3}^2 = 4e^2 + e^{-3}$$

21. The planar region forming the base of the solid is $\mathcal{R} = \{(x, y) : 0 \le x \le 4, x^{1/2} \le y \le x + 2\}$. The volume integral evaluates as follows.

$$\int_0^4 \left(\int_{x^{1/2}}^{x+2} (x + 2y)\, dy \right) dx = \int_0^4 \left((xy + y^2)\big|_{y=x^{1/2}}^{y=x+2} \right) dx$$

$$= \int_0^4 (x(x + 2 - x^{1/2}) + (x + 2)^2 - x)\, dx$$

$$= (x^3/3 + x^2/2 - 2x^{5/2}/5 + (x + 2)^3/3)\Big|_0^4 = 1288/15$$

23. The base of the solid in the xy-plane is $\mathcal{R} = \{(x, y) : 0 \le x \le 1, 1 - x \le y \le x + 1\}$. The volume integral evaluates as follows.

$$\int_0^1 \left(\int_{1-x}^{x+1} \sin(y)\, dy \right) dx = \int_0^1 \left((-\cos(y))\big|_{y=1-x}^{y=x+1} \right) dx$$

$$= \int_0^1 (\cos(x - 1) - \cos(x + 1))\, dx$$

$$= (\sin(x - 1) - \sin(x + 1))\Big|_0^1 = 2\sin(1) - \sin(2)$$

25. The region in the xy-plane is $\mathcal{R} = \{(x, y) : -1 \leq y \leq 1, y^2 \leq x \leq 2 - y^2\}$.
The volume integral evaluates as follows.

$$\int_{-1}^{1} \left(\int_{y^2}^{2-y^2} ((x + 2y^2 + 3) - (x + y)) \, dx \right) dy = \int_{-1}^{1} \left((2xy^2 + 3x - xy)\big|_{x=y^2}^{x=2-y^2} \right) dy$$

$$= \int_{-1}^{1} (2y^2 + 3 - y)(2 - 2y^2) \, dy$$

$$= (-4y^5/5 + y^4/2 - 2y^3/3 - y^2 + 6y)\Big|_{-1}^{1} = 136/15$$

27. The symmetry of the solid can be used to simplify the calculation. Its
volume is 4 times the volume of $\mathcal{U}_0$, the portion that is in the first octant.
$\mathcal{U}_0$ lies over the quarter-circle $\mathcal{R}_0 = \{(x, y) : 0 \leq y \leq 7, 0 \leq x \leq \sqrt{49 - y^2}\}$
and below the graph of $z = \sqrt{49 - y^2}$. The volume evaluates as follows.

$$4 \int_{0}^{7} \left(\int_{0}^{\sqrt{49-y^2}} \sqrt{49 - y^2} \, dx \right) dy = 4 \int_{0}^{7} \left((x\sqrt{49 - y^2})\big|_{x=0}^{x=\sqrt{49-y^2}} \right) dy$$

$$= 4 \int_{0}^{7} (49 - y^2) \, dx = 2744/3$$

29. Draw a picture and you will see that the solid lies over the triangular region
$\mathcal{R}$ in the xy-plane with vertices $(0, 0)$, $(4/5, 4/5)$, and $(4/3, -4/3)$. $\mathcal{R}$ is
neither x-simple nor y-simple so it will be divided into disjoint triangles,
both of which are y-simple:

$$\mathcal{R}_1 = \{(x, y) : 0 < x < 4/5, -x \leq y \leq x\}$$

and

$$\mathcal{R}_2 = \{(x, y) : 4/5 < x < 4/3, -x \leq y \leq 4 - 4x\}.$$

The volume is $\iint_{\mathcal{R}} f(x, y) \, dA = \iint_{\mathcal{R}_1} f(x, y) \, dA + \iint_{\mathcal{R}_2} f(x, y) \, dA$ where
$f(x, y) = 2x + 6y + 6$ (top minus bottom). We have

$$\iint_{\mathcal{R}_1} f(x, y) \, dA = \int_{0}^{4/5} \left(\int_{-x}^{x} (2x + 6y + 6) \, dy \right) dx$$

$$= \int_{0}^{4/5} \left((2xy + 3y^2 + 6y)\big|_{y=-x}^{y=x} \right) dx$$

$$= \int_{0}^{4/5} (4x^2 + 12x) \, dx = 1696/375$$

and

$$\iint_{R_2} f(x,y)\, dA = \int_{4/5}^{4/3} \left(\int_{-x}^{4-4x} (2x + 6y + 6)\, dy \right) dx$$

$$= \int_{4/5}^{4/3} \left((2xy + 3y^2 + 6y)\big|_{y=-x}^{y=4-4x} \right) dx$$

$$= \int_{4/5}^{4/3} (2x(4 - 3x) + 3(4 - 4x)^2 - 3x^2 - 18x + 24)\, dx$$

$$= (-3x^3 - 5x^2 - (4 - 4x)^3/4 + 24x)\Big|_{4/5}^{4/3} = 7616/3375\,.$$

Therefore, $\iint_{\mathcal{R}} f(x,y)\, dA = \frac{1696}{375} + \frac{7616}{3375} = \frac{4576}{675}\,.$

Further Theory and Practice

31. The circle and the line intersect at $(-3, -3)$ and $(1, 1)$. The smaller region $\mathcal{R}$ determined by the intersection is x-simple (draw a picture and solve the circle equation for x by completing the square or using the quadratic formula):

$$\mathcal{R} = \{(x, y) : -3 \le y \le 1, y \le x \le -4 - \sqrt{22 - y^2 - 4y}\,\}\,.$$

The volume of the solid is

$$\int_{-3}^{1} \left(\int_{y}^{-4-\sqrt{22-y^2-4y}} (2x + 8)\, dx \right) dy = \int_{-3}^{1} \left((x^2 + 8x)\big|_{x=y}^{x=-4-\sqrt{22-y^2-4y}} \right) dy$$

$$= \int_{-3}^{1} ((-4 - \sqrt{22 - y^2 + 4y}\,)^2 - y^2 - 32 - 8\sqrt{22 - y^2 + 4y} - 8y)\, dy$$

$$= \int_{-3}^{1} (6 - 2y^2 - 4y)\, dy = 64/3\,.$$

33. The plane $z = x + 4$ intersects the xy-plane in the line $x = -4$. This line cuts the disk in half. The right half of the disk:

$$\mathcal{R} = \{(x, y) : -1 \le y \le 1, -4 \le x \le -4 + \sqrt{1 - y^2}\,\}\,,$$

is the base of the solid. Its volume is

$$\int_{-1}^{1} \left(\int_{0}^{-4+\sqrt{1-y^2}} (x + 4)\, dx \right) dy = \int_{-1}^{1} \left((x^2/2 + 4x)\big|_{x=0}^{x=-4+\sqrt{1-y^2}} \right) dy$$

$$= \int_{-1}^{1} ((-4 + \sqrt{1 - y^2}\,)^2/2 - 8 + 4\sqrt{1 - y^2}\,)\, dy$$

$$= \int_{-1}^{1} (1/2 - y^2/2)\, dy = 2/3\,.$$

35. The paraboloids intersect in a circle: $x^2 + y^2 + 4x + 2y = 4$, or, after completing the square, $(x+2)^2 + (y+1)^2 = 9$. Therefore, the volume can be obtained by integrating top minus bottom:

$$-2(x^2 + y^2 + 4x + 2y - 4) = -2((x+2)^2 + (y+1)^2 - 9),$$

over the disk

$$\mathcal{R} = \{(x,y) : -5 \le x \le 1, -1 - \sqrt{9 - (x+2)^2} \le y \le -1 + \sqrt{9 - (x+2)^2}\,\}.$$

The computation can begin as follows.

$$V = -2 \int_{-5}^{1} \left(\int_{-1-\sqrt{9-(x+2)^2}}^{-1+\sqrt{9-(x+2)^2}} ((x+2)^2 + (y+1)^2 - 9)) \, dy \right) dx$$

$$= -2 \int_{-5}^{1} \left(((x+2)^2 y + (y+1)^3/3 - 9y)) \Big|_{y=-1-\sqrt{9-(x+2)^2}}^{y=-1+\sqrt{9-(x+2)^2}} \right) dx$$

$$= -2 \int_{-5}^{1} (2(x+2)^2 \sqrt{9-(x+2)^2} + 2(9-(x+2)^2)^{3/2}/3 - 18\sqrt{9-(x+2)^2}) \, dx$$

Factor 2 out of each term in the integrand and then make the substitution $x+2 = 3\sin(\theta)$. Note that $dx = 3\cos(\theta)\,d\theta$ and $\sqrt{9-(x+2)^2} = 3\cos(\theta)$. This yields

$$V = -4 \int_{-\pi/2}^{\pi/2} (9\sin^2(\theta) \cdot 3\cos(\theta) + 27\cos^3(\theta)/3 - 9 \cdot 3\cos(\theta)) \cdot 3\cos(\theta) \, d\theta$$

$$= -216 \int_{0}^{\pi/2} (3\sin^2(\theta)\cos^2(\theta) + \cos^4(\theta) - 3\cos^2(\theta)) \, d\theta$$

$$= -216 \int_{0}^{\pi/2} (-2\cos^4(\theta)) \, d\theta = 432 \int_{0}^{\pi/2} \cos^4(\theta) \, d\theta = 432 \cdot \frac{3\pi}{16} = 81\pi.$$

37. The solid is between the region $\mathcal{R} = \{(x,y) : 0 \le x \le 1, 0 \le y \le \sqrt{1-x^2}\,\}$ and the surface $z = xy$. Its volume can be calculated as follows.

$$V = \int_0^1 \left(\int_0^{\sqrt{1-x^2}} xy \, dy \right) dx = \int_0^1 \left((xy^2/2)\big|_{y=0}^{y=\sqrt{1-x^2}} \right) dx$$

$$= 1/2 \int_0^1 x(1-x^2) \, dx = 1/8$$

39. The two boundary surfaces intersect in the circle $x^2 + y^2 = 1$. By symmetry we can calculate the volume of the portion of the solid in the first octant and multiply by 4 (the solid lies entirely above the xy-plane). The integration can take place over $\mathcal{R} = \{(x,y) : 0 \le x \le 1, 0 \le y \le \sqrt{1-x^2}\,\}$.

The volume calculation follows. Note the substitution $x = \sin(\theta)$, $dx = \cos(\theta)\,d\theta$ in the third integral.

$$V = 4 \int_0^1 \left(\int_0^{\sqrt{1-x^2}} (1 - y^2 - x^2)\,dy \right) dx$$

$$= 4 \int_0^1 \left(((1-x^2)y - y^3/3)\big|_{y=0}^{y=\sqrt{1-x^2}} \right) dx$$

$$= 8/3 \int_0^1 (1-x^2)^{3/2}\,dx = 8/3 \int_0^{\pi/2} \cos^3(\theta) \cdot \cos(\theta)\,d\theta = \frac{8}{3} \cdot \frac{3\pi}{16} = \frac{\pi}{2}$$

41. The solid region lies below the surface and over the region $\mathcal{R} = \{(x,y) : 1 \le x \le 2, 1/x \le y \le 1\}$. The volume calculation follows.

$$V = \int_1^2 \left(\int_{1/x}^1 1/(x+y)\,dy \right) dx = \int_1^2 \left(\ln(x+y)\big|_{y=1/x}^{y=1} \right) dx$$

$$= \int_1^2 (\ln(x+1) - \ln(x+1/x))\,dx$$

$$= \int_1^2 (\ln(x+1) + \ln(x) - \ln(x^2+1))\,dx$$

$$= \int_1^2 \ln(x+1)\,dx + \int_1^2 \ln(x)\,dx - \int_1^2 \ln(x^2+1)\,dx$$

Each of these integrals are evaluated using integration-by-parts. For example, in the third one let $u = \ln(x^2 + 1)$ and $dv = dx$ to obtain

$$\int_1^2 \ln(x^2+1)\,dx = (x\ln(x^2+1))\big|_1^2 - \int_1^2 \frac{2x^2}{x^2+1}\,dx$$

$$= 2\ln(5) - \ln(2) - 2\int_1^2 \left(1 - \frac{1}{x^2+1}\right) dx$$

$$= \ln(25/2) - 2(x - \arctan(x))\big|_1^2$$

$$= \ln(25/2) - 2 + 2\arctan(2) - \pi/2.$$

The other two evaluate in a similar fashion to yield the volume formula

$$V = \ln(54/25) - 2\arctan(2) + \pi/2.$$

43. Draw a picture to see that $\mathcal{R}$, the base of the solid in the xy-plane, splits naturally into two parts. One part is in the first quadrant: $\mathcal{R}_1 = \{(x,y) : 0 \le x \le 1, x^3 \le y \le x\}$. The second part is the reflection of $\mathcal{R}_1$ through the origin. The top surface has the same symmetry so the volume of the entire solid can be obtained by doubling the volume of the part that is

above $\mathcal{R}_1$.

$$V = 2\int_0^1 \left(\int_{x^3}^x xy\,dy\right) dx = 2\int_0^1 \left((xy^2/2)\big|_{y=x^3}^{y=x}\right) dx$$

$$= \int_0^1 x(x^2 - x^6)\,dx = 1/8$$

Calculator/Computer Exercises

45. Start with a plot of the two curves in the first quadrant.

    ```
    > y1,y2 :=  x^3-3*x^2+3*x, 10*x-1-6*x^2:
      plot( [y1,y2], x=0..1.7, y=0..3.5);
    ```

    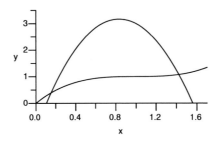

    Using this picture, the intersection points can be found using the fsolve procedure.

    ```
    > fsolve( y1=y2, x=0.05..0.2), fsolve( y1=y2, x=0.12..1.6);
    ```

    $$0.1534673051, 0.1534673051, 1.423622140$$

    For some reason `Maple` has returned two solutions for the left endpoint. The following entry names the intersection values a and b, then calculates the volume.

    ```
    > a,b := %[2..3]:
      Volume = int(int(x+y,y=y1..y2),x=a..b);
    ```

    $$Volume = 4.791820982$$

47. Start with a plot of the two curves in the upper half plane.

    ```
    > y1,y2 :=  1+x^4, 2-x^3:
      plot( [y1,y2], x=-2..2, y=0..6);
    ```

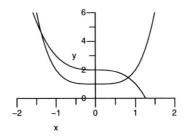

Using the picture, the intersection points are found by the fsolve procedure.

```
> a,b := fsolve( y1=y2, x=-1.5..-1),fsolve( y1=y2, x=0.5..1);
```

$$a, b := -1.380277569, 0.8191725134$$

The next entry calculates the volume.

```
> Volume = int(int(y,y=y1..y2),x=a..b);
```

$$Volume = 3.493072122$$

12.4 Polar Coordinates

Problems for Practice

1. The four points are displayed on the right. They are the vertices of a square centered at the origin.

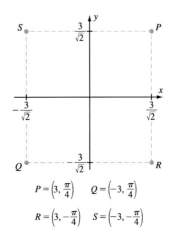

$$P = \left(3, \frac{\pi}{4}\right) \qquad Q = \left(-3, \frac{\pi}{4}\right)$$

$$R = \left(3, -\frac{\pi}{4}\right) \qquad S = \left(-3, -\frac{\pi}{4}\right)$$

3. These points are shown on the right. They overlap since they are all at the origin.

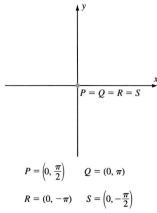

$$P = \left(0, \frac{\pi}{2}\right) \qquad Q = (0, \pi)$$
$$R = (0, -\pi) \qquad S = \left(0, -\frac{\pi}{2}\right)$$

5. These points are shown on the right. They overlap in pairs since they are all on the y-axis and the same distance from the origin.

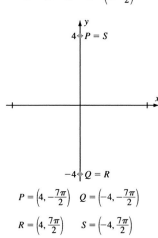

$$P = \left(4, -\frac{7\pi}{2}\right) \quad Q = \left(-4, -\frac{7\pi}{2}\right)$$
$$R = \left(4, \frac{7\pi}{2}\right) \qquad S = \left(-4, \frac{7\pi}{2}\right)$$

7. These points are shown on the right. They are the vertices of a rectangle centered at the origin.

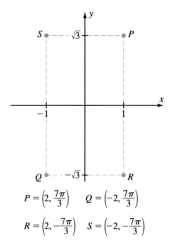

$$P = \left(2, \frac{7\pi}{3}\right) \qquad Q = \left(-2, \frac{7\pi}{3}\right)$$
$$R = \left(2, -\frac{7\pi}{3}\right) \quad S = \left(-2, -\frac{7\pi}{3}\right)$$

9. $x = 4\cos(\pi/4) = 4/\sqrt{2} = 2\sqrt{2}$, same for y.

11. $x = y = 0$

13. $x = -4\cos(-17\pi/6) = -4\cos(-3\pi + \pi/6) = 4\cos(\pi/6) = 2\sqrt{3}$
    $y = -4\sin(-17\pi/6) = -4\sin(-3\pi + \pi/6) = 4\sin(\pi/6) = 2$

15. $x = -6\cos(\pi/3) = -3$
    $y = -6\sin(\pi/3) = -3\sqrt{3}$

17. The point is in the fourth quadrant, $|r| = 4\sqrt{2}$. Using positive r, the polar coordinates are $(4\sqrt{2}, -\pi/4 + 2n\pi)$, n any integer. Using negative r, the polar coordinates are $(-4\sqrt{2}, 3\pi/4 + 2n\pi)$, n any integer.

19. The point is on the negative y-axis, $|r| = 9$. Using positive r, the polar coordinates are $(9, -\pi/2 + 2n\pi)$, n any integer. Using negative r, the polar coordinates are $(-9, \pi/2 + 2n\pi)$, n any integer.

21. The point is on the negative x-axis, $|r| = 1$. Using positive r, the polar coordinates are $(1, \pi + 2n\pi)$, n any integer. Using negative r, the polar coordinates are $(-1, 2n\pi)$, n any integer.

23. Since $r \mapsto -r$ does not change the equation there is symmetry with respect to the origin. There is also symmetry with respect to the y-axis because the equation is unchanged by the substitution $\theta \mapsto \pi - \theta$. This implies symmetry with respect to the x-axis as well.

25. There is x-axis and y-axis symmetry by default. This implies symmetry with respect to the origin.

27. There is x-axis and y-axis symmetry since the substitutions $\theta \mapsto -\theta$ and $\theta \mapsto \pi - \theta$ do not change the equation. This implies symmetry with respect to the origin.

29. The substitution $\theta \mapsto -\theta$ does not change the equation. Therefore, there is symmetry with respect to the x-axis.

31. There is no axis or origin symmetry (the polar plot is a circle of radius $1/2$ centered at $(-1/2, 1/2)$).

33. As θ increases from 0 to $\pi/2$, r increases from 0 to 1; r continues to increase from 1 to 2 as θ increases from $\pi/2$ to π. This produces the top half of the curve.

    Checking symmetry $(\theta \mapsto -\theta)$ shows that the curve is symmetric with the x-axis. See the sketch on the right.

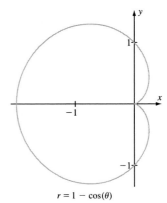

$r = 1 - \cos(\theta)$

35. As θ increases from 0 to $\pi/4$, r decreases from 3 to 0; r continues to decrease from 0 to -3 as θ increases from $\pi/4$ to $\pi/2$ yielding points in the opposite quadrant. As θ increases from $\pi/2$ to $3\pi/4$, then to π, r increases from -3 to 0 closing the loop in the opposite quadrant, then starting another half loop in quadrant two.

Checking symmetry ($\theta \mapsto -\theta$, $\theta \mapsto \pi - \theta$), shows that the curve is symmetric with the x-axis and the y-axis. See the sketch on the right.

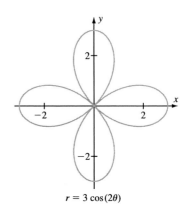

$r = 3\cos(2\theta)$

37. As θ increases from 0 to $\pi/2$, r decreases from 4 to 2; r increases from 2 back to 4 as θ increases from $\pi/2$ to π yielding the points in the second quadrant. As θ increases from π to $3\pi/2$, r increases from 4 to 6, then decreases from 6 back to 4 closing the curve in the fourth quadrant.

Checking symmetry ($\theta \mapsto \pi - \theta$), shows that the curve is symmetric with respect to the y-axis. See the sketch on the right.

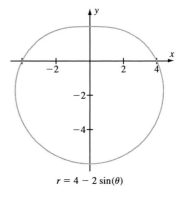

$r = 4 - 2\sin(\theta)$

39. As θ increases from 0 to $\arccos(1/3)$, r decreases from 2 to 0 tracing out the upper half of the inner loop; r decreases from 0 to -4 as θ increases from $\arccos(1/3)$ to π yielding the curve in the third and fourth quadrants that is the lower half of the outer loop. As θ increases from π to 2π, r increases from -4 to 0, first tracing out the upper half of the outer loop, then closing the curve with the lower half of the inner loop.

Checking symmetry ($\theta \mapsto -\theta$), shows that the curve is symmetric with respect to the x-axis. See the sketch on the right.

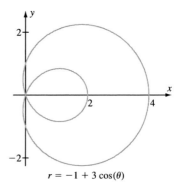

$r = -1 + 3\cos(\theta)$

41. This curve is best understood by converting to rectangular coordinates. Note that the defining equation is equivalent to $r^2 = 3r\sin(\theta) - 2r\cos(\theta)$ which, in terms of x and y, is

$$x^2 + y^2 = 3y - 2x\,.$$

Complete the square:

$$(x+1)^2 + (y-3/2)^2 = 13/4\,,$$

to recognize a circle with center at $(-1, 3/2)$ and radius $\sqrt{13}/2$.

See the sketch on the right.

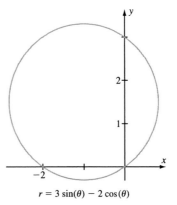

$r = 3\sin(\theta) - 2\cos(\theta)$

43. As θ increases from 0 to $\pi/2$, r decreases from 3 to 0; r increases from 0 back to 3 as θ increases from $\pi/2$ to π yielding the portion of the curve in the first and second quadrants. As θ increases from π to 2π, r increases from 3 to 6, then decreases from 6 to 3, tracing out the lower half of the curve.

Checking symmetry $(\theta \mapsto \pi - \theta)$, shows that the curve is symmetric with respect to the y-axis. See the sketch on the right.

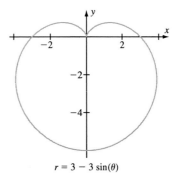

$r = 3 - 3\sin(\theta)$

Further Theory and Practice

45. (a) The point is in the first quadrant, $|r| = \sqrt{9 + \pi^2/4}$. Using positive r, the polar coordinates are $(\sqrt{9 + \pi^2/4}, \arctan(\pi/12) + 2n\pi)$, n any integer. For $r < 0$, they are $(-\sqrt{9 + \pi^2/4}, \arctan(\pi/12) + (2n+1)\pi)$, n any integer.

(b) The point is in the second quadrant, $|r| = \pi\sqrt{36 + 49\pi^2/4}$. For $r > 0$, the polar coordinates are $(\sqrt{36 + 49\pi^2/4}, -\arctan(\pi/12) + (2n+1)\pi)$, n any integer. For $r < 0$, $(-\sqrt{36 + 49\pi^2/4}, -\arctan(\pi/12) + 2n\pi)$, n any integer.

(c) The point is in the fourth quadrant, $|r| = \pi\sqrt{1033}/12$. For $r > 0$, the polar coordinates are $(\pi\sqrt{1033}/12, -\arctan(32/3) + 2n\pi)$, n any integer. For $r < 0$, $(-\pi\sqrt{1033}/12, -\arctan(32/3) + (2n+1)\pi)$, n any integer.

47. Call the points P_1 and P_2, respectively. Using rectangular coordinates,

they are $(r_1\cos(\theta_1), r_1\sin(\theta_1))$ and $(r_2\cos(\theta_2), r_2\sin(\theta_2))$. Apply the distance formula for points in the xy-plane.

$$\text{dist}(P_1, P_2) = \sqrt{(r_1\cos(\theta_1) - r_2\cos(\theta_2))^2 + (r_1\sin(\theta_1) - r_2\sin(\theta_2))^2}$$

$$= \sqrt{(r_1^2 + r_2^2)(\cos^2(\theta_1) + \sin^2(\theta_1)) - 2r_1r_2(\cos(\theta_1)\cos(\theta_2) + \sin(\theta_1)\sin(\theta_2))}$$

$$= \sqrt{r_1^2 + r_2^2 - 2r_1r_2\cos(\theta_1 - \theta_2)}$$

Using this distance formula, the polar equation for the given circle is

$$\sqrt{r^2 + 9 - 6r\cos(\theta - \pi/4)} = 1.$$

49. The curves intersect when $\cos(\theta) = 1 - \cos(\theta)$ or $\cos(\theta) = 1/2$. Therefore, $\theta = \pm\pi/3$ and $r = 1/2$. In rectangular coordinates these points are $(1/4, \pm\sqrt{3}/4)$. The curves also intersect at the origin $(0,0)$ (at different values of θ).

51. These curves intersect when $\tan(\theta) = \pm 1$. This occurs at four values of θ: $\theta = \pm\pi/4$ and $\theta = \pm 3\pi/4$. In rectangular coordinates the points of intersection are $(\pm 1/\sqrt{2}, \pm 1/\sqrt{2})$.

53. Since both curves pass through the origin, one point of intersection is $(0,0)$. More points of intersection could occur at values of θ such that $(1 + \cos(\theta))^2 = \cos(\theta)/2$, but there are none (verify by plotting their graphs in rectangular coordinates or use the quadratic formula).

Examination of the polar graphs of the equations reveals two more points of intersection, one in the second quadrant and another in the third. Since they must occur at θ values differing by π (why?) we look for solutions to the equation $(1 + \cos(\theta))^2 = \cos(\theta + \pi)/2$ or

$$\cos^2(\theta) + 2\cos(\theta) + 1 = -\cos(\theta)/2.$$

Multiply this equation by 2 and rearrange to $2\cos^2(\theta) + 5\cos(\theta) + 2 = 0$, and it factors: $(2\cos(\theta) + 1)(\cos(\theta) + 2) = 0$. This implies that the intersection occurs when $\cos(\theta) = -1/2$. Thus $r = 1/2$ also, so $x = (1/2)\cos(\theta) = -1/4$ and $y = (1/2)\sin(\theta) = \pm(1/2)\sqrt{1 - 1/4} = \pm\sqrt{3}/4$.

55. Clearly the point $(0,0)$ has this property, as does any point with rectangular coordinates of the form $(a, 0)$ with $a > 0$. Since we allow r to be negative, all points of the form $(a, 0)$ with $a < 0$ can also be included. The answer is: All points on the x-axis. There are no more.

57. The arc length is

$$L = \int_{-\pi/2}^{\pi/2} \sqrt{(e^\theta)^2 + (e^\theta)^2}\, d\theta = \sqrt{2}\int_{-\pi/2}^{\pi/2} e^\theta\, d\theta$$

$$= \sqrt{2}\,(e^{\pi/2} - e^{-\pi/2}).$$

59. Since $dr/d\theta = \sin^2(\theta/3)\cos(\theta/3)$, the arc length is

$$
\begin{aligned}
L &= \int_0^{\pi/2} \sqrt{(\sin^3(\theta/3))^2 + (\sin^2(\theta/3)\cos(\theta/3))^2}\, d\theta \\
&= \int_0^{\pi/2} \sqrt{\sin^4(\theta/3)(\sin^2(\theta/3) + \cos^2(\theta/3))}\, d\theta \\
&= \int_0^{\pi/2} \sin^2(\theta/3)\, d\theta = 1/2 \int_0^{\pi/2} (1 - \cos(2\theta/3))\, d\theta \\
&= 1/2 \bigl(\theta - (3/2)\sin(2\theta/3)\bigr)\Big|_0^{\pi/2} = 1/2\bigl(\pi/2 - (3/2)\sin(\pi/3)\bigr) \\
&= \pi/4 - 3\sqrt{3}/8\,.
\end{aligned}
$$

61. Points move outward as θ increases, tracing out ever-widening spirals.

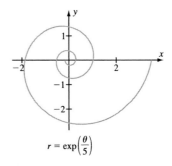

$$r = \exp\!\left(\frac{\theta}{5}\right)$$

63. The graph consists of closed loops in the sectors where $\sin(4\theta) \le 0$. For example, the loop in the first quadrant is in the sector were $\pi \le 4\theta \le 2\pi$ or $\pi/4 \le \theta \le \pi/2$. There is one loop where $r > 0$ and another reflected through the origin where $r < 0$.

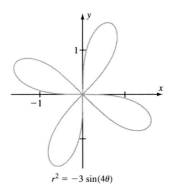

$$r^2 = -3\sin(4\theta)$$

65. The graph has one closed loop. It lies in the sector where $\sin(\theta) \geq 0$, that is, where $0 \leq \theta \leq \pi$.

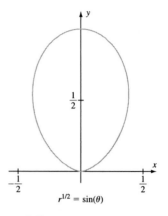

$$r^{1/2} = \sin(\theta)$$

67. The graphs are both spirals, but they possess different symmetry. The one on the left $(r = \theta^2)$ is a double spiral. One spiral is the reflection of the other in the x-axis. The graph on the right $(\theta = r^2)$ is also a double spiral but this time one spiral is the reflection of the other spiral through the origin.

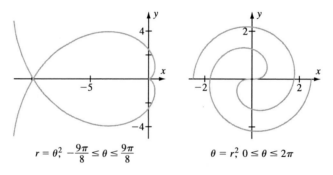

$$r = \theta^2; \ -\frac{9\pi}{8} \leq \theta \leq \frac{9\pi}{8} \qquad\qquad \theta = r^2; \ 0 \leq \theta \leq 2\pi$$

69. Observe that the polar plot (on the right below) winds around the origin, but the spiral is not uniform over the interval $0 \leq \theta \leq 2\pi$. In fact it will never be uniform, but as θ increases it will appear to become uniform as the rectangular plot of $r = \theta + \cos(\theta)$ starts to appear more like a straight line.

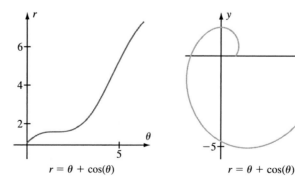

$$r = \theta + \cos(\theta) \qquad\qquad\qquad r = \theta + \cos(\theta)$$

Calculator/Computer Exercises

71. The first entry defines the function $j \mapsto P(j)$ where $P(j)$ is the plot of the
polar curve $r = \sin(j\theta)$. The display procedure is then used to display the
polar curves $r = \sin(j\theta)$ for $j = 1, 3, 5$.

```
> P := j -> plot( sin(j*theta), theta=0..2*Pi, coords=polar,
               view=[-1..1,-1..1]):
  display( array([seq(P(j), j=[1,3,5])]), axes=none );
```

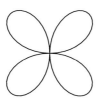  

Polar curves $r = \sin(\theta)$, $r = \sin(3\theta)$, and $r = \sin(5\theta)$.

The polar curves $r = \sin(j\theta)$ for $j = 2, 4, 6$ are displayed below.

```
> display( array([seq(P(j), j=[2,4,6])]), axes=none );
```

Polar curves $r = \sin(2\theta)$, $r = \sin(4\theta)$, and $r = \sin(6\theta)$.

If $j = 1$, then the graph of $r = \sin(k\theta)$ has more petals for all $k > j$.

If $j > 1$ is odd, then the graph of $r = \sin(k\theta)$ has more petals for all $k > j$
and for all even $k \geq (j+1)/2$.

If $j > 1$ is even, then the graph of $r = \sin(k\theta)$ has more petals for all even
$k > j$ and for all odd $k > 2j$.

73. Changing the additive constant rotates the graph in the clockwise direction
(when the constant is positive). The following entries display the polar
curves $r = \sin(\theta + j)$ for $j = 0, 1, 2, 3, 5$.

```
> P := j -> plot( sin(theta+j), theta=0..Pi, coords=polar,
               view=[-1..1,-1..1]):
  display( seq(P(j), j=[$0..3,5]) );
```

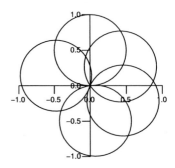

Polar curves $r = \sin(\theta + j)$, $j = 0, 1, 2, 3, 5$.

The polar curves $r = \sin(\theta + j)$ for $j = 10, 20, 30$ are displayed below.

```
> display( seq(P(10*j), j=1..3) );
```

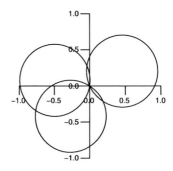

Polar curves $r = \sin(\theta + j)$, $j = 10, 20, 30$.

12.5   Integrating in Polar Coordinates

Problems for Practice

1. The polar curve $r = 1 - \sin(\theta)$ is heart-shaped, winding around the origin once for $0 \le \theta \le 2\pi$. The area enclosed is

$$\frac{1}{2} \int_0^{2\pi} (1 - \sin(\theta))^2 \, d\theta = \frac{1}{2} \int_0^{2\pi} (1 - 2\sin(\theta) + \sin^2(\theta)) \, d\theta$$

$$= \frac{1}{2} \left(2\pi + \int_0^{2\pi} \sin^2(\theta) \, d\theta \right)$$

$$= \pi + \pi/2 = 3\pi/2 \,.$$

Note. The symmetry in the graphs of $\sin(\theta)$ and $\cos(\theta)$ and the fact that

$\sin^2(\theta) + \cos^2(\theta) = 1$ imply that

$$\int_0^{2\pi} \sin^2(\theta)d\theta = \int_0^{2\pi} \cos^2(\theta)d\theta = \pi \, .$$

We will use this fact, and ones like it such as

$$\int_0^{\pi} \sin^2(\theta)d\theta = \int_0^{\pi} \cos^2(\theta)d\theta = \pi/2 \, ,$$

whenever necessary in the integrals that follow.

3. The polar curve $r = 4\sin(3\theta)$ is a 3-leafed rose that is completely swept out for $0 \le \theta \le \pi$. The area enclosed is

$$\frac{1}{2} \int_0^{\pi} (4\sin(3\theta))^2 \, d\theta = 8 \int_0^{\pi} \sin^2(3\theta) \, d\theta$$

$$= 4 \int_0^{\pi} (1 - \cos(6\theta)) \, d\theta = 4\pi \, .$$

5. The polar curve $r^2 = -\sin(\theta)$ consists of two loops. One of them is in the sector where $\sin(\theta) \le 0 \colon \pi \le \theta \le 2\pi$. The other is the reflection of the first loop through the origin. The total area enclosed is

$$2 \cdot \frac{1}{2} \int_{\pi}^{2\pi} (-\sin(\theta)) \, d\theta = - \int_{\pi}^{2\pi} \sin(\theta) \, d\theta$$

$$= \cos(\theta)\Big|_{\pi}^{2\pi} = 2 \, .$$

7. Using formula (14.12),

$$A = \frac{1}{2} \int_0^{2\pi} \left((6 - \sin(\theta))^2 - (1 + \cos(\theta))^2\right) d\theta$$

$$= \frac{1}{2} \int_0^{2\pi} (35 - 12\sin(\theta) - 2\cos(\theta) + \sin^2(\theta) - \cos^2(\theta)) \, d\theta$$

$$= \frac{1}{2} \int_0^{2\pi} (35 - 12\sin(\theta) - 2\cos(\theta) - \cos(2\theta)) \, d\theta = 35\pi \, .$$

9. Each of these polar curves is a circle passing through the origin. The circle $r = \sin(\theta)$ is swept out when $0 \le \theta \le \pi$ and the circle $r = \cos(\theta)$ is swept out when $-\pi/2 \le \theta \le \pi/2$ (draw a picture). The area of their intersection, which is in the first quadrant, is twice the area enclosed by the curve $r = \sin(\theta)$ for $0 \le \theta \le \pi/4$.

$$A = 2 \cdot \frac{1}{2} \int_0^{\pi/4} \sin^2(\theta) \, d\theta = \frac{1}{2} \int_0^{\pi/4} (1 - \cos(2\theta)) \, d\theta$$

$$= (1/2)(\theta - \sin(2\theta)/2)\Big|_0^{\pi/4} = \pi/8 - 1/4 \, .$$

11. The picture here is similar to the one described in Exercise 9, except the circle $r = \sqrt{3}\cos(\theta)$ is larger than the circle $r = \sin(\theta)$. Consequently, the intersection of the circles is not at the angle $\theta = \pi/4$ as it was there. Instead, the angle must be calculated as the first-quadrant solution for the equation $\sqrt{3}\cos(\theta) = \sin(\theta)$: $\theta = \pi/3$ (verify[1]). Consequently, the area of the region in the first quadrant that is inside of the larger circle and outside of the smaller one is

$$
\begin{aligned}
A &= \frac{1}{2}\int_0^{\pi/3} (3\cos^2(\theta) - \sin^2(\theta))\, d\theta \\
&= \frac{1}{4}\int_0^{\pi/3} (3(1+\cos(2\theta)) - (1-\cos(2\theta))\, d\theta \\
&= \frac{1}{4}\int_0^{\pi/3} (2+4\cos(2\theta))\, d\theta = \frac{1}{2}(\theta + \sin(2\theta))\Big|_0^{\pi/3} = \pi/6 + \sqrt{3}/4\,.
\end{aligned}
$$

13. Using Theorem 2,

$$
\begin{aligned}
A &= \int_{-\pi/4}^{\pi/4}\int_2^{4\cos(\theta)} r\, dr\, d\theta = \int_{-\pi/4}^{\pi/4}\left((r^2/2)\big|_2^{4\cos(\theta)}\right) d\theta \\
&= \int_{-\pi/4}^{\pi/4} (8\cos^2(\theta) - 2)\, d\theta \\
&= \int_{-\pi/4}^{\pi/4} (8(1+\cos(2\theta))/2 - 2)\, d\theta \\
&= (2\theta + 2\sin(2\theta))\Big|_{-\pi/4}^{\pi/4} = \pi + 4\,.
\end{aligned}
$$

15. The limaçon is entirely inside of the circle so the area between them is

$$
\begin{aligned}
A &= \int_0^{2\pi}\int_{3+\cos(\theta)}^5 r\, dr\, d\theta = \int_0^{2\pi}\left((r^2/2)\big|_{3+\cos(\theta)}^5\right) d\theta \\
&= 1/2\int_0^{2\pi} (25 - (3+\cos(\theta))^2)\, d\theta \\
&= 1/2\int_0^{2\pi} (16 - 6\cos(\theta) - \cos(\theta)^2)\, d\theta \\
&= 16\pi - 1/2\int_0^{2\pi} \cos(\theta)^2\, d\theta = 16\pi - \pi/2 = 31\pi/2\,.
\end{aligned}
$$

---

[1]Square both sides, replace $\sin^2(\theta)$ with $1 - \cos^2(\theta)$, and solve for θ.

17. The cardioid is entirely inside of the limaçon so the area between them is

$$A = \int_0^{2\pi} \int_{1+\cos(\theta)}^{4-\sin(\theta)} r \, dr \, d\theta = \int_0^{2\pi} \left((r^2/2) \Big|_{1+\cos(\theta)}^{4-\sin(\theta)} \right) d\theta$$

$$= 1/2 \int_0^{2\pi} \left((4-\sin(\theta))^2 - (1+\cos(\theta))^2 \right) d\theta$$

$$= 1/2 \int_0^{2\pi} \left(15 - 8\sin(\theta) + \sin^2(\theta) - 2\cos(\theta) - \cos(\theta)^2 \right) d\theta$$

$$= 15\pi + 1/2 \int_0^{2\pi} \sin(\theta)^2 \, d\theta - 1/2 \int_0^{2\pi} \cos(\theta)^2 \, d\theta = 15\pi + \pi/2 - \pi/2 = 15\pi \,.$$

19. The entire circle is traced out for $-\pi/2 \le \theta \le \pi/2$ so the integral can be evaluated as follows.

$$\int_{-\pi/2}^{\pi/2} \int_0^{\cos(\theta)} \cos(\theta) r \, dr \, d\theta = \int_{-\pi/2}^{\pi/2} \left((\cos(\theta) r^2/2) \Big|_{r=0}^{r=\cos(\theta)} \right) d\theta$$

$$= 1/2 \int_{-\pi/2}^{\pi/2} \cos^3(\theta) \, d\theta$$

$$= 1/2 \int_{-\pi/2}^{\pi/2} (1 - \sin^2(\theta)) \cos(\theta) \, d\theta$$

$$= 1/2 (\sin(\theta) - \sin^3(\theta)/3) \Big|_{-\pi/2}^{\pi/2} = 1/2(2 - 2/3) = 2/3$$

21. The leftmost petal is traced out for $3\pi/4 \le \theta \le 5\pi/4$ so the integral evaluates as follows. Be on the lookout for the use of the trigonometric identity $\cos(\alpha)\cos(\beta) = \frac{1}{2}(\cos(\alpha+\beta) + \cos(\alpha-\beta))\,.$

$$\int_{3\pi/4}^{5\pi/4} \int_0^{\cos(2\theta)} \cos(\theta) r \, dr \, d\theta = \int_{3\pi/4}^{5\pi/4} \left((\cos(\theta) r^2/2) \Big|_{r=0}^{r=\cos(2\theta)} \right) d\theta$$

$$= 1/2 \int_{3\pi/4}^{5\pi/4} \cos(\theta) \cos^2(2\theta) \, d\theta$$

$$= 1/4 \int_{3\pi/4}^{5\pi/4} \cos(\theta)(1 + \cos(4\theta)) \, d\theta$$

$$= 1/4 \int_{3\pi/4}^{5\pi/4} (\cos(\theta) + \cos(\theta)\cos(4\theta)) \, d\theta$$

$$= 1/4 \int_{3\pi/4}^{5\pi/4} \left(\cos(\theta) + \tfrac{1}{2}(\cos(5\theta) + \cos(3\theta)) \right) d\theta$$

$$= 1/4(\sin(\theta) + \sin(5\theta)/10 + \sin(3\theta)/6) \Big|_{3\pi/4}^{5\pi/4} = -4\sqrt{2}/15$$

23. The entire circle is traced out for $0 \le \theta \le \pi$ so the integral evaluates as follows.

$$\int_0^\pi \int_0^{\sin(\theta)} \sin(\theta)\, r\, dr\, d\theta = \int_0^\pi \left((\sin(\theta)\, r^2/2)\big|_{r=0}^{r=\sin(\theta)} \right) d\theta$$

$$= 1/2 \int_0^\pi \sin^3(\theta)\, d\theta$$

$$= 1/2 \int_0^\pi (1 - \cos^2(\theta)) \sin(\theta)\, d\theta$$

$$= 1/2(-\cos(\theta) + \cos^3(\theta)/3)\Big|_0^\pi = 1/2(2 - 2/3) = 2/3$$

25. The circle is entirely inside of the limaçon so the integral evaluates as follows.

$$\int_0^{2\pi} \int_3^{4-\cos(\theta)} (3r - 10)\, r\, dr\, d\theta = \int_0^{2\pi} \left((r^3 - 5r^2)\big|_3^{4-\cos(\theta)} \right) d\theta$$

$$= \int_0^{2\pi} ((4 - \cos(\theta))^3 - 5(4 - \cos(\theta))^2 + 18)\, d\theta$$

$$= \int_0^{2\pi} (2 - 8\cos(\theta) + 7\cos^2(\theta) - \cos^3(\theta))\, d\theta$$

$$= 4\pi + 7\pi = 11\pi$$

Note that both $\cos(\theta)$ and $\cos^3(\theta)$ integrate to 0 over the interval $[0, 2\pi]$.

27. The surface intersects the xy-plane in the circle $r = 2$ (polar coordinates). Moreover, the equation of the surface in polar coordinates is $z = 4 - r^2$. Therefore, the volume is

$$V = \int_0^{2\pi} \int_0^2 (4 - r^2)\, r\, dr\, d\theta = \int_0^{2\pi} \left((2r^2 - r^4/4)\big|_0^2 \right) d\theta$$

$$= \int_0^{2\pi} ((4 - \cos(\theta))^3 - 5(4 - \cos(\theta))^2 + 18)\, d\theta$$

$$= \int_0^{2\pi} 4\, d\theta = 8\pi\,.$$

29. The paraboloids intersect in a circle, $x^2 + y^2 = 4$. The region inside the circle is $\mathcal{R} = \{(r, \theta) : 0 \le \theta \le 2\pi, 0 \le r \le 2\}$ in polar coordinates. The volume between the two surfaces can be calculated with the following integral

$$V = \int_0^{2\pi} \int_0^2 ((9 - r^2) - (3r^2 - 7))\, r\, dr\, d\theta$$

$$= \int_0^{2\pi} \int_0^2 (16r - 4r^3)\, dr\, d\theta = 32\pi\,.$$

31. Using polar coordinates in the xy-plane, the top surface is $z = 5 - r^2$ and the bottom surface is $z = 2r - 3$. They intersect in a circle that lies in the plane $z = 1$ and projects to the xy-plane as the circle $r = 2$ (polar coordinates again). Therefore, the volume can be obtained as follows.

$$
V = \int_0^{2\pi} \int_0^2 \left((5 - r^2) - (2r - 3) \right) r \, dr \, d\theta
$$
$$
= \int_0^{2\pi} \int_0^2 (8r - 2r^2 - r^3) \, dr \, d\theta = 40\pi/3 \, .
$$

33. The top surface is the plane $z = 4$ and, using polar coordinates in the xy-plane, the bottom surface is $z = \sqrt{1 + r^2}$. They intersect in a circle in the plane $z = 4$ that projects to the xy-plane as the circle $r = \sqrt{15}$ (polar coordinates). Therefore, the volume can be obtained as follows.

$$
V = \int_0^{2\pi} \int_0^{\sqrt{15}} \left(4 - \sqrt{1 + r^2} \right) r \, dr \, d\theta
$$
$$
= \int_0^{2\pi} \int_0^{\sqrt{15}} (4r - r\sqrt{1 + r^2}) \, dr \, d\theta
$$
$$
= \int_0^{2\pi} \left((2r^2 - (1 + r^2)^{3/2}/3) \big|_0^{\sqrt{15}} \right) d\theta
$$
$$
= \int_0^{2\pi} 9 \, d\theta = 18\pi \, .
$$

35. The hyperboloid is in two sheets, the top: $z = \sqrt{x^2 + y^2 + 4}$, and the bottom: $z = -\sqrt{x^2 + y^2 + 4}$. The solid is bounded on the sides by the cylinder which is $\mathcal{R} = \{(r, \theta) : 0 \le \theta \le 2\pi, 0 \le r \le 2\}$ in polar coordinates. The volume between the sheets can be calculated with the following integral

$$
V = \int_0^{2\pi} \int_0^2 \left(\sqrt{r^2 + 4} - (-\sqrt{r^2 + 4}) \right) r \, dr \, d\theta
$$
$$
= \int_0^{2\pi} \int_0^2 2r \sqrt{r^2 + 4} \, dr \, d\theta
$$
$$
= \int_0^{2\pi} \left((2(r^2 + 4)^{3/2}/3) \big|_0^2 \right) d\theta
$$
$$
= 2/3 \int_0^{2\pi} (16\sqrt{2} - 8) \, d\theta = 32(2\sqrt{2} - 1)\pi/3 \, .
$$

37. The solid is bounded on the top by the surface $z = r$ (polar coordinates in the xy-plane) and on the bottom by the surface $z = -r$. The side surface

is the cylinder $x^2 + y^2 = 4$, or $r = 2$ in polar coordinates. Therefore, the volume can be obtained as follows.

$$V = \int_0^{2\pi} \int_0^2 (r - (-r))\, r\, dr\, d\theta = \int_0^{2\pi} \int_0^2 2r^2\, dr\, d\theta$$

$$= \int_0^{2\pi} \left((2r^3/3)\Big|_0^2 \right) d\theta$$

$$= \int_0^{2\pi} 16/3\, d\theta = 32\pi/3\,.$$

39. The region in the xy-plane is

$$\mathcal{R} = \{(x, y) : 0 \le y \le 2, y^2/4 \le x \le y/2\}\,.$$

The transformation $T(u, v) = (\phi(u, v), \psi(u, v))$ from the uv-plane is obtained by solving the change-of-variable equations for x and y. This yields $\phi(u, v) = 2u + 2v$ and $\psi(u, v) = 2v$. Its Jacobian is

$$J_T(u, v) = (\phi_u \psi_v - \psi_u \phi_v)(u, v) = 4\,.$$

To obtain the region $\mathcal{S}$ that corresponds to $\mathcal{R}$ note that $0 \le y \le 2$ corresponds to $0 \le v \le 1$. The second inequality in the definition of $\mathcal{R}$ implies that $v^2 \le 2u + 2v \le v$, which is equivalent to $(v^2 - 2v)/2 \le u \le -v/2$ so

$$\mathcal{S} = \{(u, v) : 0 \le v \le 2, (v^2 - 2v)/2 \le u \le -v/2\}\,.$$

Putting it all together we have

$$\int_0^2 \int_{y^2/4}^{y/2} (x - y)^2\, dx\, dy = \int_0^1 \int_{(v^2-2v)/2}^{-v/2} 4u^2 \cdot 4\, du\, dv$$

$$= \int_0^1 \left((16u^3/3)\Big|_{(v^2-2v)/2}^{-v/2} \right) dv$$

$$= 16/3 \int_0^1 (-v^3/8 - (v^2 - 2v)^3/8)\, dv$$

$$= 16/3 \int_0^1 (7v^3/8 - v^6/8 + 3v^5/4 - 3v^4/2)\, dv$$

$$= 29/210\,.$$

41. The region in the xy-plane is

$$\mathcal{R} = \{(x, y) : 1 \le y \le 5, 3 \le x \le 9\}\,.$$

The transformation $T(u, v) = (\phi(u, v), \psi(u, v))$ from the uv-plane is obtained by solving the change-of-variable equations for x and y. This yields $\phi(u, v) = 1/u$ and $\psi(u, v) = v$. Its Jacobian is

$$J_T(u, v) = (\phi_u \psi_v - \psi_u \phi_v)(u, v) = -1/u^2\,.$$

To obtain the region $\mathcal{S}$ that corresponds to $\mathcal{R}$ note that $1 \leq y \leq 5$ corresponds to $1 \leq v \leq 5$. The second inequality in the definition of $\mathcal{R}$ implies that $1/9 \leq u \leq 1/3$, so

$$\mathcal{S} = \{(u,v) : 1 \leq v \leq 5, 1/9 \leq u \leq 1/3\}.$$

Putting it all together we have

$$\int_1^5 \int_3^9 y^2/x^2 \, dx \, dy = \int_1^5 \int_{1/9}^{1/3} u^2 v^2 \cdot (-1/u^2) \, du \, dv$$

$$= -\int_1^5 \left((v^2 u) \big|_{u=1/9}^{u=1/3} \right) dv$$

$$= -2/9 \int_1^5 v^3/3 \, dv = 248/27.$$

43. The region in the xy-plane is

$$\mathcal{R} = \{(x,y) : 0 \leq x \leq 1, 3 \leq y \leq 4\}.$$

The transformation $T(u,v) = (\phi(u,v), \psi(u,v))$ from the uv-plane is obtained by solving the change-of-variable equations for x and y. This yields $\phi(u,v) = u/v$ and $\psi(u,v) = v$. Its Jacobian is

$$J_T(u,v) = (\phi_u \psi_v - \psi_u \phi_v)(u,v) = 1/v.$$

To obtain the region $\mathcal{S}$ that corresponds to $\mathcal{R}$ note that $3 \leq y \leq 4$ corresponds to $3 \leq v \leq 4$. The second inequality in the definition of $\mathcal{R}$ implies that $0 \leq u/v \leq 1$ which, because v is positive, is equivalent to $0 \leq u \leq v$ and

$$\mathcal{S} = \{(u,v) : 3 \leq v \leq 4, 0 \leq u \leq v\}.$$

Putting it all together we have

$$\int_0^1 \int_3^4 \sqrt{4-xy} \, dx \, dy = \int_3^4 \int_0^v \sqrt{4-u} \cdot (1/v) \, du \, dv$$

$$= \int_3^4 \left((-(2/3)(4-u)^{3/2}/v) \big|_{u=0}^{u=v} \right) dv$$

$$= -2/3 \int_3^4 \frac{(4-v)^{3/2} - 8}{v} \, dv.$$

One way to handle this integral is to make the substitution: $v = 4 - u^2$,

$dv = -2u\,du$. Doing so allows the calculation to proceed as follows.

$$\int_0^1 \int_3^4 \sqrt{4 - xy}\,dx\,dy = -2/3 \int_1^0 \frac{u^3 - 8}{4 - u^2} \cdot (-2u)\,du$$

$$= 4/3 \int_0^1 \frac{u^4 - 8u}{u^2 - 4}\,du$$

$$= 4/3 \int_0^1 \left(u^2 + 4 - \frac{8}{u + 2} \right)\,du$$

$$= 4/3(u^3/3 + 4u - 8\ln(u + 2))\Big|_0^1$$

$$= 52/9 + (32/3)\ln(2/3)$$

Further Theory and Practice

45. All four petals of the polar curve $r = 3\sin(2\theta)$ are inside of the polar curve $r = 4 - \cos(\theta)$. The area of the region between them can be calculated with the following integral.

$$A = \frac{1}{2} \int_0^{2\pi} \left((4 - \cos(\theta))^2 - (3\sin(2\theta))^2 \right)\,d\theta$$

$$= \frac{1}{2} \int_0^{2\pi} (16 - 8\cos(\theta) + \cos^2(\theta) - 9\sin^2(2\theta))\,d\theta$$

$$= \frac{1}{2}(32\pi + \pi - 9\pi) = 12\pi\,.$$

47. The cardiod and the circle intersect in two regions of the same size, one in the first quadrant and the other in the fourth. Draw a picture. The points of intersection are at the angles θ that satisfy the equation $1 - \cos(\theta) = \cos(\theta)$. That is, $\cos(\theta) = 1/2$, so $\theta = \pm\pi/3$. The area can be found with the following calculation. (Note the shifty move in line 3.)

$$A = 2 \cdot \left(1/2 \int_0^{\pi/3} (1 - \cos(\theta))^2\,d\theta + 1/2 \int_{\pi/3}^{\pi/2} (\cos(\theta))^2\,d\theta \right)$$

$$= \int_0^{\pi/3} (1 - 2\cos(\theta) + \cos^2(\theta))\,d\theta + \int_{\pi/3}^{\pi/2} \cos^2(\theta)\,d\theta$$

$$= (\theta - 2\sin(\theta))\Big|_0^{\pi/3} + \int_0^{\pi/2} \cos^2(\theta)\,d\theta$$

$$= \pi/3 - \sqrt{3} + \pi/4 = 7\pi/12 - \sqrt{3}$$

49. The rose has 4 petals, each one intersecting the circle in two pieces of equal area. So, there are 8 small equal pieces and (draw a picture)

$$A = 8 \times (\text{Area of One Small Piece})\,.$$

We need the smaller intersection angle in quadrant 1 and can get it by solving $\cos(2\theta) = 1/2$ to obtain $2\theta = \pi/3$, $\theta = \pi/6$.

$$A = 8 \cdot \left(1/2 \int_0^{\pi/6} (1/2)^2 \, d\theta + 1/2 \int_{\pi/6}^{\pi/4} (\cos(2\theta))^2 \, d\theta \right)$$

$$= 4 \left(\frac{1}{4} \cdot \frac{\pi}{6} + 1/2 \int_{\pi/6}^{\pi/4} (1 + \cos(4\theta)) \, d\theta \right)$$

$$= \pi/6 + 2 \int_{\pi/6}^{\pi/4} (1 + \cos(4\theta)) \, d\theta$$

$$= \pi/6 + 2(\theta + \sin(4\theta)/4) \Big|_{\pi/6}^{\pi/4} = \pi/3 - \sqrt{3}/4$$

51. Sketch the graph to see that the smaller loop is traced out for $\pi/6 \leq \theta \leq 5\pi/6$. Since the curve is symmetric with the y-axis the area can be found by doubling the area for $\pi/6 \leq \theta \leq \pi/2$.

$$A = 2 \cdot \frac{1}{2} \int_{\pi/6}^{\pi/2} (1 - 2\sin(\theta))^2 \, d\theta$$

$$= \int_{\pi/6}^{\pi/2} (1 - 4\sin(\theta) + 4\sin^2(\theta)) \, d\theta$$

$$= \int_{\pi/6}^{\pi/2} (1 - 4\sin(\theta) + 2(1 - \cos(2\theta)) \, d\theta$$

$$= (3\theta + 4\cos(\theta) - \sin(2\theta)) \Big|_{\pi/6}^{\pi/2} = \pi - 3\sqrt{3}/4$$

53. In polar coordinates the integrand is $(r^2+1)^{-1/2}$ and the integration takes place over the disk in the xy-plane having boundary $r = 2$. Therefore, the integral I can be evaluated as follows.

$$I = \int_0^{2\pi} \int_0^2 (r^2 + 1)^{-1/2} r \, dr \, d\theta$$

$$= \int_0^{2\pi} \left(((r^2 + 1)^{1/2}) \Big|_0^2 \right) d\theta$$

$$= \int_0^{2\pi} (\sqrt{5} - 1) \, d\theta = (\sqrt{5} - 1)2\pi .$$

55. In polar coordinates the integrand is $r^{2/3}$ and the integration takes place over the portion of the xy-plane described in polar coordinates as $\mathcal{R} = \{(r, \theta) : \pi \leq \theta \leq 3\pi/2, 0 \leq r \leq 3\}$. The integral I can be evaluated as

follows.

$$I = \int_\pi^{3\pi/2} \int_0^3 r^{2/3} \cdot r\, dr\, d\theta = \int_\pi^{3\pi/2} \left((3r^{8/3}/8)\big|_0^3 \right) d\theta$$

$$= \int_\pi^{3\pi/2} 3^{11/3}/8\, d\theta = 3^{11/3}\pi/16\,.$$

57. The integral I can be evaluated as follows.

$$I = \int_0^{\pi/2} \int_0^{\cos(\theta)} \theta \cdot r\, dr\, d\theta = \int_0^{\pi/2} \left((\theta r^2/2)\big|_{r=0}^{r=\cos(\theta)} \right) d\theta$$

$$= 1/2 \int_0^{\pi/2} \theta \cos^2(\theta)\, d\theta$$

Continue the calculation using integration-by-parts: $u = \theta$, $dv = \cos^2(\theta)\, d\theta$. Noting that $v = \theta/2 + \sin(2\theta)/4$, this yields

$$I = \frac{1}{2}\left((\theta^2/2 + \theta\sin(2\theta)/4)\big|_0^{\pi/2} - \int_0^{\pi/2} (\theta/2 + \sin(2\theta)/4)\, d\theta \right)$$

$$= \pi^2/16 - 1/2(\theta^2/4 - \cos(2\theta)/8)\big|_0^{\pi/2}$$

$$= \pi^2/16 - \pi^2/32 - 1/8 = \pi^2/32 - 1/8\,.$$

59. Let $\mathcal{R}$ denote the region in the xy-plane. It can be described as

$$\mathcal{R} = \{(x,y) : x > 0, y > 0, 2 \le xy \le 4, 1 \le y/x \le 2\}\,.$$

We wish to calculate its area $A = \iint_\mathcal{R} dx\, dy$. The transformation $T(u,v) = (\phi(u,v), \psi(u,v))$ from the uv-plane is obtained by solving the change-of-variable equations for x and y. This yields $\phi(u,v) = \sqrt{u/v}$ and $\psi(u,v) = \sqrt{uv}$. We leave it to you to verify that its Jacobian is

$$J_T(u,v) = (\phi_u\psi_v - \psi_u\phi_v)(u,v) = 1/(2v)\,.$$

In view of the description of $\mathcal{R}$, the region $\mathcal{S}$ that corresponds to $\mathcal{R}$ is

$$\mathcal{S} = \{(u,v) : 2 \le u \le 4, 1 \le v \le 2\}\,.$$

Putting it all together we have

$$A = \iint_\mathcal{R} dx\, dy = \int_2^4 \int_1^2 1/(2v)\, dv\, du$$

$$= 1/2 \int_2^4 \left((\ln(v))\big|_{v=1}^{v=2} \right) du$$

$$= 1/2 \int_2^4 \ln(2)\, du = \ln(2)\,.$$

61. First observe that for any value of α,

$$\lim_{R \to \infty} \int_{-R}^{R} \int_{R}^{-R} \frac{1}{(1 + x^2 + y^2)^\alpha} \, dx \, dy = \lim_{R \to \infty} \int_{0}^{2\pi} \int_{0}^{R} \frac{1}{(1 + r^2)^\alpha} \cdot r \, dr \, d\theta \, .$$

This is because if $F(R)$ denotes the rectangular double integral and $G(R)$ denotes the polar double integral, then given any positive number R,

$$G(R) \leq F(R) \leq G(\sqrt{2}\, R) \leq F(\sqrt{2}\, R) \, .$$

Now note that $G(R)$ can be calculated rather easily due to the scale factor r that must appear in polar coordinate double integrals. Assume that $\alpha \neq 1$. Then

$$\begin{aligned} G(R) &= \int_{0}^{2\pi} \int_{0}^{R} \frac{1}{(1 + r^2)^\alpha} \cdot r \, dr \, d\theta \\ &= 1/2 \int_{0}^{2\pi} \left((1 + r^2)^{-\alpha+1}/(-\alpha + 1) \right) \Big|_{r=0}^{r=R} \right) du \\ &= \frac{\pi}{\alpha - 1} \cdot \left(1 - (1 + R^2)^{1-\alpha} \right) . \end{aligned}$$

Therefore, if $\alpha > 1$ then the limit as $R \to \infty$ is $\pi/(\alpha - 1)$ and the improper integral converges. If $\alpha < 1$ then the integral diverges to ∞. The integral also diverges to ∞ for $\alpha = 1$ (verify).

Calculator/Computer Exercises

63. Start with a rectangular coordinate plot of the two curves as they would appear in the first quadrant.

```
> plot( [theta,3*cos(theta)^2], theta=0..Pi/2);
```

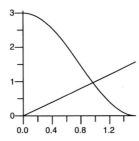

Using this picture, the intersection angle can be found using the fsolve procedure. We name it alpha.

```
> alpha := fsolve( theta=3*cos(theta)^2, theta=0.9);
```

$$\alpha := 0.9670246850$$

The next entry calculates the desired area.

```
> Area = 1/2*int( 9*cos(theta)^4 - theta^2, theta=0..alpha);
```

$$Area = 2.439314744$$

65. The double integral sets up as follows.

$$I = \int_0^{\pi/2} \int_0^\theta \frac{1}{1+\theta+r^3} \cdot r \, dr \, d\theta \, .$$

Using the code in Exercise 65 from Section 12.2, the Trapezoidal Rule approximation ($N = 4$) can proceed as follows. (The values for $c_{i,j}$ are not recalculated.)

Begin with the definition of the data used to set up the integral and the approximation.

```
> a,b,beta[1],beta[2],f :=
    0,evalf(Pi/2),unapply(0,theta),unapply(theta,theta),
              unapply(r/(1+theta+r^3),theta,r):
  N := 4:
  dx := (b-a)/N:
  for i from 0 to N
    do
       dy[i] := (beta[2](a+i*dx) - beta[1](a+i*dx))/N:
       X[i] := a + i*dx:
       for j from 0 to N
         do
            Y[i,j] := beta[1](X[i]) + j*dy[i]:
         end do
    end do:
```

Then make the approximation calculation.

```
> Approx = dx/2*add(dy[i]/2*
                add(c[i,j]*f(X[i],Y[i,j]),j=0..N),i=0..N);
```

$$Approx = 0.2422248392$$

12.6   Triple Integrals

Problems for Practice

1. $\int_2^4 \int_x^{x+1} \int_x^{x+y} (6x - 3z) \, dz \, dy \, dx = \int_2^4 \int_x^{x+1} \left((6xz - 3z^2/2) \big|_{z=x}^{z=x+y} \right) dy \, dx$

$= \int_2^4 \int_x^{x+1} (3xy - 3x^2/2) \, dy \, dx = \int_2^4 (3xy^2/2 - 3x^2y/2) \big|_{y=x}^{y=x+1} \, dx$

$= \int_2^4 (3x^2/2 - 1/2) \, dx = 27 \, .$

3. $\int_1^4 \int_{-z}^z \int_{y-z}^{y+z} (x + 3) \, dx \, dy \, dz = \int_1^4 \int_{-z}^z \left((x^2/2 + 3x) \big|_{x=y-z}^{x=y+z} \right) dx \, dy$

$$= \int_1^4 \int_{-z}^z ((y+z)^2/2 - (y-z)^2/2 + 6z) \, dy \, dz$$

$$= \int_1^4 ((y+z)^3/3 - (y-z)^3/3 + 6zy)\Big|_{y=-z}^{y=z} \, dz$$

$$= \int_1^4 12z^2 \, dx = 252 \, .$$

5. $\int_e^{e^2} \int_{1/(2y)}^{1/y} \int_1^y (4/z) \, dz \, dx \, dy = \int_e^{e^2} \int_{1/(2y)}^{1/y} \left((4\ln(z))\big|_{z=1}^{z=y} \right) dx \, dy$

$$= \int_e^{e^2} \int_{1/(2y)}^{1/y} (4\ln(y)) \, dx \, dy$$

$$= \int_e^{e^2} (2\ln(y)/y) \, dy = (\ln(y)^2)\Big|_e^{e^2} = 3 \, .$$

7. $\int_0^{\pi/2} \int_0^x \int_x^z 2\cos(y) \, dy \, dz \, dx = \int_0^{\pi/2} \int_0^x \left((2\sin(y))\big|_{y=x}^{y=z} \right) dz \, dx$

$$= \int_0^{\pi/2} \int_0^x (2\sin(z) - 2\sin(x)) \, dz \, dx$$

$$= \int_0^{\pi/2} \left((-2\cos(z) - 2z\sin(x))\big|_{z=0}^{z=x} \right) dx$$

$$= \int_0^{\pi/2} (2 - 2\cos(x) - 2x\sin(x)) \, dx$$

$$= (2x - 2\sin(x) + 2(x\cos(x) - \sin(x)))\Big|_0^{\pi/2}$$

$$= \pi - 4 \, .$$

Integration-by-parts $(u = x)$ was used to integrate $x\sin(x)$ in the penultimate step.

9. $\int_0^\pi \int_{-y}^{\sqrt{\pi}} \int_{-x}^0 4y\cos(x^2) \, dz \, dx \, dy = \int_0^\pi \int_{-y}^{\sqrt{\pi}} 4xy\cos(x^2) \, dx \, dy$

$$= \int_0^\pi \left((2y\sin(x^2))\big|_{x=-y}^{x=\sqrt{\pi}} \right) dy$$

$$= \int_0^\pi (-2y\sin(y^2)) \, dy = (\cos(y^2))\big|_0^\pi = \cos(\pi^2) - 1 \, .$$

11. $V = \iiint_{\mathcal{U}} 1 \, dV$ where $\mathcal{U}$ is the solid region inside of the circular paraboloid $z = x^2 + y^2$ and between the planes $z = 2$ and $z = 8$. This region is not z-simple. Describing it as an x-simple region

$$\mathcal{U} = \left\{ (x, y, z) : 2 \le z \le 8, -\sqrt{z} \le y \le \sqrt{z}, -\sqrt{z-y^2} \le x \le \sqrt{z-y^2} \right\},$$

so

$$V = \int_2^8 \int_{-\sqrt{z}}^{\sqrt{z}} \int_{-\sqrt{z^2-y^2}}^{\sqrt{z^2-y^2}} 1 \, dx \, dy \, dz$$

$$= \int_2^8 \int_{-\sqrt{z}}^{\sqrt{z}} 2\sqrt{z^2 - y^2} \, dy \, dz = \int_2^8 \pi z \, dz = 30\pi \, .$$

The dy integral is easy because it is the area of a circle of radius $\sqrt{z}$.

13. $V = \iiint_{\mathcal{U}} 1 \, dV$ where $\mathcal{U}$ is the solid region inside of the cylinder $y = x^2$ and between the planes $z = 1 - y$ and $z = -1 + y$. The planes intersect

in the xy-plane in the line $y = 1$. This region is z-simple.

$$\mathcal{U} = \left\{ (x, y, z) : -1 \leq x \leq 1, x^2 \leq y \leq 1, -1 + y \leq z \leq 1 - y \right\},$$

so

$$V = \int_{-1}^{1} \int_{x^2}^{1} \int_{-1+y}^{1-y} 1 \, dz \, dy \, dx = \int_{-1}^{1} \int_{x^2}^{1} (2 - 2y) \, dy \, dx$$

$$= \int_{-1}^{1} \left((2y - y^2)|_{x^2}^{1} \right) dx$$

$$= \int_{-1}^{1} (1 - 2x^2 + x^4) \, dx = 16/15$$

15. $V = \iiint_{\mathcal{U}} 1 \, dV$ where $\mathcal{U}$ is the solid region inside of the cylinder $x^2 + y^2 = 4$ and between the planes $z = x + y$ and $z = -6 - x$. This region is z-simple.

$$\mathcal{U} = \left\{ (x, y, z) : -2 \leq x \leq 2, -\sqrt{4 - x^2} \leq y \leq \sqrt{4 - x^2}, x + y \leq z \leq -6 - y \right\},$$

so

$$V = \int_{-2}^{2} \int_{-\sqrt{4-x^2}}^{\sqrt{4-x^2}} \int_{-6-y}^{x+y} 1 \, dz \, dy \, dx = \int_{-2}^{2} \int_{-\sqrt{4-x^2}}^{\sqrt{4-x^2}} (x + 2y + 6) \, dy \, dx$$

$$= \int_{-2}^{2} \left((xy + y^2 + 6y)|_{y=-\sqrt{4-x^2}}^{y=\sqrt{4-x^2}} \right) dx$$

$$= \int_{-2}^{2} (2x\sqrt{4 - x^2} + 12\sqrt{4 - x^2}) \, dx$$

$$= \int_{-\pi/2}^{\pi/2} (16 \sin(\theta) \cos(\theta) + 48 \cos^2(\theta)) \, d\theta$$

$$= (8 \sin^2(\theta)) \Big|_{-\pi/2}^{\pi/2} + 24\pi = 24\pi.$$

The last integral was obtained using the substitution $x = 2\sin(\theta)$.

17. The circular paraboloid and the plane intersect in a circle: $x^2 + y^2 + 2x + 4y = 11$, or, after completing the square, $(x + 1)^2 + (y + 2)^2 = 16$. The center of the circle is at $(-1, -2)$ and its radius is 4. Using this information, the solid $\mathcal{U}$ can be described as follows:

$$\mathcal{U} = \left\{ (x, y, z) : -5 \leq x \leq 3, \right.$$
$$-2 - \sqrt{16 - (x + 1)^2} \leq y \leq -2 + \sqrt{16 - (x + 1)^2},$$
$$\left. 2x + 4y - 2 \leq z \leq 9 - x^2 - y^2 \right\}.$$

The volume computation can begin as follows. Note the shifty rearrangement of the integrand in the third line.

$$V = \int_{-5}^{3} \int_{-2-\sqrt{16-(x+1)^2}}^{-2+\sqrt{16-(x+1)^2}} \int_{2x+4y-2}^{9-x^2-y^2} 1 \, dz \, dy \, dx$$

$$= \int_{-5}^{3} \int_{-2-\sqrt{16-(x+1)^2}}^{-2+\sqrt{16-(x+1)^2}} (11 - x^2 - y^2 - 2x - 4y) \, dy \, dx$$

$$= \int_{-5}^{3} \int_{-2-\sqrt{16-(x+1)^2}}^{-2+\sqrt{16-(x+1)^2}} (16 - (x+1)^2 - (y+2)^2) \, dy \, dx$$

$$= \int_{-5}^{3} \left(\left((16 - (x+1)^2)y - (y+2)^3/3 \right) \Big|_{y=-2-\sqrt{16-(x+1)^2}}^{y=-2+\sqrt{16-(x+1)^2}} \right) dx$$

$$= 4/3 \int_{-5}^{3} (16 - (x+1)^2)^{3/2} \, dx$$

Now make the substitution $x + 1 = 4\sin(\theta)$. Note that $dx = 4\cos(\theta) \, d\theta$ and $\sqrt{16 - (x+1)^2} = 4\cos(\theta)$. This yields

$$V = 4/3 \int_{-\pi/2}^{\pi/2} 256 \cos^4(\theta) \, d\theta = 1024/3 \int_{-\pi/2}^{\pi/2} ((1 + \cos(2\theta))/2)^2 \, d\theta$$

$$= 256/3 \int_{-\pi/2}^{\pi/2} (1 + 2\cos(2\theta) + \cos^2(2\theta)) \, d\theta$$

$$= 256/3 \int_{-\pi/2}^{\pi/2} (1 + 2\cos(2\theta) + 1/2 + \cos(4\theta)/2) \, d\theta$$

$$= 256/3 \left(3\theta/2 + \sin(2\theta) + \sin(4\theta)/8 \right) \Big|_{-\pi/2}^{\pi/2} = 128\pi.$$

19. $\int_0^2 \int_0^{3-3x/2} \int_0^{6-3x-2y} x \, dz \, dy \, dx = \int_0^2 \int_0^{3-3x/2} x(6 - 3x - 2y) \, dy \, dx$

$= \int_0^2 \left((6xy - 3x^2y - xy^2) \Big|_{y=0}^{y=3-3x/2} \right) dy \, dx = \int_0^2 (9x - 9x^2 + 9x^4/16) \, dx = 3.$

The solid can be described as

$$\mathcal{U} = \{(x,y,z) : 0 \leq x \leq 2, 0 \leq y \leq 3 - 3x/2, 0 \leq z \leq 6 - 3x - 2y\}.$$

It consists of all points in the first octant lying on and below the plane $z = 6 - 3x - 2y$.

21. $\int_0^1 \int_0^{\sqrt{1-x^2}} \int_0^{\sqrt{1-x^2-y^2}} xz \, dz \, dy \, dx = \int_0^1 \int_0^{\sqrt{1-x^2}} \left((xz^2/2) \Big|_{z=0}^{z=\sqrt{1-x^2-y^2}} \right) dy \, dx$

$$= 1/2 \int_0^1 \int_0^{\sqrt{1-x^2}} (x(1 - x^2 - y^2) \, dy \, dx$$

$$= 1/2 \int_0^1 \left((x(1-x^2)y - xy^3/3) \Big|_{y=0}^{y=\sqrt{1-x^2}} \right) dy \, dx$$

$$= 1/3 \int_0^1 (x(1-x^2)^{3/2}) \, dx = (-(1-x^2)^{5/2}/15)\big|_0^1 = 1/15 \,.$$

The solid $\mathcal{U}$ is the portion of the unit ball: $x^2 + y^2 + z^2 \le 1$, that lies in the first octant.

23. $\int_0^1 \int_0^x \int_0^{\sqrt{1-x^2}} z \, dz \, dy \, dx = \int_0^1 \int_0^x \left((z^2/2)\big|_{z=0}^{z=\sqrt{1-x^2}} \right) dy \, dx$

$\qquad\qquad = 1/2 \int_0^1 \int_0^x (1-x^2) \, dy \, dx = 1/2 \int_0^1 x(1-x^2) \, dy \, dx = 1/8$

The solid, which can be described as

$$\mathcal{U} = \left\{ (x,y,z) : 0 \le x \le 1, 0 \le y \le x, 0 \le z \le \sqrt{1-x^2} \right\},$$

is the region in the first octant bounded by the plane $y = x$ and the cylinder $x^2 + z^2 = 1$.

Further Theory and Practice

25. The solid region in Exercise 21 is y-simple. The integral can be evaluated as $\int_0^1 \int_0^{\sqrt{1-z^2}} \int_0^{\sqrt{1-x^2-z^2}} xz \, dy \, dx \, dz$ or $\int_0^1 \int_0^{\sqrt{1-x^2}} \int_0^{\sqrt{1-x^2-z^2}} xz \, dy \, dz \, dx$.

The solid region in Exercise 23 is also y-simple. The integral can be evaluated as $\int_0^1 \int_0^{\sqrt{1-z^2}} \int_0^x z \, dy \, dx \, dz$ or $\int_0^1 \int_0^{\sqrt{1-x^2}} \int_0^x z \, dy \, dz \, dx$.

27. The domain is a parallelopiped. That is, a box with parallel parallelograms for its faces.

See the picture on the right.

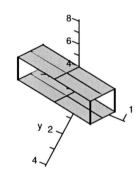

29. The domain consists of the points inside of the cylinder $x^2 + y^2 = 9$ and between the planes $z = x+y+2$ and $z = -x - y - 2$.

See the picture on the right where only the planes are displayed.

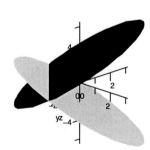

31. The domain consists of the points above the trapaziod in the xy-plane: $1 \le y \le 4$, $y \le x \le 6$, that lie between the plane $z = y$ and the cylinder $z = y^2$.

    See the picture on the right where only the bottom plane and the top cylinder are displayed.

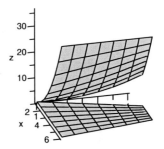

33. The circular paraboloid and the plane intersect in a circle: $x^2 + y^2 - ax - by = c$, or, after completing the square, $(x - a/2)^2 + (y - b/2)^2 = c + (a^2 + b^2)/4$. The center of the circle is at $(a/2, b/2)$ and its radius is $\sqrt{c + (a^2 + b^2)/4}$. Let $h = a/2$, $k = b/2$, and $r = \sqrt{c + (a^2 + b^2)/4} = \sqrt{c + h^2 + k^2}$. Using this information, the solid $\mathcal{U}$ can be described as follows:

$$\mathcal{U} = \big\{ (x, y, z) : h - r \le x \le h + r,$$
$$k - \sqrt{r^2 - (x - h)^2} \le y \le k + \sqrt{r^2 - (x - h)^2},$$
$$x^2 + y^2 \le z \le 2hx + 2ky + r^2 - h^2 - k^2 \big\}.$$

The volume computation can begin as follows. Note the shifty rearrangement of the integrand in the third line.

$$V = \int_{h-r}^{h+r} \int_{k-\sqrt{r^2-(x-h)^2}}^{k+\sqrt{r^2-(x-h)^2}} \int_{x^2+y^2}^{2hx+2ky+r^2-h^2-k^2} 1 \, dz \, dy \, dx$$

$$= \int_{h-r}^{h+r} \int_{k-\sqrt{r^2-(x-h)^2}}^{k+\sqrt{r^2-(x-h)^2}} (2hx + 2ky + r^2 - h^2 - k^2 - x^2 - y^2) \, dy \, dx$$

$$= \int_{h-r}^{h+r} \int_{k-\sqrt{r^2-(x-h)^2}}^{k+\sqrt{r^2-(x-h)^2}} (r^2 - (x - h)^2 - (y - k)^2) \, dy \, dx$$

$$= \int_{h-r}^{h+r} \left(((r^2 - (x - h)^2)y - (y - k)^3/3)) \Big|_{y=k-\sqrt{r^2-(x-h)^2}}^{y=k+\sqrt{r^2-(x-h)^2}} \right) dx$$

$$= 4/3 \int_{h-r}^{h+r} (r^2 - (x - h)^2)^{3/2} \, dx$$

Now make the substitution $x - h = r\sin(\theta)$. Note that $dx = r\cos(\theta) \, d\theta$ and $\sqrt{r^2 - (x - h)^2} = r\cos(\theta)$. This yields

$$V = 4/3 \int_{-\pi/2}^{\pi/2} r^4 \cos^4(\theta) \, d\theta = 4r^4/3 \int_{-\pi/2}^{\pi/2} \cos^4(\theta) \, d\theta$$

$$= \frac{4r^4}{3} \cdot \frac{3\pi}{8} = \frac{\pi}{2} \cdot \left(c + \frac{a^2 + b^2}{4} \right)^2 = \frac{\pi}{32} \cdot (a^2 + b^2 + 4c)^2.$$

The solution for Exercise 17 contains calculations that can be used to verify that $\int_{-\pi/2}^{\pi/2} \cos^4(\theta)\, d\theta = 3\pi/8$.

35. The cone and the paraboloid intersect in the circle $x^2 + y^2 = m^2$. The symmetry in the solid region over which the integration takes place and the symmetry in the integrand implies that the integral evaluates to $4I$ where I is the integral over the portion of the solid in the first octant. The first octant region is

$$\mathcal{U} = \left\{ (x,y,z) : 0 \le x \le m, 0 \le y \le \sqrt{m^2 - x^2}, x^2 + y^2 \le z \le m\sqrt{x^2 + y^2} \right\}.$$

Evaluation of the integral over this region can begin like this.

$$
\begin{aligned}
I &= \int_0^m \int_0^{\sqrt{m^2-x^2}} \int_{x^2+y^2}^{m\sqrt{x^2+y^2}} (x^2 + y^2 + z^2)\, dz\, dy\, dx \\
&= \int_0^m \int_0^{\sqrt{m^2-x^2}} \left(((x^2+y^2)z + z^3/3)\big|_{z=x^2+y^2}^{z=m\sqrt{x^2+y^2}} \right) dy\, dx \\
&= \int_0^m \int_0^{\sqrt{m^2-x^2}} ((m+m^3/3)(x^2+y^2)^{3/2} - (x^2+y^2)^2 - (x^2+y^2)^3/3)\, dy\, dx
\end{aligned}
$$

The best way to finish is to switch to polar coordinates. Doing so yields

$$
\begin{aligned}
I &= \int_0^{\pi/2} \int_0^m ((m+m^3/3)r^3 - r^4 - r^6/3)\, r\, dr\, d\theta \\
&= \int_0^{\pi/2} ((m+m^3/3)r^5/5 - r^6/6 - r^8/24)\big|_0^m\, d\theta \\
&= \int_0^{\pi/2} (m^8/40 + m^6/30)\, d\theta = \frac{\pi}{240} m^6 (3m^2 + 1).
\end{aligned}
$$

The value of the integral over the entire region is $4I = \dfrac{\pi}{60} m^6 (3m^2 + 1)$.

37. $\int_0^1 \int_x^1 \int_0^x z \sin(y^4)\, dz\, dy\, dx = \int_0^1 \int_x^1 \left((\sin(y^4)z^2/2)\big|_{z=0}^{z=x} \right) dy\, dx$

$$= 1/2 \int_0^1 \int_x^1 x^2 \sin(y^4)\, dy\, dx$$

In order to continue it will be convenient to change the order of integration in the xy-plane. Doing so yields

$$
\begin{aligned}
\int_0^1 \int_x^1 \int_0^x z \sin(y^4)\, dz\, dy\, dx &= 1/2 \int_0^1 \int_0^y x^2 \sin(y^4)\, dx\, dy \\
&= 1/2 \int_0^1 \left((x^3 \sin(y^4)/3)\big|_{x=0}^{x=y} \right) dy \\
&= 1/6 \int_0^1 y^3 \sin(y^4)\, dy \\
&= 1/6(-\cos(y^4)/4)\big|_0^1 = (1 - \cos(1))/24.
\end{aligned}
$$

39. Draw a picture to see that the solid region is

$$\mathcal{U} = \left\{ (x,y,z) : 0 \le x \le 2, 0 \le y \le 4 - x^2, 0 \le z \le 2 + 7x \right\}.$$

Therefore,

$$\iiint_{\mathcal{U}} (xy + z)\, dV = \int_0^2 \int_0^{4-x^2} \int_0^{2+7x} (xy + z)\, dz\, dy\, dx$$

$$= \int_0^2 \int_0^{4-x^2} \left((xyz + z^2/2)\big|_{z=0}^{z=2+7x} \right) dy\, dx$$

$$= \int_0^2 \int_0^{4-x^2} (xy(2 + 7x) + (2 + 7x)^2/2)\, dy\, dx$$

$$= 1/2 \int_0^2 \left(((2 + 7x)(xy^2 + (2 + 7x)y)) \big|_{y=0}^{y=4-x^2} \right) dx$$

$$= 1/2 \int_0^2 (2 + 7x)(4 - x^2)(x(4 - x^2) + (2 + 7x))\, dx$$

$$= 1/2 \int_0^2 (7x^6 + 2x^5 - 105x^4 - 44x^3 + 304x^2 + 144x + 16)\, dx$$

$$= 216\,.$$

41. (a) As a z-simple solid

$$\int_0^1 \int_x^1 \int_0^1 f(x, y, z)\, dz\, dy\, dx\,, \qquad \int_0^1 \int_0^y \int_0^1 f(x, y, z)\, dz\, dx\, dy$$

(b) As a y-simple solid

$$\int_0^1 \int_0^1 \int_x^1 f(x, y, z)\, dy\, dz\, dx\,, \qquad \int_0^1 \int_0^1 \int_x^1 f(x, y, z)\, dy\, dx\, dz$$

(c) As an x-simple solid

$$\int_0^1 \int_0^1 \int_0^y f(x, y, z)\, dx\, dy\, dz\,, \qquad \int_0^1 \int_0^1 \int_0^y f(x, y, z)\, dx\, dz\, dy$$

43. (a) As a y-simple solid

$$\int_{-1}^0 \int_2^{37} \int_{x+1}^{\sqrt{x+1}} y\sqrt{1 - x}\, dy\, dz\, dx\,, \qquad \int_2^{37} \int_{-1}^0 \int_{x+1}^{\sqrt{x+1}} y\sqrt{1 - x}\, dy\, dx\, dz$$

(b) As a z-simple solid

$$\int_{-1}^0 \int_{x+1}^{\sqrt{x+1}} \int_2^{37} y\sqrt{1 - x}\, dz\, dy\, dx\,, \qquad \int_0^1 \int_{y^2-1}^{y-1} \int_2^{37} y\sqrt{1 - x}\, dz\, dx\, dy$$

(c) As an x-simple solid

$$\int_2^{37} \int_0^1 \int_{y^2-1}^{y-1} y\sqrt{1-x}\, dx\, dy\, dz\,, \quad \int_0^1 \int_2^{37} \int_{y^2-1}^{y-1} y\sqrt{1-x}\, dx\, dz\, dy$$

45. The solid $\mathcal{U}$ can be described as

$$\mathcal{U} = \left\{ (x,y,z) : 0 \le x \le a, 0 \le y \le b(1 - \sqrt{x/a}\,)^2, 0 \le z \le c(1 - \sqrt{x/a} - \sqrt{y/b}\,)^2 \right\}.$$

Therefore, its volume is given by the following integral.

$$V = \int_0^a \int_0^{b(1-\sqrt{x/a}\,)^2} \int_0^{c(1-\sqrt{x/a}-\sqrt{y/b}\,)^2} 1\, dz\, dy\, dx$$

$$= \int_0^a \int_0^{b(1-\sqrt{x/a}\,)^2} c(1 - \sqrt{x/a} - \sqrt{y/b}\,)^2\, dy\, dx$$

In order to continue make a change-of-variables: $u = \sqrt{x/a}$, $v = \sqrt{y/b}$. This transforms the region

$$\mathcal{R} = \left\{ (x,y) : 0 \le x \le a, 0 \le y \le b(1 - \sqrt{x/a}\,)^2 \right\}$$

in the xy-plane to the region

$$\mathcal{S} = \left\{ (u,v) : 0 \le u \le 1, 0 \le v \le 1 - u \right\}$$

in the uv-plane.

The transformation from the uv-plane to the xy-plane is $T = (\phi, \psi)$ where $\phi(u,v) = au^2$ and $\psi(u,v) = bv^2$. Its Jacobian is

$$J_T(u,v) = (\phi_u \psi_v - \psi_u \phi_v)(u,v) = 4abuv\,.$$

Consequently,

$$V = \int_0^1 \int_0^{1-u} c(1 - u - v)^2 \cdot 4abuv\, dv\, du$$

$$= 4abc \int_0^1 \int_0^{1-u} uv(1 - u - v)^2\, dv\, du = abc/90\,.$$

The second integral, $I = \iiint_{\mathcal{U}} xyz\, dV$, can also be evaluated using the change-of-variable technique. To begin,

$$I = \int_0^a \int_0^{b(1-\sqrt{x/a}\,)^2} \int_0^{c(1-\sqrt{x/a}-\sqrt{y/b}\,)^2} xyz\, dz\, dy\, dx$$

$$= 1/2 \int_0^a \int_0^{b(1-\sqrt{x/a}\,)^2} xyc^2(1 - \sqrt{x/a} - \sqrt{y/b}\,)^4\, dy\, dx\,.$$

After changing variables,

$$I = 1/2 \int_0^1 \int_0^{1-u} abu^2 v^2 c^2 (1 - u - v)^4 \cdot 4abuv \, dv \, du$$

$$= 2a^2 b^2 c^2 \int_0^1 \int_0^{1-u} u^3 v^3 (1 - u - v)^4 \, dv \, du = \frac{a^2 b^2 c^2}{277200}.$$

Calculator/Computer Exercises

47. Start with a plot of the two curves in the xy-plane.

```
> plot( [sin(x),1/(x^2+1)], x=0..Pi);
```

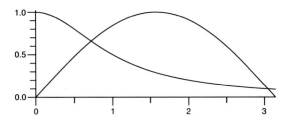

Using this picture as a guide, the fsolve procedure is used to obtain the x-coordinates of the intersection points. They are named a and b. This is followed by the integral calculation.

```
> a,b := fsolve( sin(x)=1/(x^2+1), x=1),
        fsolve( sin(x)=1/(x^2+1), x=3):
  Int(Int(Int( z, z=0..x), y=1/(x^2+1)..sin(x)), x=a..b) =
  int(int(int( z, z=0..x), y=1/(x^2+1)..sin(x)), x=a..b);
```

$$\int_{0.7194212963}^{3.044030166} \int_{\frac{1}{x^2+1}}^{\sin(x)} \int_0^x z \, dz \, dy \, dx = 2.033269733$$

12.7   Physical Applications

1. $\mathcal{M} = \int_0^2 \int_{x+1}^{2x+3} (3x + 4y) \, dy \, dx = \int_0^2 (3xy + 2y^2) \big|_{y=x+1}^{y=2x+3} \, dx$

$\qquad = \int_0^2 (9x^2 + 26x + 16) \, dx = 108$

3. $\mathcal{M} = \int_0^1 \int_{1-x^2}^{2(1-x^2)} 12xy \, dy \, dx = \int_0^1 6xy^2) \big|_{y=1-x^2}^{y=2(1-x^2)} \, dx$

$\qquad = \int_0^1 (18x - 36x^3 + 18x^5) \, dx = 3$

5. $\mathcal{M}_{x=0} = \int_0^1 \int_{1-x^3}^{2(1-x^3)} x \cdot 12xy \, dy \, dx = \int_0^1 6x^2 y^2) \big|_{y=1-x^3}^{y=2(1-x^3)} \, dx$

$\qquad = \int_0^1 (18x^2 - 36x^5 + 18x^8) \, dx = 2$

7. $\mathcal{M}_{y=0} = \int_1^2 \int_x^{5-x} y \cdot 12x\, dy\, dx = \int_1^2 (6xy^2)\big|_{y=x}^{y=5-x}\, dx$

$\qquad = \int_1^2 (150x - 60x^2)\, dx = 85$

9. $\mathcal{M}_{y=0} = \int_1^4 \int_{\sqrt{y}}^{y^2} y \cdot 10\, dx\, dy = \int_1^4 (10xy)\big|_{x=\sqrt{y}}^{x=y^2}\, dy$

$\qquad = \int_1^4 (10y^3 - 10y^{3/2})\, dy = \frac{1027}{2}$

11. $\mathcal{M}_{x=0} = \int_{-3}^{-1} \int_2^4 x(-x/y)\, dy\, dx = -\int_{-3}^{-1} \left((x^2 \ln(y))\big|_{y=2}^{y=4}\right) dx$

$\qquad = -\ln(2) \int_{-3}^{-1} x^2\, dx = -26\ln(2)/3$

13. $\mathcal{M} = \int_0^1 \int_x^{\sqrt{x}} (x+6)\, dy\, dx = \int_0^1 (x+6)(\sqrt{x} - x)\, dx = 16/15$

$\qquad \mathcal{M}_{x=0} = \int_0^1 \int_x^{\sqrt{x}} x(x+6)\, dy\, dx = \int_0^1 x(x+6)(\sqrt{x} - x)\, dx = 61/140$

$\qquad \mathcal{M}_{y=0} = \int_0^1 \int_x^{\sqrt{x}} y(x+6)\, dy\, dx = \int_0^1 (x+6)(x - x^2)/2\, dx = 13/24$

$$\bar{x} = \frac{61/140}{16/15} = \frac{183}{448}\,,\ \bar{y} = \frac{13/24}{16/15} = \frac{65}{128}$$

15. $\mathcal{M} = \int_{-1}^1 \int_2^{y+3} x(y+1)\, dx\, dy = 1/2 \int_{-1}^1 ((y+3)^2 - 4)(y+1)\, dy = 22/3$

$\qquad \mathcal{M}_{x=0} = \int_{-1}^1 \int_2^{y+3} x^2(y+1)\, dx\, dy = 1/3 \int_{-1}^1 ((y+3)^3 - 8)(y+1)\, dy = 104/5$

$\qquad \mathcal{M}_{y=0} = \int_{-1}^1 \int_2^{y+3} xy(y+1)\, dx\, dy = 1/2 \int_{-1}^1 ((y+3)^2 - 4)y(y+1)\, dy$

$\qquad\qquad = 58/15$

$$\bar{x} = \frac{104/5}{22/3} = \frac{156}{55}\,,\ \bar{y} = \frac{58/15}{22/3} = \frac{29}{55}$$

17. $\mathcal{M} = \int_1^4 \int_1^{2/\sqrt{y}} 12/\sqrt{y}\, dx\, dy = \int_1^4 (24/y - 12/\sqrt{y})\, dy = 48\ln(2) - 24$

$\qquad \mathcal{M}_{x=0} = \int_1^4 \int_1^{2/\sqrt{y}} x \cdot (12/\sqrt{y})\, dx\, dy = \int_1^4 (24/y^{3/2} - 6/\sqrt{y})\, dy = 12$

$\qquad \mathcal{M}_{y=0} = \int_1^4 \int_1^{2/\sqrt{y}} y \cdot (12/\sqrt{y})\, dx\, dy = \int_1^4 (24 - 12y)\, dy = 16$

$$\bar{x} = \frac{1}{2}k\,,\ \bar{y} = \frac{2}{3}k \quad \text{where} \quad k = (2\ln(2) - 1)^{-1}$$

19. $\mathcal{M} = \int_0^1 \int_0^2 \int_0^{xy} (x+1)\, dz\, dy\, dx = \int_0^1 \int_0^2 (x+1)xy\, dy\, dx$

$\qquad = \int_0^1 2(x+1)x\, dx = 5/3$

21. $\mathcal{M} = \int_0^2 \int_z^3 \int_y^{y+1} (x+z)\, dx\, dy\, dz = \int_0^2 \int_z^3 (((y+1)^2 - y^2)/2 + zy)\, dy\, dz$

$\qquad = \int_0^2 \int_z^3 (y + \frac{1}{2} + zy)\, dy\, dz = \int_0^2 \left(\frac{1}{2}y^2 + \frac{1}{2}y + \frac{1}{2}zy^2\right)\big|_{y=z}^{y=3}\, dz$

$\qquad = \int_0^2 \left(6 + \frac{5}{2}z - \frac{3}{2}z^2\right) dz = 13$

23. $\mathcal{M} = \int_0^1 \int_{x^2}^x \int_y^1 12x \, dz \, dy \, dx = \int_0^1 \int_{x^2}^x 12x(1-y) \, dy \, dx$

$\qquad = \int_0^1 (12xy - 6xy^2)\big|_{y=x^2}^{y=x} \, dx$

$\qquad = \int_0^1 (-18x^3 + 6x^5 + 12x^2) \, dx = \frac{1}{2}$

25. $\mathcal{M} = \int_{-1}^1 \int_{-\sqrt{1-x^2}}^{\sqrt{1-x^2}} \int_0^{12} z \, dz \, dy \, dx = \int_{-1}^1 \int_{-\sqrt{1-x^2}}^{\sqrt{1-x^2}} \left(\frac{1}{2}z^2\right)\big|_0^{12} \, dy \, dx$

$\qquad = 72 \int_{-1}^1 \int_{-\sqrt{1-x^2}}^{\sqrt{1-x^2}} dy \, dx = 72\pi$ See footnote.[2]

$\mathcal{M}_{yz} = \int_{-1}^1 \int_{-\sqrt{1-x^2}}^{\sqrt{1-x^2}} \int_0^{12} x \cdot z \, dz \, dy \, dx = \int_{-1}^1 \int_{-\sqrt{1-x^2}}^{\sqrt{1-x^2}} \left(\frac{1}{2}xz^2\right)\big|_0^{12} \, dy \, dx$

$\qquad = 72 \int_{-1}^1 \int_{-\sqrt{1-x^2}}^{\sqrt{1-x^2}} x \, dy \, dx = 72 \int_{-1}^1 2x\sqrt{1-x^2} \, dx = 0$ (odd)

$\mathcal{M}_{xz} = \int_{-1}^1 \int_{-\sqrt{1-x^2}}^{\sqrt{1-x^2}} \int_0^{12} y \cdot z \, dz \, dy \, dx = \int_{-1}^1 \int_{-\sqrt{1-x^2}}^{\sqrt{1-x^2}} \left(\frac{1}{2}yz^2\right)\big|_0^{12} \, dy \, dx$

$\qquad = 72 \int_{-1}^1 \int_{-\sqrt{1-x^2}}^{\sqrt{1-x^2}} y \, dy \, dx = 0$ (odd)

$\mathcal{M}_{xy} = \int_{-1}^1 \int_{-\sqrt{1-x^2}}^{\sqrt{1-x^2}} \int_0^{12} z \cdot z \, dz \, dy \, dx = \frac{12^3}{3} \int_{-1}^1 \int_{-\sqrt{1-x^2}}^{\sqrt{1-x^2}} dy \, dx$

$\qquad = 576\pi$ (See footnote 2.)

$$\bar{x} = 0 \, , \ \bar{y} = 0 \, , \ \bar{z} = \frac{576}{72} = 8$$

27. $\mathcal{M} = \int_{-1}^0 \int_x^1 \int_{x+y}^2 (x+3) \, dz \, dy \, dx = \int_{-1}^0 \int_x^1 (x+3)(2-x-y) \, dy \, dx$

$\qquad = \int_{-1}^0 \left(((x+3)((2-x)y - y^2/2))\big|_{y=x}^{y=1}\right) dx$

$\qquad = \int_{-1}^0 ((x+3)((2-x)(1-x) - (1-x^2)/2) \, dx = 67/8$

$\mathcal{M}_{x=0} = \int_{-1}^0 \int_x^1 \int_{x+y}^2 x(x+3) \, dz \, dy \, dx = \int_{-1}^0 \int_x^1 x(x+3)(2-x-y) \, dy \, dx$

$\qquad = \int_{-1}^0 \left((x(x+3)((2-x)y - y^2/2))\big|_{y=x}^{y=1}\right) dx$

$\qquad = \int_{-1}^0 (x(x+3)((2-x)(1-x) - (1-x^2)/2) \, dx = -193/40$

$\mathcal{M}_{y=0} = \int_{-1}^0 \int_x^1 \int_{x+y}^2 y(x+3) \, dz \, dy \, dx = \int_{-1}^0 \int_x^1 y(x+3)(2-x-y) \, dy \, dx$

$\qquad = \int_{-1}^0 \left(((x+3)((1-x/2)y^2 - y^3/3))\big|_{y=x}^{y=1}\right) dx$

$\qquad = \int_{-1}^0 ((x+3)((1-x/2)(1-x^2) - (1-x^3)/3) \, dx = 25/24$

$\mathcal{M}_{z=0} = \int_{-1}^0 \int_x^1 \int_{x+y}^2 z(x+3) \, dz \, dy \, dx = \int_{-1}^0 \int_x^1 (2-(x+y)^2/2)(x+3) \, dy \, dx$

$\qquad = \int_{-1}^0 \left(((2y - (x+y)^3/6)(x+3))\big|_{y=x}^{y=1}\right) dx$

$\qquad = \int_{-1}^0 ((2(1-x) - ((x+1)^3 - 8x^3)/6)(x+3)) \, dx = 389/60$

$$\bar{x} = \frac{-193/40}{67/8} = -\frac{193}{335} \, , \ \bar{y} = \frac{25/24}{67/8} = \frac{25}{201} \, , \ \bar{z} = \frac{389/60}{67/8} = \frac{778}{1005}$$

29. $\mathcal{M} = \int_1^2 \int_0^z \int_x^z (y-x) \, dy \, dx \, dz = \int_1^2 \int_0^z \left((y^2/2 - xy)\big|_{y=x}^{y=z}\right) dx \, dz$

---

[2]The double integral is the area of a unit circle.

$$= \int_1^2 \int_0^z ((z^2 - x^2)/2 - x(z - x)) \, dx \, dz$$

$$= \int_1^2 \left((xz^2/2 + x^3/6 - x^2 z/2) \big|_{x=0}^{x=z} \right) dz$$

$$= \int_1^2 z^3/6 \, dz = 5/8$$

$$\mathcal{M}_{yz} = \int_1^2 \int_0^z \int_x^z x(y - x) \, dy \, dx \, dz = \int_1^2 \int_0^z \left((x(y^2/2 - xy)) \big|_{y=x}^{y=z} \right) dx \, dz$$

$$= \int_1^2 \int_0^z (x(z^2 - x^2)/2 - x^2(z - x)) \, dx \, dz$$

$$= \int_1^2 \left((x^4/8 + z^2 x^2/4 - zx^3/3) \big|_{x=0}^{x=z} \right) dz$$

$$= \int_1^2 z^4/24 \, dz = 31/120$$

$$\mathcal{M}_{xz} = \int_1^2 \int_0^z \int_x^z y(y - x) \, dy \, dx \, dz = \int_1^2 \int_0^z \left((y^3/3 - xy^2/2) \big|_{y=x}^{y=z} \right) dx \, dz$$

$$= \int_1^2 \int_0^z ((z^3 - x^3)/3 - x(z^2 - x^2))/2) \, dx \, dz$$

$$= \int_1^2 (xz^3/3 + x^4/24 - x^2 z^2/4) \big|_{x=0}^{x=z} \, dz$$

$$= \int_1^2 z^4/8 \, dz = 31/40$$

$$\mathcal{M}_{xy} = \int_1^2 \int_0^z \int_x^z z(y - x) \, dy \, dx \, dz = \int_1^2 \int_0^z \left((z(y^2/2 - xy)) \big|_{y=x}^{y=z} \right) dx \, dz$$

$$= \int_1^2 \int_0^z (z(z^2 - x^2)/2 - xz(z - x)) \, dx \, dz$$

$$= \int_1^2 \left((xz^3/2 - x^2 z^2/2 + x^3 z/6) \big|_{x=0}^{x=z} \right) dz$$

$$= \int_1^2 z^4/6 \, dz = 31/30$$

$$\bar{x} = \frac{31/120}{5/8} = \frac{31}{75}, \ \bar{y} = \frac{31/40}{5/8} = \frac{31}{25}, \ \bar{z} = \frac{31/30}{5/8} = \frac{31}{75}$$

31. $I_{x=0} = \int_0^2 \int_0^{4-x^2} x^2 \cdot (x + 1) \, dy \, dx = \int_0^2 x^2(x + 1)y \big|_{y=0}^{y=4-x^2} \, dx$

$$= \int_0^2 x^2(x + 1)(4 - x^2) \, dx = \frac{48}{5}$$

$I_{y=0} = \int_0^2 \int_0^{4-x^2} y^2 \cdot (x + 1) \, dy \, dx = \int_0^2 (x + 1) \cdot \frac{1}{3} y^3 \big|_{y=0}^{y=4-x^2} \, dx$

$$= \int_0^2 \frac{1}{3}(x + 1)(4 - x^2)^3 \, dx = \frac{1056}{35}$$

33. $I_{x=0} = \int_0^1 \int_{1-x}^{2(1-x)} x^2 \cdot (x + y) \, dy \, dx = \int_0^1 \left(x^3 y + \frac{1}{2} x^2 y^2 \right) \big|_{y=1-x}^{y=2(1-x)} \, dx$

$$= \int_0^1 \left(\frac{3}{2} x^2 - 2x^3 + \frac{1}{2} x^4 \right) dx = \frac{1}{10}$$

$I_{y=0} = \int_0^1 \int_{1-x}^{2(1-x)} y^2 \cdot (x + y) \, dy \, dx = \int_0^1 \left(\frac{1}{3} xy^3 + \frac{1}{4} y^4 \right) \big|_{y=1-x}^{y=2(1-x)} \, dx$

$$= \int_0^1 \left(\frac{17}{12} x^4 - 8x^3 + \frac{31}{2} x^2 - \frac{38}{3} x + \frac{15}{4} \right) dx = \frac{13}{15}$$

35. $I_0 = \int_0^1 \int_0^2 (x^2 + y^2) \cdot 6x \, dy \, dx = \int_0^1 \left(6x^3 y + 2xy^3 \right) \big|_{y=0}^{y=2} \, dx$

$$= \int_0^1 \left(12x^3 + 16x \right) dx = 11$$

37. $I_0 = \int_0^1 \int_1^{2-x} (x^2 + y^2) \cdot \frac{6y}{x^2 + y^2} \, dy \, dx = \int_0^1 \left(3y^2 \right) \big|_{y=1}^{y=2-x} \, dx$

$$= \int_0^1 \left(3x^2 - 12x + 9 \right) dx = 4$$

39. $I_x = \int_3^5 \int_1^4 \int_1^2 (y^2 + z^2) xy \, dz \, dy \, dx = \int_3^5 \int_1^4 \left(((y^2 z + z^3/3) xy) \big|_{z=1}^{z=2} \right) dy \, dx$

$= \int_3^5 \int_1^4 (7xy/3 + xy^3) \, dy \, dx$

$= \int_3^5 \left((7xy^2/6 + xy^4/4) \big|_{y=1}^{y=4} \right) dx$

$= \int_3^5 (325x/4) \, dx = 650$

$I_y = \int_3^5 \int_1^4 \int_1^2 (x^2 + z^2) xy \, dz \, dy \, dx = \int_3^5 \int_1^4 \left(((x^2 z + z^3/3) xy) \big|_{z=1}^{z=2} \right) dy \, dx$

$= \int_3^5 \int_1^4 (7xy/3 + x^3 y) \, dy \, dx$

$= \int_3^5 \left((7xy^2/6 + x^3 y^2/2) \big|_{y=1}^{y=4} \right) dx$

$= 1/2 \int_3^5 (15x^3 + 35x) \, dx = 1160$

$I_z = \int_3^5 \int_1^4 \int_1^2 (x^2 + y^2) xy \, dz \, dy \, dx = \int_3^5 \int_1^4 \left(((x^2 + y^2) xyz) \big|_{z=1}^{z=2} \right) dy \, dx$

$= \int_3^5 \int_1^4 (x^2 + y^2) xy \, dy \, dx$

$= \int_3^5 \left((xy^4/4 + x^3 y^2/2) \big|_{y=1}^{y=4} \right) dx$

$= \int_3^5 \left(15x^3/2 + 255x/4 \right) dx$

$= \frac{1}{8} (15x^4 + 255x^2) \big|_3^5 = 1530$

41. $I_x = \int_0^3 \int_1^2 \int_{x+y}^{x+y+1} 4(y^2 + z^2) \, dz \, dy \, dx = 4 \int_0^3 \int_1^2 (y^2 z + z^3/3) \big|_{z=x+y}^{z=x+y+1} dy \, dx$

$= (4/3) \int_0^3 \int_1^2 (6y^2 + 3x^2 + 6xy + 3x + 3y + 1) \, dy \, dx$

$= 2 \int_0^3 (2x^2 + 8x + 13) \, dx = 186$

$I_y = \int_0^3 \int_1^2 \int_{x+y}^{x+y+1} 4(x^2 + z^2) \, dz \, dy \, dx = 4 \int_0^3 \int_1^2 (x^2 z + z^3/3) \big|_{z=x+y}^{z=x+y+1} dy \, dx$

$= (4/3) \int_0^3 \int_1^2 (6x^2 + 3y^2 + 6xy + 3x + 3y + 1) \, dy \, dx$

$= (2/3) \int_0^3 (12x^2 + 24x + 25) \, dx = 194$

$I_z = \int_0^3 \int_1^2 \int_{x+y}^{x+y+1} 4(x^2 + y^2) \, dz \, dy \, dx = 4 \int_0^3 \int_1^2 ((x^2 + y^2) z) \big|_{z=x+y}^{z=x+y+1} dy \, dx$

$= 4 \int_0^3 \int_1^2 (x^2 + y^2) \, dy \, dx = (4/3) \int_0^3 (3x^2 + 7) \, dx = 64$

43. $\mathcal{M} = \int_0^\pi \int_0^{\sin(x)} 8y \, dy \, dx = \int_0^\pi 4y^2 \big|_{y=0}^{y=\sin(x)} dx = \int_0^\pi 4\sin^2(x) \, dx = 2\pi$

$\mathcal{M}_{x=0} = \int_0^\pi \int_0^{\sin(x)} x \cdot 8y \, dy \, dx = \int_0^\pi 4xy^2 \big|_{y=0}^{y=\sin(x)} dx$

$= \int_0^\pi 4x \sin^2(x) \, dx = \pi^2 \text{ (Integration by parts } u = 2x)$

$\mathcal{M}_{y=0} = \int_0^\pi \int_0^{\sin(x)} y \cdot 8y \, dy \, dx = \int_0^\pi \frac{8}{3} y^3 \big|_{y=0}^{y=\sin(x)} dx$

$= \int_0^\pi \frac{8}{3} \sin^3(x) \, dx = \frac{32}{9} \text{ (Split off } \sin(x) \text{ and convert } \sin^2(x) \text{ to cosines)}$

$$\bar{x} = \frac{\pi^2}{2\pi} = \frac{\pi}{2} \, , \ \bar{y} = \frac{32/9}{2\pi} = \frac{16}{9\pi}$$

45. $\mathcal{M} = \int_0^1 \int_0^{e^x} 8x \, dy \, dx = \int_0^1 8xy \big|_{y=0}^{y=e^x} \, dx = \int_0^1 8xe^x \, dx$ (integration by parts, $u = x$)

$$= 8(xe^x - e^x)\big|_0^1 = 8$$

$\mathcal{M}_{x=0} = \int_0^1 \int_0^{e^x} x \cdot 8x \, dy \, dx = \int_0^1 8x^2 y \big|_{y=0}^{y=e^x} \, dx = \int_0^1 8x^2 e^x \, dx$ (parts, twice)

$$= 8(2 - 2x + x^2)e^x \big|_0^1 = 8e - 16$$

$\mathcal{M}_{y=0} = \int_0^1 \int_0^{e^x} y \cdot 8x \, dy \, dx = \int_0^1 4xy^2 \big|_{y=0}^{y=e^x} \, dx = \int_0^1 4xe^{2x} \, dx$ (parts, $u = 2x$)

$$= (2x - 1)e^{2x} \big|_0^1 = e^2 + 1$$

$$\bar{x} = \frac{8e - 16}{8} = e - 2 \,, \; \bar{y} = \frac{e^2 + 1}{8}$$

Further Theory and Practice

47. $\mathcal{R} = \{(x, y) : y \geq 0, 15 \leq x^2 + y^2 \leq 16\}$

49. Consider a small piece of the solid at the point (x, y, z) with volume ΔV and mass $\Delta m \approx \delta(x, y, z)\Delta V$ where $\delta(x, y, z)$ is the mass density of the solid at (x, y, z). Since its distance to the x-axis is $\sqrt{y^2 + z^2}$, its speed when it rotates is $v(x, y, z) = \omega\sqrt{y^2 + z^2}$. Consequently, the kinetic energy of the piece is $\frac{1}{2}(\Delta m)v^2 \approx \frac{1}{2}\delta(x, y, z)\Delta V \omega^2(y^2 + z^2)$. Summing over the entire solid $\mathcal{U}$ yields the following approximation for the total kinetic energy:

$$KE_{\mathcal{U}} \approx \sum \tfrac{1}{2}\omega^2(y^2 + z^2)\delta(x, y, z)\, \Delta V \,.$$

If follows that

$$KE_{\mathcal{U}} = \iiint_{\mathcal{U}} \tfrac{1}{2}\omega^2(y^2 + z^2)\delta(x, y, z)\, dV$$

$$= \tfrac{1}{2}\omega^2 \iiint_{\mathcal{U}} (y^2 + z^2)\delta(x, y, z)\, dV$$

$$= \tfrac{1}{2}I_x\omega^2 \,.$$

A point mass $M_{\mathcal{U}}$ at a distance ρ from the axis that rotates with angular velocity ω has kinetic energy $\frac{1}{2}M_{\mathcal{U}}(\omega\rho)^2$. Therefore, the radius ρ must satisfy the equation $\frac{1}{2}M_{\mathcal{U}}\omega^2\rho^2 = \frac{1}{2}I_x\omega^2$. Consequently, $\rho = \sqrt{I_x/M_{\mathcal{U}}}$.

51. The area of the disk is πr^2 and its centroid traverses a circle of radius R having circumference $2\pi R$. Therefore, the volume of the solid is

$$V = \pi r^2 \cdot 2\pi R = 2\pi^2 r^2 R \,.$$

Calculator/Computer Exercises

53. The following plot displays the region.

```
> y1,y2 := exp(x), 1+x+x^2:
  plot( [y1,y2], x = 0..2.5, 0..7 );
```

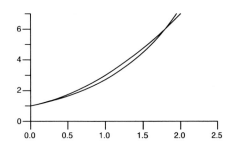

Using this picture as a guide, the fsolve procedure is used to obtain the
x-coordinate of the intersection point near $x = 2$. It is named b. This is
followed by the definition of the density function and the calculation of
the mass of the region, M.

```
> b := fsolve( y1 = y2, x=2);
  delta := 1+x:  M := int(int( delta, y=y1..y2), x=0..b);
```

$$b := 1.793282133$$

$$M := 0.6631249413$$

The calculation of the first moments and the center of mass follows.

```
> Mx0 := int(int( x*delta, y=y1..y2), x=0..b):
  My0 := int(int( y*delta, y=y1..y2), x=0..b):
  cm   := [Mx0/M,My0/M];
```

$$cm := [1.168646533, 3.522136312]$$

55. The density function is defined, followed by the calculation of the mass,
    the first moments, and the center of mass.

```
> delta := 2 + sin(x^2) + exp(-y^4):
  M    := evalf(Int(Int(   delta, y=0..1), x=0..1)):
  Mx0  := evalf(Int(Int( x*delta, y=0..1), x=0..1)):
  My0  := evalf(Int(Int( y*delta, y=0..1), x=0..1)):
  cm   := [Mx0/M,My0/M];
```

$$cm := [0.5236805593, 0.4844673310]$$

12.8   Other Coordinate Systems

Problems for Practice

1. The distance from the origin to the projection of P to the xy-plane is $\sqrt{12+4} = 4$, so $r = 4$. The angle in the xy-plane is $\theta = \arctan(1/\sqrt{3}) = \pi/6$. Therefore, the cylindrical coordinates are $(4, \pi/6, 5)$.

3. The distance from the origin to the projection of P to the xy-plane is $\sqrt{4+4} = 2\sqrt{2}$, so $r = 2\sqrt{2}$. The angle in the xy-plane is $\theta = \arctan(-1) + \pi = 3\pi/4$. Therefore, the cylindrical coordinates are $(2\sqrt{2}, 3\pi/4, -1)$.

5. The distance from the origin to the projection of P to the xy-plane is 3, so $r = 3$. The angle in the xy-plane is $\theta = 3\pi/2$. Therefore, the cylindrical coordinates are $(3, 3\pi/2, -2)$.

7. The distance from the origin to the projection of P to the xy-plane is $\sqrt{1+3} = 2$, so $r = 2$. The angle in the xy-plane is $\theta = \arctan(-\sqrt{3}) + 2\pi = 5\pi/3$. Therefore, the cylindrical coordinates are $(2, 5\pi/3, -2)$.

9. The distance from P to the origin is $\sqrt{1+1+2} = 2$, so $\rho = 2$. The angle from the positive z-axis is $\phi = \arccos(\sqrt{2}/2) = \pi/4$. Since $\theta = \pi/4$ also, the spherical coordinates are $(2, \pi/4, \pi/4)$.

11. The distance from P to the origin is $4\sqrt{2}$, so $\rho = 4\sqrt{2}$. The angle from the positive z-axis is $\phi = \arccos(4/(4\sqrt{2})) = \arccos(\sqrt{2}/2) = \pi/4$. Since $\theta = 0$, the spherical coordinates are $(4\sqrt{2}, \pi/4, 0)$.

13. The distance from P to the origin is $\sqrt{1+1+6} = 2\sqrt{2}$, so $\rho = 2\sqrt{2}$. The angle from the positive z-axis is $\phi = \arccos(\sqrt{6}/(2\sqrt{2})) = \arccos(\sqrt{3}/2) = \pi/6$. Since $\theta = \pi/4$, the spherical coordinates are $(2\sqrt{2}, \pi/6, \pi/4)$.

15. The distance from P to the origin is $4\sqrt{1+1+6} = 8\sqrt{2}$, so $\rho = 8\sqrt{2}$. The angle from the positive z-axis is $\phi = \arccos(-4\sqrt{6}/(8\sqrt{2})) = \arccos(-\sqrt{3}/2) = 5\pi/6$. Since $\theta = 3\pi/4$, the spherical coordinates are $(8\sqrt{2}, 5\pi/6, 3\pi/4)$.

17. Since $\rho = 3$ and $\phi = \pi/3$, $r = 3\sin(\pi/3) = 3\sqrt{3}/2$ and $z = 3\cos(\pi/3) = 3/2$. Therefore, the cylindrical coordinates of P are $(3\sqrt{3}/2, 2\pi/3, 3/2)$.

    The rectangular coordinates x and y are $x = r\cos(\theta) = (3\sqrt{3}/2)\cos(2\pi/3) = -3\sqrt{3}/4$ and $y = r\sin(\theta) = (3\sqrt{3}/2)\sin(2\pi/3) = (3\sqrt{3}/2)(\sqrt{3}/2) = 9/4$. The rectangular coordinates of P are $(-3\sqrt{3}/4, 9/4, 3/2)$.

19. The distance from the origin to the projection of P to the xy-plane is $\sqrt{4+4} = 2\sqrt{2}$, so $r = 2\sqrt{2}$. The angle in the xy-plane is $\theta = 7\pi/4$. Therefore, the cylindrical coordinates are $(2\sqrt{2}, 7\pi/4, -2\sqrt{2})$.

    The distance from P to the origin is $2\sqrt{1+1+2} = 4$, so $\rho = 4$. The angle from the positive z-axis is $\phi = \arccos(-2\sqrt{2}/4) = \arccos(-\sqrt{2}/2) = 3\pi/4$. Since $\theta = 7\pi/4$, the spherical coordinates are $(4, 3\pi/4, 7\pi/4)$.

21. Since $r = 2$ and $\theta = -3\pi/4$, $x = 2\cos(-3\pi/4) = -2/\sqrt{2} = -\sqrt{2}$ and $y = 2\sin(-3\pi/4) = -2/\sqrt{2} = -\sqrt{2}$. Therefore, the rectangular coordinates of P are $(-\sqrt{2}, -\sqrt{2}, -2)$.

The distance from P to the origin is $\sqrt{2+2+4} = 2\sqrt{2}$, so $\rho = 2\sqrt{2}$. The angle from the positive z-axis is $\phi = \arccos(-2/(2\sqrt{2})) = \arccos(-1/\sqrt{2}) = 3\pi/4$. Using the convention that $0 \le \theta < 2\pi$, the angle θ is $5\pi/4$ so the spherical coordinates of P are $(2\sqrt{2}, 3\pi/4, 5\pi/4)$.

23. The distance from the origin to the projection of P to the xy-plane is 3, so $r = 3$. The angle in the xy-plane is $\theta = 0$. Therefore, the cylindrical coordinates are $(3, 0, -3)$. Note that these are the same as the rectangular coordinates.

    The distance from P to the origin is $3\sqrt{2}$, so $\rho = 3\sqrt{2}$. The angle from the positive z-axis is $\phi = \arccos(-3/(3\sqrt{2})) = \arccos(-1/\sqrt{2}) = 3\pi/4$. Since $\theta = 0$, the spherical coordinates are $(3\sqrt{2}, 3\pi/4, 0)$.

25. The paraboloid and the plane intersect in the circle $x^2 + y^2 = 4$. The description of the solid in cylindrical coordinates is

$$\mathcal{U} = \left\{ (r, \theta, z) : 0 \le \theta \le 2\pi, 0 \le r \le 2, -4 \le z \le -r^2 \right\}.$$

Since $f(r\cos(\theta), r\sin(\theta), z) = r^2$, the integral evaluates as follows.

$$\int_0^{2\pi} \int_0^2 \int_{-4}^{-r^2} r^2 \cdot r \, dz \, dr \, d\theta = \int_0^{2\pi} \int_0^2 r^3(4 - r^2) \, dr \, d\theta$$
$$= \int_0^{2\pi} 16/3 \, d\theta = 32\pi/3$$

27. In cylindrical coordinates, the solid is

$$\mathcal{U} = \left\{ (r, \theta, z) : 0 \le \theta \le 2\pi, 0 \le r \le 2, -3 \le z \le r^2 + 1 \right\}.$$

Since $f(r\cos(\theta), r\sin(\theta), z) = r$, the integral evaluates as follows.

$$\int_0^{2\pi} \int_0^2 \int_{-3}^{r^2+1} r \cdot r \, dz \, dr \, d\theta = \int_0^{2\pi} \int_0^2 r^2(r^2 + 4) \, dr \, d\theta$$
$$= \int_0^{2\pi} 256/15 \, d\theta = 512\pi/15$$

29. The description of the solid in cylindrical coordinates is

$$\mathcal{U} = \left\{ (r, \theta, z) : 0 \le \theta \le 2\pi, \sqrt{2} \le r \le 2, -2 \le z \le 8 \right\}.$$

Since $f(r\cos(\theta), r\sin(\theta), z) = 1/r^2$, the integral evaluates as follows.

$$\int_0^{2\pi} \int_{\sqrt{2}}^2 \int_{-2}^8 (1/r^2) \cdot r \, dz \, dr \, d\theta = \int_0^{2\pi} \int_{\sqrt{2}}^2 (10/r) \, dr \, d\theta$$
$$= \int_0^{2\pi} 5\ln(2) \, d\theta = 10\pi \ln(2)$$

31. The description of the solid in spherical coordinates is

$$\mathcal{U} = \left\{ (\rho, \phi, \theta) : 0 \le \rho \le 3, 0 \le \phi \le \pi, 0 \le \theta \le 2\pi \right\}.$$

Therefore, the integral evaluates as follows.

$$\int_0^3 \int_0^\pi \int_0^{2\pi} 5 \cdot \rho^2 \sin(\phi) \, d\theta \, d\phi \, d\rho = 10\pi \int_0^3 \int_0^\pi \rho^2 \sin(\phi) \, d\phi \, d\rho$$

$$= 20\pi \int_0^3 \rho^2 \, d\rho = 180\pi$$

33. The description of the solid in spherical coordinates is

$$\mathcal{U} = \left\{ (\rho, \phi, \theta) : 2 \le \rho \le 3, 0 \le \phi \le \pi, 0 \le \theta \le 2\pi \right\}.$$

Since $f(\rho \sin(\phi) \cos(\theta), \rho \sin(\phi) \sin(\theta), \rho \cos(\phi)) = \rho$, the integral evaluates as follows.

$$\int_2^3 \int_0^\pi \int_0^{2\pi} \rho \cdot \rho^2 \sin(\phi) \, d\theta \, d\phi \, d\rho = 2\pi \int_2^3 \int_0^\pi \rho^3 \sin(\phi) \, d\phi \, d\rho$$

$$= 4\pi \int_2^3 \rho^3 \, d\rho = 65\pi$$

35. The cone and the sphere intersect at the angle $\phi = \pi/4$. Therefore, the description of the solid in spherical coordinates is

$$\mathcal{U} = \left\{ (\rho, \phi, \theta) : 0 \le \rho \le 2, \pi/4 \le \phi \le \pi, 0 \le \theta \le 2\pi \right\}.$$

Since $f(\rho \sin(\phi) \cos(\theta), \rho \sin(\phi) \sin(\theta), \rho \cos(\phi)) = \rho \cos(\phi)$, the integral evaluates as follows.

$$\int_0^2 \int_{\pi/4}^\pi \int_0^{2\pi} \rho \cos(\phi) \cdot \rho^2 \sin(\phi) \, d\theta \, d\phi \, d\rho = 2\pi \int_0^2 \int_{\pi/4}^\pi \rho^3 \sin(\phi) \cos(\phi) \, d\phi \, d\rho$$

$$= -\pi/2 \int_0^2 \rho^3 \, d\rho = -2\pi$$

37. The description of the solid in spherical coordinates is

$$\mathcal{U} = \left\{ (\rho, \phi, \theta) : \sqrt{2} \le \rho \le \sqrt{3}, 0 \le \phi \le \pi, 0 \le \theta \le 2\pi \right\}.$$

Since $f(\rho \sin(\phi) \cos(\theta), \rho \sin(\phi) \sin(\theta), \rho \cos(\phi)) = 1/\rho$, the integral evaluates as follows.

$$\int_{\sqrt{2}}^{\sqrt{3}} \int_0^\pi \int_0^{2\pi} (1/\rho) \cdot \rho^2 \sin(\phi) \, d\theta \, d\phi \, d\rho = 2\pi \int_{\sqrt{2}}^{\sqrt{3}} \int_0^\pi \rho \sin(\phi) \, d\phi \, d\rho$$

$$= 4\pi \int_{\sqrt{2}}^{\sqrt{3}} \rho \, d\rho = 2\pi$$

Further Theory and Practice

39. The description of the solid region in cylindrical coordinates is

$$\mathcal{U} = \left\{ (r, \theta, z) : 0 \le \theta \le 2\pi, 0 \le r \le 2, -r \le z \le r \right\}.$$

Taking advantage of the symmetry of the region, the volume can be calculated as follows.

$$V = \iiint_{\mathcal{U}} 1\, dV = 2 \int_0^{2\pi} \int_0^2 \int_0^r r\, dz\, dr\, d\theta$$

$$= 2 \int_0^{2\pi} \int_0^2 r^2\, dr\, d\theta$$

$$= 16/3 \int_0^{2\pi} d\theta = 32\pi/3$$

41. Since $\rho = 3$ and $\phi = 2$, $r = 3\sin(2)$ and $z = 3\cos(2)$. Therefore, the cylindrical coordinates of P are $(3\sin(2), 5, 3\cos(2))$.

The rectangular coordinates x and y are $x = r\cos(\theta) = 3\sin(2)\cos(5)$ and $y = r\sin(\theta) = 3\sin(2)\sin(5)$. The rectangular coordinates of P are $(3\sin(2)\cos(5), 3\sin(2)\sin(5), 3\cos(2))$.

43. Since the z-coordinates are the same in both systems, these points are the ones whose projection to the xy-plane have the same polar coordinates. Therefore, the projection must be on the x-axis. Thus the points are the ones in the xz-plane.

45. The points satisfy the equation $\rho^2 = \rho\cos(\phi)$ or $x^2 + y^2 + z^2 = z$. In the xz-plane this is $x^2 + z^2 = z$ and the trace of the surface in that plane is the circle passing through the origin having the line segment from $(0,0,0)$ to $(0,0,1)$ as a diameter. The surface is obtained by rotating this circle about the z-axis.

See the picture on the right.

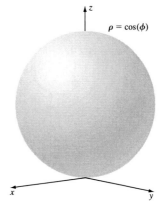

47. The points satisfy the equation

$$z = 2 + x^2 + y^2 \,.$$

In the xz-plane this is $z = 2 + x^2$ and the trace of the surface in that plane is a parabola opening upward with vertex at $(0, 0, 2)$. The surface is obtained by rotating this parabola about the z-axis.

See the picture on the right.

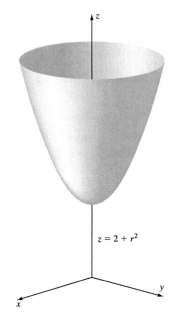

$z = 2 + r^2$

Calculator/Computer Exercises

49. Start with a picture of the region in the xy-plane.

```
> r1,r2 := sin(theta)^3,3-2*cos(theta):
  plot( [r1,r2], theta=0..Pi/2, coords=polar);
```

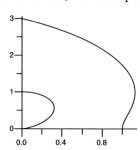

The next entry sets up and evaluates the integral (cylindrical coordinates).

```
> Int(Int(Int( r^(2/3)*r, z=0..r*sin(theta)),
                      r=r1..r2),
                      theta=0..Pi/2) =
    evalf( Int(Int(Int( r^(2/3)*r, z=0..r*sin(theta)),
                          r=r1..r2),
                          theta=0..Pi/2) );
```

$$\int_0^{\frac{1}{2}\pi} \int_{\sin(\theta)^3}^{3-2\cos(\theta)} \int_0^{r\sin(\theta)} r^{5/3} \, dz \, dr \, d\theta = 4.797453714$$

51. The spherical cap can be described in cylindrical coordinates as

$$\mathcal{U} = \{(r, \theta, z) : 0 \le \theta \le 2\pi, 0 \le r \le \sqrt{2a - a^2}, 1 - a \le z \le \sqrt{1 - r^2}\}.$$

Therefore, its volume is $V = \int_0^{2\pi} \int_0^{\sqrt{2a-a^2}} \int_{1-a}^{\sqrt{1-r^2}} r \, dz \, dr \, d\theta$. Since the volume of the unit ball is $\frac{4}{3}\pi$, the constant a must satisfy the equation

$$\int_0^{2\pi} \int_0^{\sqrt{2a-a^2}} \int_{1-a}^{\sqrt{1-r^2}} r \, dz \, dr \, d\theta = \frac{1}{3}\pi.$$

Under the assumption that $0 < a < 1$, *Maple* evaluates this to

$$\pi a^2 - \frac{1}{3}\pi a^3 = \frac{1}{3}\pi.$$

Using *fsolve*, the solutions to this equation are -0.5320888862, 0.6527036445, and 2.879385243, so $a \approx 0.6527036445$. It turns out that the *solve* procedure can generate the exact solutions and $a = 1 + \sqrt{3}\sin(\pi/9) - \cos(\pi/9)$.

Chapter 13

Vector Calculus

13.1 Vector Fields

Problems for Practice

1. This is a constant vector field. All vectors are parallel and have the same length and direction.

   See the picture on the right.

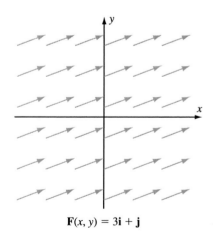

$$\mathbf{F}(x, y) = 3\mathbf{i} + \mathbf{j}$$

3. The vectors in this field point downward. The streamlines appear to follow parabolic paths.

   See the picture on the right.

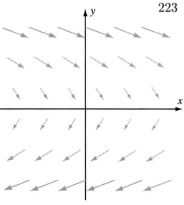

$$\mathbf{F}(x, y) = 2y\mathbf{i} - 3\mathbf{j}$$

5. The vectors in this field generate streamlines that spiral around the origin in what appear to be closed paths.

   See the picture on the right.

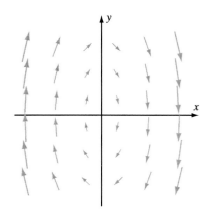

$$\mathbf{F}(x, y) = y\mathbf{i} - 5x\mathbf{j}$$

7. The vectors in this field point to the right and downward except for the ones on the coordinate axes.

   See the picture on the right.

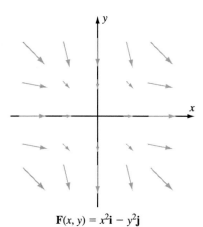

$$\mathbf{F}(x, y) = x^2\mathbf{i} - y^2\mathbf{j}$$

9. The direction of the field at any point $P = (x, y, z)$ is the $\mathbf{j}$ vector. Since the magnitude of the field is $\|\mathbf{F}(P)\| = \alpha\sqrt{x^2 + z^2}$, $\mathbf{F}(x, y, z) = \alpha\sqrt{x^2 + z^2}\,\mathbf{j}$. Its domain is $-4 < x < 4$, $0 < y < 10$.

11. The direction of the field at a point $P = (x, y, z)$ is $(1/\sqrt{x^2 + y^2})(y\,\mathbf{i} - x\,\mathbf{j})$. Since its magnitude is $\alpha\sqrt{x^2 + y^2}$ the field is

$$\mathbf{F}(P) = \|\mathbf{F}(P)\|\,\mathbf{dir}(\mathbf{F}(P)) = \alpha\sqrt{x^2 + y^2}\left(\frac{1}{\sqrt{x^2 + y^2}}(y\,\mathbf{i} - x\,\mathbf{j})\right)$$

$$= \alpha(y\,\mathbf{i} - x\,\mathbf{j}).$$

The domain of the field is $x^2 + y^2 < 16$, $0 < z < 5$.

13. $\nabla u(x, y) = \langle y + 3, x - 2y \rangle$

15. $\nabla u(x, y) = \langle 3x^2/(x^3 - y^2), -2y/(x^3 - y^2) \rangle$

17. $\nabla u(x, y, z) = \langle 2xy^3 z, 3x^2 y^2 z, x^2 y^3 \rangle$

19. $\nabla u(x, y) = ((x - y)\ln(z))^{-1}\,\mathbf{i} - ((x - y)\ln(z))^{-1}\,\mathbf{j} - \ln(x - y)(z\ln^2(z))^{-1}\,\mathbf{k}$

21. We seek $(x, y) \mapsto u(x, y)$ such that $\langle u_x, u_y \rangle = \langle 1, 1 \rangle$. In scalar form: $u_x = 1$ and $u_y = 1$.

$$u_x = 1 \implies u(x, y) = x + \phi(y)$$
$$u_y = 1 \implies \phi'(y) = 1$$

This implies that $\phi(y) = y + C$, and $u(x, y) = x + y + C$.

23. We seek $(x, y) \mapsto u(x, y)$ such that $\langle u_x, u_y \rangle = \langle y, x \rangle$. In scalar form: $u_x = y$ and $u_y = x$.

$$u_x = y \implies u(x, y) = xy + \phi(y)$$
$$u_y = x \implies \phi'(y) = 0$$

This implies that $\phi(y) = C$, and $u(x, y) = xy + C$.

25. We seek $(x, y) \mapsto u(x, y)$ such that $\langle u_x, u_y \rangle = \langle y^3 + 2xy, 3xy^2 + x^2 \rangle$. In scalar form: $u_x = y^3 + 2xy$ and $u_y = 3xy^2 + x^2$.

$$u_x = y^3 + 2xy \implies u(x, y) = xy^3 + x^2 y + \phi(y)$$
$$u_y = 3xy^2 + x^2 \implies 3xy^2 + x^2 + \phi'(y) = 3xy^2 + x^2$$

This implies that $\phi'(y) = 0$ so we may take $\phi(y) = C$, a constant, and $u(x, y) = xy^3 + x^2 y + C$.

27. We seek $(x, y, z) \mapsto u(x, y, z)$ such that

$$\langle u_x, u_y, u_z \rangle = \langle 2x, 1, 0 \rangle.$$

In scalar form: $u_x = 2x$, $u_y = 1$, and $u_z = 0$.

$$u_x = 2x \implies u(x, y, z) = x^2 + \phi(y, z)$$
$$u_y = 1 \implies \phi_y(y, z) = 1$$

This implies that $\phi(y, z) = y + \psi(z)$, and $u(x, y, z) = x^2 + y + \psi(z)$. The third scalar equation implies that $\psi'(z) = 0$. Therefore $\psi(z) = C$, a constant. Consequently, $u(x, y, z) = x^2 + y + C$.

29. We seek $(x, y, z) \mapsto u(x, y, z)$ such that

$$\langle u_x, u_y, u_z \rangle = \langle 2xyz^3, x^2z^3 + 5, 3x^2yz^2 \rangle.$$

In scalar form: $u_x = 2xyz^3$, $u_y = x^2z^3 + 5$, and $u_z = 3x^2yz^2$.

$$u_x = 2xyz^3 \implies u(x, y, z) = x^2yz^3 + \phi(y, z)$$
$$u_y = x^2z^3 + 5 \implies x^2z^3 + \phi_y(y, z) = x^2z^3 + 5$$

This implies that $\phi_y(y, z) = 5$ so $\phi(y, z) = 5y + \psi(z)$, and $u(x, y, z) = x^2yz^3 + 5y + \psi(z)$. The third scalar equation implies that $3x^2yz^2 + \psi'(z) = 3x^2yz^2$. Therefore $\psi'(z) = 0$ and $\psi(z) = C$, a constant. Consequently, $u(x, y, z) = x^2yz^3 + 5y + C$.

31. We seek $(x, y, z) \mapsto u(x, y, z)$ such that

$$\langle u_x, u_y, u_z \rangle = \langle z^3 - 2xy^2, z^2 - 2yx^2, 3xz^2 + 2yz \rangle.$$

In scalar form: $u_x = z^3 - 2xy^2$, $u_y = z^2 - 2yx^2$, and $u_z = 3xz^2 + 2yz$.

$$u_x = z^3 - 2xy^2 \implies u(x, y, z) = xz^3 - x^2y^2 + \phi(y, z)$$
$$u_y = z^2 - 2yx^2 \implies -2x^2y + \phi_y(y, z) = z^2 - 2yx^2$$

This implies that $\phi_y(y, z) = z^2$ so $\phi(y, z) = yz^2 + \psi(z)$, and $u(x, y, z) = xz^3 - x^2y^2 + yz^2 + \psi(z)$. The third scalar equation implies that $3xz^2 + 2yz + \psi'(z) = 3xz^2 + 2yz$. Therefore $\psi'(z) = 0$ and $\psi(z) = C$, a constant. Consequently, $u(x, y, z) = xz^3 - x^2y^2 + yz^2 + C$.

33. The component functions of $\mathbf{r}(t) = \langle x(t), y(t) \rangle$ must satisfy the equations $x'(t) = 1$ and $y'(t) = 1$. Therefore, $x(t) = t + C_1$ and $y(t) = t + C_2$, and $\mathbf{r}(t) = \langle t + C_1, t + C_2 \rangle$.

35. The component functions of $\mathbf{r}(t) = \langle x(t), y(t) \rangle$ must satisfy the equations $x'(t) = x(t)$ and $y'(t) = 2$. Therefore, $x(t) = C_1e^t$ and $y(t) = 2t + C_2$, and $\mathbf{r}(t) = \langle C_1e^t, 2t + C_2 \rangle$.

37. The component functions of $\mathbf{r}(t) = \langle x(t), y(t) \rangle$ must satisfy the equations $x'(t) = 2y(t)$ and $y'(t) = 1$. Therefore, $y(t) = t + C_1$ and, substituting this into the differential equation for $x(t)$: $x'(t) = 2(t + C_1)$. Therefore, $x(t) = t^2 + 2C_1t + C_2$, and $\mathbf{r}(t) = \langle t^2 + 2C_1t + C_2, t + C_1 \rangle$.

Further Theory and Practice

39. We seek $(x, y) \mapsto u(x, y)$ such that $\langle u_x, u_y \rangle = \langle e^{2x} + 2xe^{2x}, 1 \rangle$. In scalar form: $u_x = e^{2x} + 2xe^{2x}$ and $u_y = 1$.

$$u_x = e^{2x} + 2xe^{2x} \implies u(x, y) = xe^{2x} + \phi(y)$$
$$u_y = 1 \implies \phi'(y) = 1$$

This implies that $\phi(y) = y + C$, and $u(x,y) = xe^{2x} + y + C$.

41. We seek $(x,y) \mapsto u(x,y)$ with $\langle u_x, u_y \rangle = \langle y^2 \cos(xy), \sin(xy) + xy\cos(xy) \rangle$.
    In scalar form: $u_x = y^2 \cos(xy)$ and $u_y = \sin(xy) + xy\cos(xy)$.

$$u_x = y^2 \cos(xy) \implies u(x,y) = y\sin(xy) + \phi(y)$$
$$u_y = 1 \implies xy\cos(xy) + \sin(xy) + \phi'(y) = \sin(xy) + xy\cos(xy)$$

This implies that $\phi'(y) = 0$. Therefore, $\phi(y) = C$, a constant, and $u(x,y) = y\sin(xy) + C$.

43. We seek $(x,y,z) \mapsto u(x,y,z)$ such that

$$\langle u_x, u_y, u_z \rangle = \langle z(y+z)^{-1}, -xz(y+z)^{-2}, xy(y+z)^{-2} \rangle.$$

In scalar form: $u_x = z(y+z)^{-1}$, $u_y = -xz(y+z)^{-2}$, and $u_z = xy(y+z)^{-2}$.

$$u_x = z(y+z)^{-1} \implies u(x,y,z) = xz(y+z)^{-1} + \phi(y,z)$$
$$u_y = -xz(y+z)^{-2} \implies -xz(y+z)^{-2} + \phi_y(y,z) = -xz(y+z)^{-2}$$

This implies that $\phi_y(y,z) = 0$ so $\phi(y,z) = \psi(z)$, and

$$u(x,y,z) = xz(y+z)^{-1} + \psi(z).$$

The third scalar equation implies that $-xz(y+z)^{-2} + x(y+z)^{-1} + \psi'(z) = xy(y+z)^{-2}$. Therefore

$$\psi'(z) = xz(y+z)^{-2} + xy(y+z)^{-2} - x(y+z)^{-1}$$
$$= x(y+z)^{-2}(z + y - (y+z)) = 0$$

and $\psi(z) = C$, a constant. Consequently, $u(x,y,z) = xz(y+z)^{-1} + C$.

45. We seek $(x,y,z) \mapsto u(x,y,z)$ with $\langle u_x, u_y, u_z \rangle = \langle x^{y-1}y, x^y\ln(x), \ln(z) \rangle$.
    In scalar form: $u_x = x^{y-1}y$, $u_y = x^y\ln(x)$, and $u_z = \ln(z)$.

$$u_x = x^{y-1}y \implies u(x,y,z) = x^y + \phi(y,z)$$
$$u_y = x^y\ln(x) \implies x^y\ln(x) + \phi_y(y,z) = x^y\ln(x)$$

This implies that $\phi_y(y,z) = 0$ so $\phi(y,z) = \psi(z)$, and $u(x,y,z) = x^y + \psi(z)$. The third scalar equation implies that $\psi'(z) = \ln(z)$. Therefore $\psi(z) = z\ln(z) - z + C$, and $u(x,y,z) = x^y + z\ln(z) - z + C$.

47. If there were a continuously differentiable function $(x,y) \mapsto u(x,y)$ such that $\nabla u(x,y) = \langle y, 2x \rangle$, then $u_x(x,y) = y$ and $u_y(x,y) = 2x$. From this it would follow that $u_{xy}(x,y) = 1$ and $u_{yx}(x,y) = 2$, contradicting the fact that $u_{xy}(x,y) = u_{yx}(x,y)$.

49. Since $r(x, y, z) = \sqrt{x^2 + y^2 + z^2}$,

$$\nabla(1/r)(x, y, z) = \frac{\partial}{\partial x}(x^2 + y^2 + z^2)^{-1/2}\,\mathbf{i}$$
$$+ \frac{\partial}{\partial y}(x^2 + y^2 + z^2)^{-1/2}\,\mathbf{j}$$
$$+ \frac{\partial}{\partial z}(x^2 + y^2 + z^2)^{-1/2}\,\mathbf{k}$$
$$= -x(x^2 + y^2 + z^2)^{-3/2}\,\mathbf{i}$$
$$- y(x^2 + y^2 + z^2)^{-3/2}\,\mathbf{j}$$
$$- z(x^2 + y^2 + z^2)^{-3/2}\,\mathbf{k}$$
$$= -r^{-3}(x\,\mathbf{i} + y\,\mathbf{j} + z\,\mathbf{k}).$$

51. The vectors in this field are the negatives of the gradient vectors for the function h:

$$\mathbf{F}(x, y) = -\nabla h(x, y)$$
$$= (2x - 3y)\,\mathbf{i} + (-3x - 4y)\,\mathbf{j}.$$

See the picture on the right.

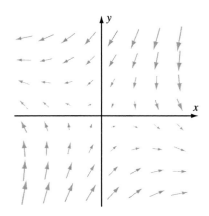

53. The component functions of $\mathbf{r}(t) = \langle x(t), y(t) \rangle$ must satisfy the equations $x'(t) = 3x(t) + y(t)$ and $y'(t) = 2y(t)$. Differentiate the first equation: $x''(t) = 3x'(t) + y'(t)$, and use the second equation to substitute for $y'(t)$:

$$x''(t) = 3x'(t) + 2y(t).$$

Now use the first differential equation to eliminate $y(t)$ yielding

$$x''(t) = 3x'(t) + 2(x'(t) - 3x(t)).$$

Therefore, the function $t \mapsto x(t)$ must be a solution for the differential equation

$$x''(t) - 5x'(t) + 6x(t) = 0.$$

Using the solution formula ($r_1 = 2$, $r_2 = 3$), $x(t) = Ae^{2t} + Be^{3t}$. To find $y(t)$ substitute this into the first differential equation:

$$y(t) = x'(t) - 3x(t)$$
$$= 2Ae^{2t} + 3Be^{3t} - 3(Ae^{2t} + Be^{3t})$$
$$= -Ae^{2t}.$$

55. The component functions of $\mathbf{r}(t) = \langle x(t), y(t) \rangle$ must satisfy the equations $x'(t) = 3x(t) - 4y(t)$ and $y'(t) = -x(t)$. Differentiate the first equation: $x''(t) = 3x'(t) - 4y'(t)$, and use the second equation to substitute for $y'(t)$:

$$x''(t) = 3x'(t) - 4(-x(t)).$$

Therefore, the function $t \mapsto x(t)$ must be a solution for the differential equation

$$x''(t) - 3x'(t) - 4x(t) = 0.$$

Using the solution formula ($r_1 = 4$, $r_2 = -1$), $x(t) = Ae^{4t} + Be^{-t}$. To find $y(t)$ substitute this into the first differential equation:

$$
\begin{aligned}
y(t) &= \frac{1}{4}\left(3x(t) - x'(t)\right) \\
&= \frac{3}{4}\left(Ae^{4t} + Be^{-t}\right) - \frac{1}{4}\left(4Ae^{4t} - Be^{-t}\right) \\
&= -\frac{1}{4}Ae^{4t} + Be^{-t}.
\end{aligned}
$$

57. A suitable vector field $\mathbf{F}(x, y)$ can be defined in terms of the position vector $\mathbf{r}(x, y) = x\mathbf{i} + y\mathbf{j}$. Observe that $\mathbf{F}(x, y) = -y\mathbf{i} + x\mathbf{j}$ is a vector field that is everywhere perpendicular to $\mathbf{r}(x, y)$ which implies that it is tangent to $\mathcal{C}$. Moreover, since $\mathcal{C}$ is the unit circle, $\mathbf{F}(x, y)$ will be a unit vector on $\mathcal{C}$.

The vector field $\mathbf{F}(x, y) = y\mathbf{i} - x\mathbf{j}$ works just as well.

13.2   Line Integrals

Problems for Practice

1. Observe that $\mathbf{F}(\mathbf{r}(t)) = \langle \cos(t)\sin(t), \sin^2(t) \rangle$ and $\mathbf{r}'(t) = \langle -\sin(t), \cos(t) \rangle$. Therefore, $\mathbf{F}(\mathbf{r}(t)) \cdot \mathbf{r}'(t) = -\cos(t)\sin^2(t) + \sin^2(t)\cos(t) = 0$, and

$$
\begin{aligned}
\int_{\mathcal{C}} \mathbf{F} \cdot d\mathbf{r} &= \int_0^{\pi/3} \mathbf{F}(\mathbf{r}(t)) \cdot \mathbf{r}'(t)\, dt \\
&= \int_0^{\pi/3} 0\, dt = 0.
\end{aligned}
$$

3. Observe that $\mathbf{F}(\mathbf{r}(t)) = \langle \sin(t^2), -\cos(t^3) \rangle$ and $\mathbf{r}'(t) = \langle 2t, 3t^2 \rangle$. There-

fore, $\mathbf{F}(\mathbf{r}(t)) \cdot \mathbf{r}'(t) = 2t\sin(t^2) - 3t^2\cos(t^3)$, and

$$\int_C \mathbf{F} \cdot d\mathbf{r} = \int_0^{\sqrt{\pi}} \mathbf{F}(\mathbf{r}(t)) \cdot \mathbf{r}'(t)\, dt$$

$$= \int_0^{\sqrt{\pi}} (2t\sin(t^2) - 3t^2\cos(t^3))\, dt$$

$$= (-\cos(t^2) - \sin(t^3))\Big|_0^{\sqrt{\pi}} = 2 - \sin(\pi^{3/2})\,.$$

5. Observe that $\mathbf{F}(\mathbf{r}(t)) = \langle t^{-1}, t^{-2} \rangle$ and $\mathbf{r}'(t) = \langle (1/2)t^{-1/2}, (-3/2)t^{-5/2} \rangle$. Therefore, $\mathbf{F}(\mathbf{r}(t)) \cdot \mathbf{r}'(t) = (1/2)t^{-3/2} - (3/2)t^{-9/2}$, and

$$\int_C \mathbf{F} \cdot d\mathbf{r} = \int_1^4 \mathbf{F}(\mathbf{r}(t)) \cdot \mathbf{r}'(t)\, dt$$

$$= \int_1^4 ((1/2)t^{-3/2} - (3/2)t^{-9/2})\, dt$$

$$= (-t^{-1/2} + (3/7)x^{-7/2})\Big|_1^4 = 67/896\,.$$

7. Observe that $\mathbf{F}(\mathbf{r}(t)) = \langle (t^2+1)^{-1}, -t^3 \rangle$ and $\mathbf{r}'(t) = \langle 1, 2t \rangle$. Therefore, $\mathbf{F}(\mathbf{r}(t)) \cdot \mathbf{r}'(t) = (t^2+1)^{-1} - 2t^4$, and

$$\int_C \mathbf{F} \cdot d\mathbf{r} = \int_{-4}^{-1} \mathbf{F}(\mathbf{r}(t)) \cdot \mathbf{r}'(t)\, dt$$

$$= \int_{-4}^{-1} ((t^2+1)^{-1} - 2t^4)\, dt$$

$$= (\arctan(t) - (2/5)t^5)\Big|_{-4}^{-1} = \arctan(4) - 2046/5 - \pi/4\,.$$

9. Observe that $\mathbf{F}(\mathbf{r}(t)) = \langle t^4, -t^3, t^3 \rangle$ and $\mathbf{r}'(t) = \langle 1, 2t, 3t^2 \rangle$. Therefore, $\mathbf{F}(\mathbf{r}(t)) \cdot \mathbf{r}'(t) = t^4 - 2t^4 + 3t^5 = 3t^5 - t^4$, and

$$\int_C \mathbf{F} \cdot d\mathbf{r} = \int_{-1}^1 \mathbf{F}(\mathbf{r}(t)) \cdot \mathbf{r}'(t)\, dt$$

$$= \int_{-1}^1 (3t^5 - t^4)\, dt = -2/5\,.$$

11. Observe that $\mathbf{F}(\mathbf{r}(t)) = \langle \cos(t), \sin(t), \cos(t) \rangle$ and $\mathbf{r}'(t) = \langle -\sin(t), 1, 1 \rangle$.

Therefore, $\mathbf{F}(\mathbf{r}(t)) \cdot \mathbf{r}'(t) = -\cos(t)\sin(t) + \sin(t) + \cos(t)$, and

$$\int_{\mathcal{C}} \mathbf{F} \cdot d\mathbf{r} = \int_{\pi/4}^{\pi/2} \mathbf{F}(\mathbf{r}(t)) \cdot \mathbf{r}'(t)\, dt$$

$$= \int_{\pi/4}^{\pi/2} (-\cos(t)\sin(t) + \sin(t) + \cos(t))\, dt$$

$$= ((1/2)\cos^2(t) - \cos(t) + \sin(t))\Big|_{\pi/4}^{\pi/2} = 3/4\,.$$

13. Observe that $\mathbf{F}(\mathbf{r}(t)) = \langle \cos^2(t), 0, 1 \rangle$ and $\mathbf{r}'(t) = \langle 1, 0, -2\cos(t)\sin(t) \rangle$. Therefore, $\mathbf{F}(\mathbf{r}(t)) \cdot \mathbf{r}'(t) = \cos^2(t) - 2\cos(t)\sin(t)$, and

$$\int_{\mathcal{C}} \mathbf{F} \cdot d\mathbf{r} = \int_0^{\pi} \mathbf{F}(\mathbf{r}(t)) \cdot \mathbf{r}'(t)\, dt$$

$$= \int_0^{\pi} (\cos^2(t) - 2\cos(t)\sin(t))\, dt$$

$$= \int_0^{\pi} ((1/2)(1 + \cos(2t)) - 2\cos(t)\sin(t))\, dt$$

$$= ((1/2)(t + \sin(2t)/2) + \cos^2(t))\Big|_0^{\pi} = \pi/2\,.$$

15. Observe that $\mathbf{F}(\mathbf{r}(t)) = \langle 0, 0, \sin(t) \rangle$ and $\mathbf{r}'(t) = \langle \cos(t), -\sin(t), 1 \rangle$. Therefore, $\mathbf{F}(\mathbf{r}(t)) \cdot \mathbf{r}'(t) = \sin(t)$, and

$$\int_{\mathcal{C}} \mathbf{F} \cdot d\mathbf{r} = \int_0^{\pi} \mathbf{F}(\mathbf{r}(t)) \cdot \mathbf{r}'(t)\, dt$$

$$= \int_0^{\pi} \sin(t)\, dt = 2\,.$$

17. Observe that $\mathbf{F}(\mathbf{r}(t)) = \langle t^{3/4}, t^{3/2} \rangle$ and $\mathbf{r}'(t) = \langle (1/2)t^{-1/2}, (1/4)t^{-3/4} \rangle$. Therefore, $\mathbf{F}(\mathbf{r}(t)) \cdot \mathbf{r}'(t) = (1/2)t^{1/4} + (1/4)t^{3/4}$, and the work performed is

$$\int_1^{16} \mathbf{F}(\mathbf{r}(t)) \cdot \mathbf{r}'(t)\, dt = \int_1^{16} \left((1/2)t^{1/4} + (1/4)t^{3/4} \right) dt$$

$$= ((2/5)t^{5/4} + (1/7)t^{7/4})\Big|_1^{4} = 1069/35\,.$$

19. Since $\mathbf{F}(\mathbf{r}(t)) = \langle (1+t^2)^2, t^4, (2+\sin(\pi t))^2 \rangle$ and $\mathbf{r}'(t) = \langle 2t, 2t, \pi\cos(\pi t) \rangle$, $\mathbf{F}(\mathbf{r}(t)) \cdot \mathbf{r}'(t) = 2t(1+t^2)^2 + 2t^5 + \pi\cos(\pi t)(2+\sin(\pi t))^2$. Therefore, the work performed is

$$\int_0^1 \mathbf{F}(\mathbf{r}(t)) \cdot \mathbf{r}'(t)\, dt = \int_0^1 \left(2t(1+t^2)^2 + 2t^5 + \pi\cos(\pi t)(2+\sin(\pi t))^2 \right) dt$$

$$= ((1/3)(1+t^2)^3 + (1/2)t^6 + (1/3)(2+\sin(\pi t))^3)\Big|_0^1 = 8/3\,.$$

21. Using formula (15.20) with $x(t) = t^2$, $y(t) = t^3$, and $z(t) = t^4$,

$$\int_C \mathbf{F} \cdot d\mathbf{r} = \int_C xyz \, dx$$
$$= \int_0^1 \left((t^2 \cdot t^3 \cdot t^4) \frac{dx}{dt} \right) dt$$
$$= \int_0^1 t^9 (2t) \, dt = 2/11 \,.$$

23. Using formula (15.20) with $x(t) = t^2$, $y(t) = t^3$, and $z(t) = t^4$,

$$\int_C \mathbf{F} \cdot d\mathbf{r} = \int_C xz \, dx + yz \, dy + xy \, dz$$
$$= \int_0^1 \left((t^2 \cdot t^4) \frac{dx}{dt} + (t^3 \cdot t^4) \frac{dy}{dt} + (t^2 \cdot t^3) \frac{dz}{dt} \right) dt$$
$$= \int_0^1 \left(t^6 (2t) + t^7 (3t^2) + t^5 (4t^3) \right) dt$$
$$= \int_0^1 (2t^7 + 3t^9 + 4t^8) \, dt = 179/180 \,.$$

25. The curve consists of three line segments. Let C_1 be the segment from P_0 to P_1. It can be parametrized with $\mathbf{r}_1(t) = \overrightarrow{OP_0} + t(\overrightarrow{P_0 P_1}) = \langle 2t, 1 \rangle$, $0 \leq t \leq 1$. Thus $x(t) = 2t$ and $y(t) = 1$ so

$$\int_{C_1} x \, dy = \int_0^1 \left(2t \frac{dy}{dt} \right) dt = \int_0^1 (2t \cdot 0) \, dt = 0 \,.$$

The segment C_2 from P_1 to P_2 can be parametrized with $\mathbf{r}_2(t) = \overrightarrow{OP_1} + t(\overrightarrow{P_1 P_2}) = \langle 2+t, 1+3t \rangle$, $0 \leq t \leq 1$. Thus $x(t) = 2+t$, $y(t) = 1+3t$, and

$$\int_{C_2} x \, dy = \int_0^1 \left((2+t) \frac{dy}{dt} \right) dt = \int_0^1 (2+t) \cdot 3 \, dt = 15/2 \,.$$

The segment C_3 from P_2 to P_0 is parametrized with $\mathbf{r}_3(t) = \overrightarrow{OP_2} + t(\overrightarrow{P_2 P_0}) = \langle 3 - 3t, 4 - 3t \rangle$, $0 \leq t \leq 1$. Thus $x(t) = 3 - 3t$, $y(t) = 4 - 3t$, and

$$\int_{C_2} x \, dy = \int_0^1 \left((3 - 3t) \frac{dy}{dt} \right) dt = \int_0^1 (3 - 3t) \cdot (-3) \, dt = -9/2 \,.$$

The integral over the closed curve $C = C_1 + C_2 + C_3$ is

$$\oint_C x \, dy = \int_{C_1} x \, dy + \int_{C_2} x \, dy + \int_{C_3} x \, dy = 3 \,.$$

27. Using the parametrizations defined in Exercise 25:

$$\mathbf{r}_1(t) = \langle 2t, 1 \rangle, \; \mathbf{r}_2(t) = \langle 2 + t, 1 + 3t \rangle, \; \text{and} \; \mathbf{r}_3(t) = \langle 3 - 3t, 4 - 3t \rangle,$$

(each one for $0 \le t \le 1$), the integrals over the three line segments evaluate as follows:

$$\int_{C_1} (2x + y)\,dx + y\,dy = \int_0^1 ((4t + 1) \cdot 2 + 1 \cdot 0)\,dt = 6$$

$$\int_{C_2} (2x + y)\,dx + y\,dy = \int_0^1 ((5 + 5t) \cdot 1 + (1 + 3t) \cdot 3)\,dt = 15$$

$$\int_{C_3} (2x + y)\,dx + y\,dy = \int_0^1 ((10 - 9t) \cdot (-3) + (4 - 3t) \cdot (-3))\,dt = -24$$

Therefore, $\oint_C (2x + y)\,dx + y\,dy = 6 + 15 - 24 = -3$.

29. Since $\|\mathbf{r}'(t)\| = \|\langle \cos(t), -\sin(t) \rangle\| = \sqrt{\cos^2(t) + \sin^2(t)} = 1$, this is the arc length parametrization and

$$\int_C f\,ds = \int_0^{2\pi} (\sin^2(t) + \cos^2(t))\,dt = 2\pi.$$

31. Since $\|\mathbf{r}'(t)\| = \|\langle 2t, 4t^3 \rangle\| = 2t\sqrt{1 + 4t^4}$, this is not the arc length parametrization. Noting that $f(\mathbf{r}(t)) = 2t^2$,

$$\int_C f\,ds = \int_1^2 f(\mathbf{r}(t))\,\|\mathbf{r}'(t)\|\,dt$$

$$= \int_1^2 2t^2 \cdot 2t\sqrt{1 + 4t^4}\,dt$$

$$= (1/6)(1 + 4t^4)^{3/2}\Big|_1^2 = (65\sqrt{65} - 5\sqrt{5})/6.$$

Problems for Practice

33. The path C_1

$\mathbf{F}(\mathbf{r}_1(t)) = \langle 3, -2 \rangle$ and $\mathbf{r}_1'(t) = \langle -2\sin(2t), -2/\pi \rangle$. Therefore, $\mathbf{F}(\mathbf{r}_1(t)) \cdot \mathbf{r}_1'(t) = -6\sin(2t) + 4/\pi$, and

$$\int_{C_1} \mathbf{F} \cdot d\mathbf{r} = \int_0^\pi (-6\sin(2t) + 4/\pi)\,dt$$

$$= (3\cos(2t) + 4t/\pi)\Big|_0^\pi = 4.$$

The path C_2

$\mathbf{F}(\mathbf{r}_2(t)) = \langle 3, -2 \rangle$ and $\mathbf{r}_2'(t) = \langle 2\pi \cos(2\pi t), -\pi \sin(\pi t) \rangle$. Therefore, $\mathbf{F}(\mathbf{r}_2(t)) \cdot$
$\mathbf{r}_2'(t) = 6\pi \cos(2\pi t) + 2\pi \sin(\pi t)$, and

$$\int_{\mathcal{C}_2} \mathbf{F} \cdot d\mathbf{r} = \int_0^1 (6\pi \cos(2\pi t) + 2\pi \sin(\pi t))\, dt$$
$$= (3\sin(2\pi t) - 2\cos(\pi t))\Big|_0^1 = 4\,.$$

The path $\mathcal{C}_3$

$\mathbf{F}(\mathbf{r}_3(t)) = \langle 3, -2 \rangle$ and $\mathbf{r}_3'(t) = \langle 0, -2 \rangle$. Therefore, $\mathbf{F}(\mathbf{r}_3(t)) \cdot \mathbf{r}_2'(t) = 4$,
and

$$\int_{\mathcal{C}_3} \mathbf{F} \cdot d\mathbf{r} = \int_0^1 4\, dt = 4\,.$$

35. The path $\mathcal{C}_1$

$\mathbf{F}(\mathbf{r}_1(t)) = \langle \cos(2t), 1 - 2t/\pi \rangle$ and $\mathbf{r}_1'(t) = \langle -2\sin(2t), -2/\pi \rangle$. Therefore,
$\mathbf{F}(\mathbf{r}_1(t)) \cdot \mathbf{r}_1'(t) = -2\cos(2t)\sin(2t) - 2/\pi + 4t/\pi^2$, and

$$\int_{\mathcal{C}_1} \mathbf{F} \cdot d\mathbf{r} = \int_0^\pi (-2\cos(2t)\sin(2t) - 2/\pi + 4t/\pi^2)\, dt$$
$$= (\cos(2t) - 2t/\pi + 2t^2/\pi^2)\Big|_0^\pi = 0\,.$$

The path $\mathcal{C}_2$

$\mathbf{F}(\mathbf{r}_2(t)) = \langle 1 + \sin(2\pi t), \cos(\pi t) \rangle$ and $\mathbf{r}_2'(t) = \langle 2\pi \cos(2\pi t), -\pi \sin(\pi t) \rangle$.
Therefore, $\mathbf{F}(\mathbf{r}_2(t)) \cdot \mathbf{r}_2'(t) = 2\pi(1 + \sin(2\pi t))\cos(2\pi t) - \pi \cos(\pi t)\sin(\pi t)$,
and

$$\int_{\mathcal{C}_2} \mathbf{F} \cdot d\mathbf{r} = \int_0^1 2\pi(1 + \sin(2\pi t))\cos(2\pi t) - \pi \cos(\pi t)\sin(\pi t))\, dt$$
$$= \left(\sin(2\pi t) + \sin^2(2\pi t) - \sin^2(\pi t)\right)\Big|_0^1 = 0\,.$$

The path $\mathcal{C}_3$

$\mathbf{F}(\mathbf{r}_3(t)) = \langle 1, 1 - 2t \rangle$ and $\mathbf{r}_3'(t) = \langle 0, -2 \rangle$. Therefore, $\mathbf{F}(\mathbf{r}_3(t)) \cdot \mathbf{r}_3'(t) = -2 + 4t$, and

$$\int_{\mathcal{C}_3} \mathbf{F} \cdot d\mathbf{r} = \int_0^1 (-2 + 4t)\, dt = (-2t + 2t^2)\Big|_0^1 = 0\,.$$

37. The path $\mathcal{C}_1$

$\mathbf{F}(\mathbf{r}_1(t)) = \langle 2(1 - 2t/\pi), \cos(2t)) \rangle$ and $\mathbf{r}_1'(t) = \langle -2\sin(2t), -2/\pi \rangle$. Therefore, $\mathbf{F}(\mathbf{r}_1(t)) \cdot \mathbf{r}_1'(t) = -2(2 - 4t/\pi)\sin(2t) - 2\cos(2t)/\pi$, and

$$\int_{\mathcal{C}_1} \mathbf{F} \cdot d\mathbf{r} = \int_0^\pi (-2(2 - 4t/\pi)\sin(2t) - 2\cos(2t)/\pi)\, dt$$

$$= (2\cos(2t) + 2(\sin(2t) - t\cos(2t)/\pi - \sin(2t)/\pi)\Big|_0^\pi = -4\,.$$

The path $\mathcal{C}_2$

$\mathbf{F}(\mathbf{r}_2(t)) = \langle 2\cos(\pi t), 1 + \sin(2\pi t) \rangle$ and $\mathbf{r}_2'(t) = \langle 2\pi\cos(2\pi t), -\pi\sin(\pi t) \rangle$. Therefore, $\mathbf{F}(\mathbf{r}_2(t)) \cdot \mathbf{r}_2'(t) = 4\pi\cos(\pi t)\cos(2\pi t) - \pi\sin(\pi t)(1 + \sin(2\pi t))$, and

$$\int_{\mathcal{C}_2} \mathbf{F} \cdot d\mathbf{r} = \int_0^1 (4\pi\cos(\pi t)\cos(2\pi t) - \pi\sin(\pi t)(1 + \sin(2\pi t)))\, dt$$

$$= \left(\frac{3}{2}\sin(\pi t) + \cos(\pi t) + \frac{5}{6}\sin(3\pi t)\right)\Big|_0^1 = -2 \;\; (\text{\textit{Maple} was used here}).$$

The path $\mathcal{C}_3$

$\mathbf{F}(\mathbf{r}_3(t)) = \langle 2 - 4t, 1 \rangle$ and $\mathbf{r}_3'(t) = \langle 0, -2 \rangle$. Therefore, $\mathbf{F}(\mathbf{r}_3(t)) \cdot \mathbf{r}_3'(t) = -2$, and

$$\int_{\mathcal{C}_3} \mathbf{F} \cdot d\mathbf{r} = \int_0^1 -2\, dt = -2\,.$$

39. Parametrize the circle with $x(t) = a\cos(t)$, $y(t) = a\sin(t)$, $0 \le t \le 2\pi$. Then

$$\oint_C \frac{-y\, dx + x\, dy}{x^2 + y^2} = \frac{1}{a^2} \int_0^{2\pi} (-a\sin(t) \cdot (-a\sin(t)) + a\cos(t) \cdot a\cos(t))\, dt$$

$$= \int_0^{2\pi} (\sin^2(t) + \cos^2(t))\, dt = 2\pi\,.$$

41. Let $\mathcal{C}_1$ be the arc and $\mathcal{C}_2$ be the line segment. Using the parametrization for the circle in Exercise 39 ($a = \sqrt{2}$ and $-\pi/4 \le t \le \pi/4$):

$$\oint_{\mathcal{C}_1} \frac{-y\, dx + x\, dy}{x^2 + y^2} = \int_{-\pi/4}^{\pi/4} 1\, dt = \pi/2\,.$$

Using the parametrization $\mathbf{r}(t) = \langle 1, -t \rangle$ with $-1 \le t \le 1$ for the line:

$$\oint_{\mathcal{C}_2} \frac{-y\, dx + x\, dy}{x^2 + y^2} = \int_{-1}^1 \frac{1 \cdot (-dt)}{1 + t^2}$$

$$= (-\arctan(t))\Big|_{-1}^1 = -\pi/2\,.$$

Add these to see that the integral over the closed curve is 0.

43. The curve $\mathcal{C}$ can be parametrized with $\mathbf{r}(t) = \langle \cos(t), \sin(t), 1 + \cos(t) \rangle$ where $0 \le t \le 2\pi$. That is, $x(t) = \cos(t)$, $y(t) = \sin(t)$, and $z(t) = 1 + \cos(t)$. The calculation uses the identity $\cos(t)\sin(t) = (1/2)\sin(2t)$ to obtain the third line.

$$
\begin{aligned}
\int_{\mathcal{C}_1} xyz\,dz &= \int_0^{2\pi} \cos(t)\sin(t)(1 + \cos(t))(-\sin(t)\,dt) \\
&= -\int_0^{2\pi} (\cos(t)\sin^2(t) + \cos^2(t)\sin^2(t))\,dt \\
&= -\int_0^{2\pi} (\cos(t)\sin^2(t) + (1/4)\sin^2(2t))\,dt \\
&= -(1/3)\sin^3(t)\Big|_0^{2\pi} - 1/4\int_0^{2\pi}\sin^2(2t)\,dt \\
&= 0 - 1/4 \cdot \pi = -\pi/4.
\end{aligned}
$$

Calculator/Computer Exercises

45. Begin with a picture of the paths and the calculation of a and b.

```
> y := 4 + 2*x - 3*x^4:
  plot( y, x=-1..1.5, -2..5);
  a,b := fsolve( y=0, x=-1..-0.8),
         fsolve( y=0, x=1..2);
```

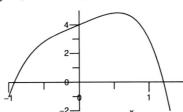

$$a, b := -0.9209526657, 1.209440530$$

The line integral can be calculated by ignoring the dx part because these contributions will cancel. Note also that $dy = 0$ on the line segment from $(a, 0)$ to $(b, 0)$. Consequently, we have

```
> Int( x^2*diff(y,x), x=b..a) =
  evalf(int( x^2*diff(y,x), x=b..a));
```

$$\int_{1.209440530}^{-.9209526657} x^2(2 - 12x^3)\,dx = 3.339056409$$

47. Note that $f(\mathbf{r}(t)) = te^{t^2}$ and $ds = ||\mathbf{r}'(t)||\,dt = \sqrt{1 + 4t^2}\,dt$. Therefore,

```
> Int( t*exp(t^2)*t*sqrt(1+4*t^2), t=0..1) =
  evalf(Int( t*exp(t^2)*sqrt(1+4*t^2), t=0..1));
```

$$\int_0^1 t^2 e^{t^2} \sqrt{1 + 4t^2}\, dt = 1.541224356$$

49. The curve can be parametrized using $x(\theta) = \theta \cos(\theta)$, $y(\theta) = \theta \sin(\theta)$, $0 \leq \theta \leq 2\pi$. Observe that $f(x(\theta), y(\theta)) = \theta^2 \cos^2(\theta)$ and

$$\begin{aligned}
ds &= ||\langle x'(\theta), y'(\theta) \rangle||\, d\theta \\
&= \sqrt{(-\theta \sin(\theta) + \cos(\theta))^2 + (\theta \cos(\theta) + \sin(\theta))^2}\, d\theta \\
&= \sqrt{\theta^2 + 1}\, d\theta
\end{aligned}$$

We have

```
> Int(theta^2*cos(theta)^2*sqrt(theta^2+1), theta=0..2*Pi) =
  evalf(Int(theta^2*cos(theta)^2*sqrt(theta^2+1),
                                        theta=0..2*Pi));
```

$$\int_0^{2\pi} \theta^2 \cos(\theta)^2 \sqrt{\theta^2 + 1}\, d\theta = 214.4199297$$

13.3   Conservative Vector Fields and Path-Independence

Problems for Practice

1. This is a closed field on Q because $M_y(x, y) = 0 = N_x(x, y)$ at all points (x, y) in Q. Consequently, there is a function $(x, y) \mapsto u(x, y)$ on Q such that $u_x(x, y) = x + \pi$ and $u_y(x, y) = 1$.

   The first equation implies that $u(x, y) = x^2/2 + \pi x + \phi(y)$ and the second equation will be satisfied if ϕ is defined so that $\phi'(y) = 1$. That is, $\phi(y) = y + C$. All potential functions have the form

   $$V(x, y) = -u(x, y) = -x^2/2 - \pi x - y + C.$$

3. This is a closed field on Q because $M_y(x, y) = 3y^2 = N_x(x, y)$ at all points (x, y) in Q. Consequently, there is a function $(x, y) \mapsto u(x, y)$ on Q such that $u_x(x, y) = y^3$ and $u_y(x, y) = 3y^2 x - 7y^6$.

   The first equation implies that $u(x, y) = xy^3 + \phi(y)$ and the second equation will be satisfied if ϕ is defined so that $3xy^2 + \phi'(y) = 3y^2 x - 7y^6$. That is, $\phi'(y) = -7y^6$ and $\phi(y) = -y^7 + C$. All potential functions have the form
   $$V(x, y) = -u(x, y) = -xy^3 + y^7 + C.$$

5. This is a not a closed field because $M_y(x, y) = 3y^2 - 2xy$ and $N_x(x, y) = -2xy^2 - 1$.

7. This is a closed field on Q because $M_y(x, y) = 0 = N_x(x, y)$ at all points (x, y) in Q. Consequently, there is a function $(x, y) \mapsto u(x, y)$ such that $u_x(x, y) = (x - 2)^{-2}$ and $u_y(x, y) = (y - 1/2)^2$.

   The first equation implies that $u(x, y) = -(x-2)^{-1} + \phi(y)$ and the second equation will be satisfied if ϕ is defined so that $\phi'(y) = (y - 1/2)^2$. That is, $\phi(y) = (y - 1/2)^3/3 + C$. All potential functions have the form

$$V(x, y) = -u(x, y) = (x - 2)^{-1} - (y - 1/2)^3/3 + C.$$

9. This is a closed field on Q because $M_y(x, y) = -2x \sin(x^2 - y) = N_x(x, y)$ at all points (x, y) in Q. Consequently, there is a function $(x, y) \mapsto u(x, y)$ such that $u_x(x, y) = 2x \sin(x^2 - y)$ and $u_y(x, y) = -\sin(x^2 - y)$.

   The first equation implies that $u(x, y) = -\cos(x^2 - y) + \phi(y)$ and the second equation will be satisfied if ϕ is defined so that $-\sin(x^2 - y) + \phi'(y) = -\sin(x^2 - y)$. That is, $\phi(y) = 0$ and $\phi(y) = C$. All potential functions have the form

$$V(x, y) = -u(x, y) = \cos(x^2 - y) + C.$$

11. Although it is true that $M_y(x, y) = (y^2 - x^2)/(x^2 + y^2)^2 = N_x(x, y)$ almost everywhere, this field is not closed on Q. This is because it is not defined at $(0, 0)$.

13. This is a closed field on Q. The required conditions are easily seen to be satisfied:

$$M_y(x, y, z) = z = N_x(x, y, z)$$

$$M_z(x, y, z) = y = R_x(x, y, z)$$

$$N_z(x, y, z) = x = R_y(x, y, z)$$

   at all points (x, y, z) in Q. Consequently, there is a function $(x, y, z) \mapsto u(x, y, z)$ on Q such that $u_x(x, y, z) = yz + 1$, $u_y(x, y, z) = xz + 2$, and $u_z(x, y, z) = xy + 3$.

   The first equation implies that $u(x, y, z) = xyz + x + \phi(y, z)$ and the second equation will be satisfied if ϕ is defined so that $xz + \phi_y(y, z) = xz + 2$. That is, $\phi'(y) = 2$, so $\phi(x, y) = 2y + \psi(z)$ and $u(x, y, z) = xyz + x + 2y + \psi(z)$. To satisfy the third equation choose ψ so that $xy + \psi'(z) = xy + 3$. That is, $\psi'(z) = 3$, so $\psi(z) = 3z + C$, a constant. All potential functions have the form

$$V(x, y) = -u(x, y, z) = -xyz - x - 2y - 3z + C.$$

15. This field is not closed. $N_z(x, y, z) = 1/(2 + x + y)$ but $R_y(x, y, z) = 1 + 1/(2 + x + y)$.

17. This is a closed field on Q. The required conditions are satisfied:

$$M_y(x,y,z) = \frac{2z^2}{(x+y-4)^3} = N_x(x,y,z)$$

$$M_z(x,y,z) = -\frac{2z}{(x+y-4)^2} = R_x(x,y,z)$$

$$N_z(x,y,z) = -\frac{2z}{(x+y-4)^2} = R_y(x,y,z)$$

at all points (x,y,z) in Q. Consequently, there is a function $(x,y,z) \mapsto u(x,y,z)$ on Q such that $u_x(x,y,z) = -z^2/(x+y-4)^2$, $u_y(x,y,z) = -z^2/(x+y-4)^2$, and $u_z(x,y,z) = 2z/(x+y-4)$.

The first equation implies that $u(x,y,z) = z^2/(x+y-4)+\phi(y,z)$ and the second equation will be satisfied if ϕ is defined so that $-z^2/(x+y-4)^2 + \phi_y(y,z) = -z^2/(x+y-4)^2$. That is, $\phi_y(y,z) = 0$, so $\phi(x,y) = \psi(z)$ and $u(x,y,z) = z^2/(x+y-4) + \psi(z)$. To satisfy the third equation choose ψ so that $2z/(x+y-4) + \psi'(z) = 2z/(x+y-4)$. That is, $\psi'(z) = 0$ so $\psi(z) = C$, a constant. All potential functions have the form

$$V(x,y) = -u(x,y,z) = -z^2/(x+y-4) + C.$$

19. This field is not closed. $N_z(x,y,z) = -\sin(y)$ but $R_y(x,y,z) = \sin(x)$.

21. This field is not closed on the unit cube. Although the partial derivative conditions are satisfied at those points in the cube where $\mathbf{F}$ is defined, the field is not defined at all points in Q (the point $(\pi/4, \pi/4, 0)$ for example).

23. This field is continuously differentiable on $\mathcal{G} = \{(x,y) : xy > -5\}$, a simply-connected planar region containing P_0 and P_1 (draw a picture). Because $M_y(x,y) = -10x/(5+xy)^3 = N_x(x,y)$ at all points (x,y) in $\mathcal{G}$ (verify), the field is closed. Consequently, there is a function $(x,y) \mapsto u(x,y)$ defined on $\mathcal{G}$ such that $u_x(x,y) = 5/(5+xy)^2$ and $u_y(x,y) = -x^2/(5+xy)^2$.

The first equation implies that $u(x,y) = -5y^{-1}/(5+xy) + \phi(y)$ and the second equation will be satisfied if ϕ is defined so that

$$\frac{5xy^{-1}}{(5+xy)^2} + \frac{5y^{-2}}{5+xy} + \phi'(y) = \frac{-x^2}{(5+xy)^2}.$$

This simplifies to $\phi'(y) = -y^{-2}$ (verify!). Consequently, $\phi(y) = y^{-1} + C$ and

$$u(x,y) = -\frac{5}{y(5+xy)} + \frac{1}{y} + C = \frac{x}{5+xy} + C.\,{}^1$$

Consequently, $\int_{\mathcal{C}} \mathbf{F} \cdot d\mathbf{r} = u(P_1) - u(P_0) = -1/36$.

---

[1]Simplification of the formula for $u(x,y)$ is to be expected here. This is because the region $\mathcal{G}$ includes both coordinate axes. In particular, u must be defined on the x-axis where $y = 0$.

25. This field is continuously differentiable on the entire (x, y)-plane. Because $M_y(x, y) = 2y + yxe^{xy} + e^{xy} = N_x(x, y)$, the field is closed. Consequently, there is a function $(x, y) \mapsto u(x, y)$ defined everywhere such that $u_x(x, y) = y^2 + ye^{xy}$ and $u_y(x, y) = 2xy + xe^{xy}$.

The first equation implies that $u(x, y) = xy^2 + e^{xy} + \phi(y)$ and the second equation will be satisfied if ϕ is defined so that $2xy + xe^{xy} + \phi'(y) = 2xy + xe^{xy}$. Consequently, $\phi'(y) = 0$ and $\phi(y) = C$, a constant. Therefore, $u(x, y) = xy^2 + e^{xy} + C$.

Using formula (15.23), $\int_C \mathbf{F} \cdot d\mathbf{r} = u(P_1) - u(P_0) = e^{-1} - 2$.

27. This field is continuously differentiable at all points. Because $M_y(x, y, z) = e^{y+2z} = N_x(x, y, z)$, $M_z(x, y, z) = 2e^{y+2z} = R_x(x, y, z)$, and $N_z(x, y, z) = 2xe^{x+2y} = R_y(x, y, z)$, the field is closed. Consequently, there is a function $(x, y, z) \mapsto u(x, y, z)$ defined everywhere such that $u_x(x, y, z) = e^{y+2z}$ and $u_y(x, y, z) = xe^{y+2z}$, and $u_z(x, y, z) = 2xe^{y+2z}$.

The first equation implies that $u(x, y, z) = xe^{y+2z} + \phi(y, z)$ and the second equation will be satisfied if ϕ is defined so that $xe^{y+2z} + \phi'(y) = xe^{y+2z}$. Consequently, $\phi'(y) = 0$ and $\phi(y) = \psi(z)$. Therefore, $u(x, y, z) = xe^{y+2z} + \psi(z)$. The third condition will be satisfied if $2xe^{y+2z} + \psi'(z) = 2xe^{y+2z}$ so $\psi'(z) = 0$ and $\psi(z) = C$. We conclude that $u(x, y, z) = xe^{y+2z} + C$.

Using formula (15.23), $\int_C \mathbf{F} \cdot d\mathbf{r} = u(P_1) - u(P_0) = -e^4$.

29. Using the following parametrizations:

$$\mathbf{r}_1(t) = \langle 0, t \rangle, \ \mathbf{r}_2(t) = \langle t, 1 \rangle, \ \text{and} \ \mathbf{r}_3(t) = \langle 1 - t, 1 - t \rangle,$$

(each one for $0 \le t \le 1$), the integrals over the three line segments evaluate as follows:

$$\int_{\overrightarrow{P_0 P_1}} \mathbf{F} \cdot d\mathbf{r} = \int_0^1 \left((t^2 + 2t) \cdot 0 + 0 \cdot 1 \right) \, dt = 0$$

$$\int_{\overrightarrow{P_1 P_2}} \mathbf{F} \cdot d\mathbf{r} = \int_0^1 (3 \cdot 1 + 4t \cdot 0) \, dt = 3$$

$$\int_{\overrightarrow{P_2 P_0}} \mathbf{F} \cdot d\mathbf{r} = \int_0^1 \left((1-t)^2 + 2(1-t) \right) \cdot (-1) + \left(2(1-t)^2 + 2(1-t) \cdot (-1) \right) \, dt$$

$$= \int_0^1 \left(-3(1-t)^2 - 4 + 4t \right) \, dt$$

$$= \left. \left((1-t)^3 - 4t + 2t^2 \right) \right|_0^1 = -3$$

These sum to 0.

31. Using the following parametrizations:

$$\mathbf{r}_1(t) = \langle 0, t, 1 \rangle, \ \mathbf{r}_2(t) = \langle t, 1, 1 \rangle, \ \text{and} \ \mathbf{r}_3(t) = \langle 1 - t, 1 - t, 1 \rangle,$$

(each one for $0 \le t \le 1$), the integrals over the three line segments evaluate as follows:

$$\int_{\overrightarrow{P_0 P_1}} \mathbf{F} \cdot d\mathbf{r} = \int_0^1 \left((t^2) \cdot 0 + 0 \cdot 1 + 2 \cdot 0 \right) dt = 0$$

$$\int_{\overrightarrow{P_1 P_2}} \mathbf{F} \cdot d\mathbf{r} = \int_0^1 \left(2t \cdot 1 + t^2 \cdot 0 + 2 \cdot 0 \right) dt = 1$$

$$\int_{\overrightarrow{P_2 P_0}} \mathbf{F} \cdot d\mathbf{r} = \int_0^1 \left(2(1-t)^2 \cdot (-1) + (1-t)^2 \cdot (-1) + 2 \cdot 0 \right) dt$$

$$= \int_0^1 -3(1-t)^2 \, dt$$

$$= (1-t)^3 \Big|_0^1 = -1$$

These sum to 0.

Further Theory and Practice

33. A straightforward (but long) calculation will show that

$$\mathbf{F}(\mathbf{r}(t)) \cdot \mathbf{r}'(t) = 2688t^6 - 5376t^5 + 4800t^4 - 2752t^3 + 936t^2 - 168t + 20 \,.$$

Therefore,

$$\int_{\mathcal{C}} \mathbf{F} \cdot d\mathbf{r} = \int_0^1 \mathbf{F}(\mathbf{r}(t)) \cdot \mathbf{r}'(t) \, dt$$

$$= \int_0^1 (2688t^6 - 5376t^5 + 4800t^4 - 2752t^3 + 936t^2 - 168t + 20) \, dt$$

$$= 8 \,.$$

35. Observe that when $\mathbf{r} = \langle x, y, z \rangle$ (i.e. $\overrightarrow{OP}$) and $r = \|\mathbf{r}\| = \sqrt{x^2 + y^2 + z^2}$, then $\nabla(1/r) = -r^{-3}\mathbf{r}$. Consequently, $u(x,y,z) = Gm/r$ has the property that $\nabla u(x,y,z) = -(Gmr^{-3})\mathbf{r}$ and $V(x,y,z) = Gm/r$ is a potential function for $\mathbf{F}$.

Using this potential function

$$W_P = \int_{\overrightarrow{P_0 P}} \mathbf{F} \cdot d\mathbf{r} = u(P) - u(P_0) = Gm/r - Gm/r_0 \,,$$

and $\lim_{P \to \infty} W_P = -Gm/r_0$.

Note that $-Gm/r_0$ is the work done by the gravitional force.
The work done by a force that pulls against gravity is Gm/r_0.

37. The region $\mathcal{G}$ is *not* simply-connected if either $-1 < c < d < 1$ or $c \le -1 < 1 \le d$. Therefore, it is simply connected for all pairs (c,d) with $c \le -1 < d < 1$ or $d \ge 1 > c > -1$ (always assuming that $c < d$).

39. See the picture on the right. Both regions are simply-connected, but their washer-shaped intersection has a hole in the middle.

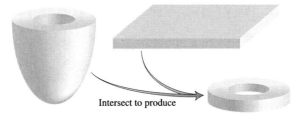

Intersect to produce

Calculator/Computer Exercises

41. The first entry defines the functions M and N and verifies that $\partial M/\partial y = \partial N/\partial x$.

```
> M,N := sin(Pi*x*y^2)/x, 2*sin(Pi*x*y^2)/y:
  diff(M,y), diff(N,x);
```

$$2\cos(\pi xy^2)\pi y,\ 2\cos(\pi xy^2)\pi y$$

The next entry obtains a potential function for the field, valid in the first quadrant. The function Si is a special function defined as follows: $\mathrm{Si}(t) = \int_0^t (\sin(\tau)/\tau)\,d\tau$.

```
> u := unapply(int(M,x),x,y);
  check, diff(u(x,y),x), diff(u(x,y),y);
```

$$u := (x,y) \to \mathrm{Si}(\pi xy^2)$$

$$check,\ \frac{\sin(\pi xy^2)}{x},\ \frac{2\sin(\pi xy^2)}{y}$$

Here is the evaluation of the integral, both exactly and approximately.

```
> integral = u(1/2,1) - u(1,1/2);
  evalf(%);
```

$$integral = \mathrm{Si}\left(\frac{1}{2}\pi\right) - \mathrm{Si}\left(\frac{1}{4}\pi\right)$$

$$integral = 0.6117862868$$

13.4 Divergence, Gradient, and Curl

Problems for Practice

1. $\mathrm{div}(\mathbf{F} = \frac{\partial}{\partial x}(3x^2y^3) - \frac{\partial}{\partial y}(xy^4) = 6xy^3 - 4xy^3 = 2xy^3$

3. $\mathrm{div}(\mathbf{F} = \frac{\partial}{\partial x}(x^2\sin(y)) + \frac{\partial}{\partial y}(x^2 + \sin(y)) = 2x\sin(y) + \cos(y)$

5. $\mathrm{div}(\mathbf{F}) = \frac{\partial}{\partial x}(\cos(x)) + \frac{\partial}{\partial y}(-\sin^2(xy)) = -\sin(x) - 2x\sin(xy)\cos(xy)$

7. $\mathrm{div}(\mathbf{F}) = \frac{\partial}{\partial x}(x/(x+y)) + \frac{\partial}{\partial y}(-2y/(x-y)) = y/(x+y)^2 - 2x/(x-y)^2$

9. $\operatorname{div}(\mathbf{F}) = \frac{\partial}{\partial x}(\tan(x/y)) + \frac{\partial}{\partial y}(\cot(y/x)) = (1/y)\sec^2(x/y) - (1/x)\csc^2(y/x)$

11. $\operatorname{div}(\mathbf{F}) = \frac{\partial}{\partial x}(2y - 3z) + \frac{\partial}{\partial y}(4z - 5x) + \frac{\partial}{\partial z}(6x + 7z) = 0 + 0 + 7 = 7$

13. $\operatorname{div}(\mathbf{F}) = \frac{\partial}{\partial x}(x + 2y + 3z) + \frac{\partial}{\partial y}(2(x + 2y + 3z)) + \frac{\partial}{\partial z}(3(x + 2y + 3z))$
    $= 1 + 4 + 9 = 14$

15. $\operatorname{div}(\mathbf{F}) = \frac{\partial}{\partial x}(x^2) + \frac{\partial}{\partial y}(y^2) + \frac{\partial}{\partial z}(z^2) = 2x + 2y + 2z$

17. $\operatorname{div}(\mathbf{F}) = \frac{\partial}{\partial x}(e^{x-z}) + \frac{\partial}{\partial y}(e^{z-y}) - \frac{\partial}{\partial z}(e^{y-x}) = e^{x-z} - e^{z-y}$

19. $\operatorname{div}(\mathbf{F}) = \frac{\partial}{\partial x}(\cos(xy)) - \frac{\partial}{\partial y}(\sin(yz)) + \frac{\partial}{\partial z}(\cos(y)\sin(x))$
    $= -y\sin(xy) - z\cos(yz)$

21. $\mathbf{curl}(\mathbf{F}) = \left(\frac{\partial}{\partial y}(xyz) - \frac{\partial}{\partial z}(-y^3)\right)\mathbf{i} + \left(\frac{\partial}{\partial z}(zx) - \frac{\partial}{\partial x}(xyz)\right)\mathbf{j}$
    $+ \left(\frac{\partial}{\partial x}(-y^3) - \frac{\partial}{\partial y}(zx)\right)\mathbf{k}$
    $= xz\,\mathbf{i} + (x - yz)\,\mathbf{j}$

23. $\mathbf{curl}(\mathbf{F}) = \left(\frac{\partial}{\partial y}(xy) - \frac{\partial}{\partial z}(-e^{yz})\right)\mathbf{i} + \left(\frac{\partial}{\partial z}\ln(x + z) - \frac{\partial}{\partial x}(xy)\right)\mathbf{j}$
    $+ \left(\frac{\partial}{\partial x}(-e^{yz}) - \frac{\partial}{\partial y}\ln(x + z)\right)\mathbf{k}$
    $= (x + ye^{yz})\,\mathbf{i} + (1/(x + z) - y)\,\mathbf{j}$

25. $\mathbf{curl}(\mathbf{F}) = \left(\frac{\partial}{\partial y}(-\sin(xy)) - \frac{\partial}{\partial z}\cos(xy)\right)\mathbf{i} + \left(\frac{\partial}{\partial z}\tan(xy) - \frac{\partial}{\partial x}(-\sin(xy))\right)\mathbf{j}$
    $+ \left(\frac{\partial}{\partial x}\cos(xy) - \frac{\partial}{\partial y}\tan(xy)\right)\mathbf{k}$
    $= -x\cos(xy)\,\mathbf{i} + y\cos(xy)\,\mathbf{j} - (y\sin(xy) + x\sec^2(xy))\,\mathbf{k}$

27. Using the determinant formula,

$$\nabla \times \mathbf{F} = \det\left(\begin{bmatrix} \mathbf{i} & \mathbf{j} & \mathbf{k} \\ \frac{\partial}{\partial x} & \frac{\partial}{\partial y} & \frac{\partial}{\partial z} \\ y^2z & 2xyz & xy^2 \end{bmatrix}\right)$$
$$= \langle 2xy - 2xy, y^2 - y^2, 2yz - 2yz \rangle = \langle 0, 0, 0 \rangle$$

The field is closed.

29. Using the determinant formula,

$$\nabla \times \mathbf{F} = \det\left(\begin{bmatrix} \mathbf{i} & \mathbf{j} & \mathbf{k} \\ \frac{\partial}{\partial x} & \frac{\partial}{\partial y} & \frac{\partial}{\partial z} \\ \cos(yz) & -xz\sin(yz) & -xy\sin(yz) \end{bmatrix}\right)$$
$$= \langle -xyz\sin(yz) - x\sin(yz) - (-xyz\sin(yz) - x\sin(yz)),$$
$$- y\sin(yz) - (-y\sin(yz)), -z\sin(yz) - (-z\sin(yz)) \rangle$$
$$= \langle 0, 0, 0 \rangle$$

The field is closed.

31. Using the determinant formula,

$$\nabla \times \mathbf{F} = \det\left(\begin{bmatrix} \mathbf{i} & \mathbf{j} & \mathbf{k} \\ \frac{\partial}{\partial x} & \frac{\partial}{\partial y} & \frac{\partial}{\partial z} \\ x\cos(y) & -z\sin(x) & xy\sin(z) \end{bmatrix}\right)$$
$$= \langle x\sin(z) + \sin(x), -y\sin(z), -z\cos(x) + x\sin(y) \rangle$$

The field is not closed.

33. $\text{div}(\mathbf{grad}(u)) = \text{div}(2x\,\mathbf{i} - 2y\,\mathbf{j}) = 2 - 2 = 0$

35. $\text{div}(\mathbf{grad}(u)) = \text{div}((1/(z - y^2))\,\mathbf{i} + (2xy/(z - y^2)^2)\,\mathbf{j} - (x/(z - y^2)^2)\,\mathbf{k})$
$$= 0 + 2x(3y^2 + z)/(z - y^2)^3 + 2x/(z - y^2)^3$$
$$= 2x(3y^2 + z + 1)/(z - y^2)^3$$

37. $\text{div}(\mathbf{grad}(u)) = \text{div}((y + 2z)\,\mathbf{i} + x\,\mathbf{j} + 2x\,\mathbf{k}) = 0 + 0 + 0 = 0$

39. $\mathbf{grad}(\text{div}(\mathbf{F})) = \mathbf{grad}(-\sin(x) + \cos(y)) = \langle -\cos(x), -\sin(y) \rangle$

41. $\mathbf{grad}(\text{div}(\mathbf{F})) = \mathbf{grad}(yz + z^2 + x) = \langle 1, z, y + 2z \rangle$

Further Theory and Practice

43. $\nabla \times (\nabla \times \mathbf{F}) = \det\left(\begin{bmatrix} \mathbf{i} & \mathbf{j} & \mathbf{k} \\ \frac{\partial}{\partial x} & \frac{\partial}{\partial y} & \frac{\partial}{\partial z} \\ xz - xy & xy - yz & yz - xz \end{bmatrix}\right)$
$$= \langle z + y, x + z, y + x \rangle$$

45. $\nabla \times (\nabla \times \mathbf{F}) = \det\left(\begin{bmatrix} \mathbf{i} & \mathbf{j} & \mathbf{k} \\ \frac{\partial}{\partial x} & \frac{\partial}{\partial y} & \frac{\partial}{\partial z} \\ -2y - 1/y & -2x & x/y^2 \end{bmatrix}\right)$
$$= -y^{-3}\langle 2x, y, y \rangle$$

47. Product rule for the gradient

$$\nabla(gh) = \left\langle \frac{\partial(gh)}{\partial x}, \frac{\partial(gh)}{\partial y}, \frac{\partial(gh)}{\partial z} \right\rangle$$
$$= \left\langle g\frac{\partial h}{\partial x} + h\frac{\partial g}{\partial x}, g\frac{\partial h}{\partial y} + h\frac{\partial g}{\partial y}, g\frac{\partial h}{\partial z} + h\frac{\partial g}{\partial z} \right\rangle$$
$$= \left\langle g\frac{\partial h}{\partial x}, g\frac{\partial h}{\partial y}, g\frac{\partial h}{\partial z} \right\rangle + \left\langle h\frac{\partial g}{\partial x}, h\frac{\partial g}{\partial y}, h\frac{\partial g}{\partial z} \right\rangle$$
$$= g\nabla(h) + h\nabla(g)$$

49. Product rule for the curl

Let $\mathbf{F} = \langle M, N, R \rangle$. Then

$$\nabla \times (g\mathbf{F}) = \left\langle \frac{\partial(gR)}{\partial y} - \frac{\partial(gN)}{\partial z}, \frac{\partial(gM)}{\partial z} - \frac{\partial(gR)}{\partial x}, \frac{\partial(gN)}{\partial x} - \frac{\partial(gM)}{\partial y} \right\rangle$$

$$= \left\langle g\frac{\partial R}{\partial y} + R\frac{\partial g}{\partial y} - g\frac{\partial N}{\partial z} - N\frac{\partial g}{\partial z}, \right.$$

$$g\frac{\partial M}{\partial z} + M\frac{\partial g}{\partial z} - g\frac{\partial R}{\partial x} - R\frac{\partial g}{\partial x},$$

$$\left. g\frac{\partial N}{\partial x} + N\frac{\partial g}{\partial x} - g\frac{\partial M}{\partial y} - M\frac{\partial g}{\partial y} \right\rangle$$

$$= \left\langle R\frac{\partial g}{\partial y} - N\frac{\partial g}{\partial z}, M\frac{\partial g}{\partial z} - R\frac{\partial g}{\partial x}, N\frac{\partial g}{\partial x} - M\frac{\partial g}{\partial y} \right\rangle$$

$$+ \left\langle g\frac{\partial R}{\partial y} - g\frac{\partial N}{\partial z}, g\frac{\partial M}{\partial z} - g\frac{\partial R}{\partial x}, g\frac{\partial N}{\partial x} - g\frac{\partial M}{\partial y} \right\rangle$$

$$= \nabla g \times \mathbf{F} + g(\nabla \times \mathbf{F}).$$

51. Using the product rule for the curl (Exercise 49) we have

$$\nabla \times (g\nabla h - h\nabla g) = \nabla \times (g\nabla h) - \nabla \times (h\nabla g)$$

$$= \nabla g \times \nabla h + g(\nabla \times \nabla h) - (\nabla h \times \nabla g + h(\nabla \times \nabla g))$$

$$= 2\nabla g \times \nabla h$$

Note that $\nabla \times \nabla h = \nabla \times \nabla g = \vec{0}$ (Theorem 1 c).

53. Let $||\mathbf{r}|| = r$ and observe that $\nabla(1/r) = -r^3\mathbf{r}$ (Section 13.1, Exercise 49). Consequently, $\mathbf{F} = \lambda\nabla(1/r)$ and

$$\nabla \times \mathbf{F} = \nabla \times (\lambda\nabla(1/r)) = \lambda(\nabla \times \nabla(1/r)) = \mathbf{0}$$

by Theorem 1 c.

55. The function u is harmonic because $\nabla u = \langle 3x^2 - 3y^2, -6xy \rangle$ and $\nabla \cdot \nabla u = 6x - 6x = 0$. Observe that $u(2,1) = 3$ and

$$\frac{1}{2\pi} \int_0^{2\pi} u(2 + r\cos(t), 1 + r\sin(t))\, dt$$

$$= \frac{1}{2\pi} \int_0^{2\pi} ((2 + r\cos(t))^3 - 3(2 + r\cos(t))(1 + r\sin(t))^2 + 1)\, dt$$

$$= \frac{1}{2\pi} \int_0^{2\pi} (r^3\cos^3(t) + 6r^2\cos^2(t) + (9r - 6r^2\sin(t) - 3r^3\sin^2(t))\cos(t) + 3 - 12r\sin(t) - 6r^2\sin^2(t))\, dt$$

$$= \frac{1}{2\pi}\left(3\pi r^2 + 6\pi - 3\pi r^2\right) = 3.$$

Note that only three terms in the last integrand yield non-zero values when integrated over the interval $[0, 2\pi]$, the second one, the last one, and the constant term 3.

57. Let $\omega = \langle a, b, c \rangle$ and $\mathbf{r} = \langle x, y, z \rangle$. Since $\omega \times \mathbf{r} = \langle bz - cy, cx - az, ay - bx \rangle$,

$$\nabla \times (\omega \times \mathbf{r}) =$$

$$\left\langle \frac{\partial(ay - bx)}{\partial y} - \frac{\partial(cx - az)}{\partial z}, \frac{\partial(bz - cy)}{\partial z} - \frac{\partial(ay - bx)}{\partial x}, \frac{\partial(cx - az)}{\partial x} - \frac{\partial(bz - cy)}{\partial y} \right\rangle$$

$$= \langle 2a, 2b, 2c \rangle = 2\omega.$$

59. Since V is a potential function for $\nabla \times \mathbf{F}$, $\nabla \times \mathbf{F} = -\nabla V$. Therefore, according to Theorem 1 b,

$$\nabla \cdot (\nabla V) = -\nabla \cdot (\nabla \times \mathbf{F}) = 0,$$

and V is harmonic. Moreover, starting with a rearrangement of the identity in Example 9 and using Theorem 1 c, we have

$$\triangle \mathbf{F} = \mathbf{grad}(\operatorname{div}(\mathbf{F})) - \mathbf{curl}(\mathbf{curl}(\mathbf{F}))$$
$$= \mathbf{grad}(\operatorname{div}(\mathbf{F})) + \mathbf{curl}(\mathbf{grad}(V))$$
$$= \mathbf{grad}(\operatorname{div}(\mathbf{F})).$$

Calculator/Computer Exercises

61. Define u, evaluate it at $(1, 3)$, then evaluate the integral.

```
> u := (x,y) -> exp(x)*cos(y): u(1,3) = evalf(u(1,3));
  1/(2*Pi)*Int(u(1+cos(theta),3+sin(theta)),theta=0..2*Pi) =
    evalf(1/(2*Pi)*Int(u(1+cos(theta),3+sin(theta)),theta=0..2*Pi));
```

$$e \cos(3) = -2.691078613$$

$$\frac{1}{2} \frac{\int_0^{2\pi} e^{1+\cos(\theta)} \cos(3 + \sin(\theta)) \, d\theta}{\pi} = -2.691078614$$

13.5 Green's Theorem

Problems for Practice

1. The curve $\mathcal{C}$ is the circle $x^2 + y^2 = 1$ traversed in the counter-clockwise direction. Since $\mathbf{F}(\mathbf{r}(t)) = \langle \cos^2(t), -3\sin(t) \rangle$ and $\mathbf{r}'(t) = \langle \cos(t), \sin(t) \rangle$, $\mathbf{F}(\mathbf{r}(t)) \cdot \mathbf{r}'(t) = \cos^3(t) - 3\sin^2(t)$. Therefore,

$$\oint_{\mathcal{C}} \mathbf{F} \cdot d\mathbf{r} = \int_{\pi}^{3\pi} (\cos^3(t) - 3\sin^2(t)) \, dt$$

$$= \int_{\pi}^{3\pi} ((1 - \sin^2(t)) \cos(t) - (3/2)(1 - \cos(2t))) \, dt$$

$$= (\sin(t) - (1/3) \sin^3(t) - (3/2)(t - \sin(2t)/2)) \Big|_{\pi}^{3\pi}$$

$$= -3\pi.$$

Note that this result can be obtained more quickly by observing that $\int_\pi^{3\pi} \cos^3(t)\,dt = 0$ because of the shape of the graph of $\cos^3(t)$. Moreover, $\int_\pi^{3\pi} \sin^2(t)\,dt = \pi$ because of the fact that $\sin^2(t) + \cos^2(t) = 1$. We will take advantage of these, and similar, observations whenever possible.

For the right hand side of Green's Theorem, let $\mathcal{R}$ denote the region enclosed by the circle. Then

$$\iint_\mathcal{R} \left(\frac{\partial N}{\partial x} - \frac{\partial M}{\partial y} \right) dA = \iint_\mathcal{R} \left(\frac{\partial}{\partial x}(-3x) - \frac{\partial}{\partial y}(y^2) \right) dA$$
$$= \iint_\mathcal{R} (-3 - 2y)\,dA$$
$$= \iint_\mathcal{R} -3\,dA = -3\pi\,.$$

Note that $\iint_\mathcal{R} y\,dA = 0$ by symmetry.

3. The curve $\mathcal{C}$ is the circle $x^2 + y^2 = 1$ traversed in the counter-clockwise direction. Since $\mathbf{F}(\mathbf{r}(t)) = \langle \sin(2t) - 2\cos(2t), 3\cos(2t) - 4\sin(2t) \rangle$ and $\mathbf{r}'(t) = \langle -2\sin(2t), 2\cos(2t) \rangle$, $\mathbf{F}(\mathbf{r}(t)){\cdot}\mathbf{r}'(t) = -2\sin^2(2t) + 4\cos(2t)\sin(2t) + 6\cos^2(2t) - 8\sin(2t)\cos(2t)$. Therefore,

$$\oint_\mathcal{C} \mathbf{F} \cdot d\mathbf{r} = \int_0^\pi (-2\sin^2(2t) - 4\cos(2t)\sin(2t) + 6\cos^2(2t))\,dt$$
$$= -\pi + \cos^2(2t)\Big|_0^\pi + 3\pi = 2\pi\,.$$

For the right hand side of Green's Theorem, let $\mathcal{R}$ denote the region enclosed by the circle. Then

$$\iint_\mathcal{R} \left(\frac{\partial N}{\partial x} - \frac{\partial M}{\partial y} \right) dA = \iint_\mathcal{R} \left(\frac{\partial}{\partial x}(3x - 4y) - \frac{\partial}{\partial y}(y - 2x) \right) dA$$
$$= \iint_\mathcal{R} (3 - 1)\,dA = 2\pi\,.$$

5. The curve $\mathcal{C}$ is the circle $x^2 + y^2 = 1$ traversed in the counter-clockwise direction. Since $\mathbf{F}(\mathbf{r}(t)) = \langle -\cos(t), 2\sin(t) \rangle$ and $\mathbf{r}'(t) = \langle \cos(t), \sin(t) \rangle$, $\mathbf{F}(\mathbf{r}(t)) \cdot \mathbf{r}'(t) = -\cos^2(t) + 2\sin^2(t)$. Therefore,

$$\oint_\mathcal{C} \mathbf{F} \cdot d\mathbf{r} = \int_{-\pi}^\pi (-\cos^2(t) + 2\sin^2(t))\,dt$$
$$= -\pi + 2\pi = \pi\,.$$

For the right hand side of Green's Theorem, let $\mathcal{R}$ denote the region enclosed by the circle. Then

$$\iint_{\mathcal{R}} \left(\frac{\partial N}{\partial x} - \frac{\partial M}{\partial y} \right) dA = \iint_{\mathcal{R}} \left(\frac{\partial}{\partial x}(2x) - \frac{\partial}{\partial y}(y) \right) dA$$

$$= \iint_{\mathcal{R}} (2 - 1) \, dA = \pi \, .$$

7. The curve $\mathcal{C}$ is the circle $x^2 + y^2 = 1$ traversed in the counter-clockwise direction. Therefore, letting $\mathcal{R}$ denote the region enclosed by the circle,

$$\oint_{\mathcal{C}} \mathbf{F} \cdot d\mathbf{r} = \iint_{\mathcal{R}} \left(\frac{\partial}{\partial x}(-x^2) - \frac{\partial}{\partial y}(yx) \right) dA$$

$$= \iint_{\mathcal{R}} (-3x) \, dA = 0 \quad \text{(by symmetry)}.$$

9. The curve $\mathcal{C}$ is the circle $x^2 + y^2 = 1$ traversed in the counter-clockwise direction. Therefore, letting $\mathcal{R}$ denote the region enclosed by the circle,

$$\oint_{\mathcal{C}} \mathbf{F} \cdot d\mathbf{r} = \iint_{\mathcal{R}} \left(\frac{\partial}{\partial x}(x^3) - \frac{\partial}{\partial y}(-y^2) \right) dA$$

$$= \iint_{\mathcal{R}} (3x^2 + 2y) \, dA$$

$$= \int_0^{2\pi} \int_0^1 3r^2 \cos^2(\theta) \cdot r \, dr \, d\theta$$

$$= 3/4 \int_0^{2\pi} \cos^2(\theta) \, d\theta = 3\pi/4 \, .$$

Note that $\iint_{\mathcal{R}} y \, dA = 0$ by symmetry.

11. Let $\mathcal{C}$ denote the boundary curve and $\mathcal{R}$ the enclosed rectangle,

$$\oint_{\mathcal{C}} \mathbf{F} \cdot d\mathbf{r} = \iint_{\mathcal{R}} \left(\frac{\partial}{\partial x}(-y^2 x^3) - \frac{\partial}{\partial y}(xy^2) \right) dA$$

$$= \int_{-1}^1 \int_{-2}^2 (-3y^2 x^2 - 2xy) \, dy \, dx$$

$$= \int_{-1}^1 (-16x^2) \, dx = -32/3 \, .$$

13. The square is $\mathcal{R} = \{(x, y) : -1 \leq x \leq 1, -1 \leq y \leq 1\}$. Therefore, letting

$\mathcal{C}$ denote its boundary curve,

$$\oint_{\mathcal{C}} \mathbf{F} \cdot d\mathbf{r} = \iint_{\mathcal{R}} \left(\frac{\partial}{\partial x}(-xy) - \frac{\partial}{\partial y} \ln(3+y) \right) dA$$

$$= \int_{-1}^{1} \int_{-1}^{1} (-y - (3+y)^{-1}) \, dy \, dx$$

$$= \int_{-1}^{1} (-\ln(2)) \, dx = -2\ln(2) .$$

15. The planar region is $\mathcal{R} = \{(x,y) : -1 \le x \le 5, x^2 \le y \le 4x+5\}$. Therefore, letting $\mathcal{C}$ denote its boundary curve,

$$\oint_{\mathcal{C}} \mathbf{F} \cdot d\mathbf{r} = \iint_{\mathcal{R}} \left(\frac{\partial}{\partial x}(xy) - \frac{\partial}{\partial y}(x^2 y) \right) dA$$

$$= \int_{-1}^{5} \int_{x^2}^{4x+5} (y - x^2) \, dy \, dx$$

$$= \int_{-1}^{5} (x^4/2 - 4x^3 + 3x^2 + 20x + 25/2) \, dx = 648/5 .$$

17. Let $P = (2,5)$, $Q = (3,-6)$, and $R = (4,1)$. This is counter-clockwise around the triangular region $\mathcal{R}$, so the boundary curve $\mathcal{C}$ consists of $\overline{PQ}$ followed by $\overline{QR}$ and then $\overline{RP}$. Their parametrizations, in order, are

$$\mathbf{r}_1(t) = \langle 2+t, 5-11t \rangle, \ \mathbf{r}_2(t) = \langle 3+t, -6+7t \rangle, \ \mathbf{r}_3(t) = \langle 4-2t, 1+4t \rangle,$$

each one over the interval $0 \le t \le 1$. The area calculation can begin like this:

$$(\text{area of } \mathcal{R}) = -\frac{1}{2} \oint_{\mathcal{C}} y \, dx - x \, dy$$

$$= -\frac{1}{2} \int_{0}^{1} \left((5-11t) \cdot 1 - (2+t) \cdot (-11) \right) dt$$

$$- \frac{1}{2} \int_{0}^{1} \left((-6+7t) \cdot 1 - (3+t) \cdot 7 \right) dt$$

$$- \frac{1}{2} \int_{0}^{1} \left((1+4t) \cdot (-2) - (4-2t) \cdot 4 \right) dt ,$$

and conclude with the evaluation of each integral.

$$(\text{area of } \mathcal{R}) = -\frac{1}{2} \int_{0}^{1} 27 \, dt - \frac{1}{2} \int_{0}^{1} (-27) \, dt - \frac{1}{2} \int_{0}^{1} (-18) \, dt = 9$$

19. The region is $\mathcal{R} = \{(x,y) : -3 \le x \le 1, 4x \le y \le -2x^2 + 6\}$, so the boundary curve $\mathcal{C}$ consists of the line from $(-3,-12)$ to $(1,4)$ followed

by the parabola from $x = 1$ back to $x = -3$. Their parametrizations, in order, are

$$\mathbf{r}_1(t) = \langle -3 + 4t, -12 + 16t \rangle, \ 0 \le t \le 1$$

and

$$\mathbf{r}_2(t) = \langle t, -2t^2 + 6 \rangle, \ 1 \ge t \ge -3.$$

The area calculation can begin like this:

$$(\text{area of } \mathcal{R}) = -\frac{1}{2} \oint_{\mathcal{C}} y \, dx - x \, dy$$

$$= -\frac{1}{2} \int_0^1 \left((-12 + 16t) \cdot 4 - (-3 + 4t) \cdot 16 \right) dt$$

$$- \frac{1}{2} \int_1^{-3} \left((-2t^2 + 6) \cdot 1 - t \cdot (-4t) \right) dt,$$

and conclude with the evaluation of each integral.

$$(\text{area of } \mathcal{R}) = -\frac{1}{2} \int_0^1 0 \, dt - \frac{1}{2} \int_1^{-3} (2t^2 + 6) \, dt = 64/3$$

21. The region is $\mathcal{R} = \{(x, y) : -1 \le x \le 1/2, (x + 1)/\sqrt{3} \le y \le \sqrt{1 - x^2}\}$, so the boundary curve $\mathcal{C}$ consists of the line from $(-1, 0)$ to $(1/2, \sqrt{3}/2)$ followed by the top part of the circle from $x = 1/2$ back to $x = -1$. Their parametrizations, in order, are

$$\mathbf{r}_1(t) = \langle -1 + 3t/2, 0 + \sqrt{3}\, t/2 \rangle, \ 0 \le t \le 1$$

and

$$\mathbf{r}_2(t) = \langle \cos(t), \sin(t) \rangle, \ \pi/3 \ge t \ge \pi.$$

The area calculation can begin like this:

$$(\text{area of } \mathcal{R}) = -\frac{1}{2} \oint_{\mathcal{C}} y \, dx - x \, dy$$

$$= -\frac{1}{2} \int_0^1 \left((\sqrt{3}\, t/2) \cdot (3/2) - (-1 + 3t/2) \cdot (\sqrt{3}/2) \right) dt$$

$$- \frac{1}{2} \int_{\pi/3}^{\pi} \left(\sin(t) \cdot (-\sin(t)) - \cos(t) \cdot \cos(t) \right) dt,$$

and conclude with the evaluation of each integral.

$$(\text{area of } \mathcal{R}) = -\frac{1}{2} \int_0^1 \sqrt{3}/2 \, dt - \frac{1}{2} \int_{\pi/3}^{\pi} (-1) \, dt = \pi/3 - \sqrt{3}/4$$

23. The region is $\mathcal{R} = \{(x,y) : 0 \leq y \leq 1, -y \leq x \leq y\}$. If $\mathcal{C}$ denotes its positively oriented boundary curve, then according to formula 15.39,

$$\oint_{\mathcal{C}} \mathbf{F} \cdot \mathbf{n}\, ds = \iint_{\mathcal{R}} \operatorname{div}(\mathbf{F})\, dA$$
$$= \iint_{\mathcal{R}} \left(\frac{\partial M}{\partial x} + \frac{\partial N}{\partial y} \right) dA$$
$$= \int_0^1 \int_{-y}^{y} y^2\, dx\, dy$$
$$= \int_0^1 2y^3\, dy = 1/2\,.$$

25. The region is $\mathcal{R} = \{(x,y) : 0 \leq y \leq 1, 0 \leq x \leq y\}$. If $\mathcal{C}$ denotes its positively oriented boundary curve, then according to formula 15.39,

$$\oint_{\mathcal{C}} \mathbf{F} \cdot \mathbf{n}\, ds = \iint_{\mathcal{R}} \operatorname{div}(\mathbf{F})\, dA$$
$$= \iint_{\mathcal{R}} \left(\frac{\partial M}{\partial x} + \frac{\partial N}{\partial y} \right) dA$$
$$= \int_0^1 \int_0^{y} (y^{7/2} - x^{5/2})\, dx\, dy$$
$$= \int_0^1 \left((xy^{7/2} - (2/7)x^{7/2})\big|_{x=0}^{x=y} \right) dy$$
$$= \int_0^1 (y^{9/2} - (2/7)y^{7/2})\, dy = 82/693\,.$$

Further Theory and Practice

27. The boundary curve $\mathcal{C}$ can be parametrized as follows:

$$\mathbf{r}(t) = \langle (2 + 2\cos(\theta))\cos(\theta), (2 + 2\cos(\theta))\sin(\theta) \rangle\,, \ 0 \leq \theta \leq 2\pi\,.$$

Therefore, the area calculation goes like this.

$$\text{(area of } \mathcal{R}) = -\frac{1}{2} \oint_{\mathcal{C}} y\, dx - x\, dy$$
$$= -\frac{1}{2} \int_0^{2\pi} ((2 + 2\cos(\theta))\sin(\theta) \cdot (-2\sin(\theta) - 4\cos(\theta)\sin(\theta))$$
$$\qquad - (2 + 2\cos(\theta))\cos(\theta) \cdot (2\cos(\theta) + 2\cos^2(\theta) - 2\sin^2(\theta)))\, d\theta$$
$$= -\frac{1}{2} \int_0^{2\pi} (4\cos(\theta) - 12\cos^3(\theta) - 8\cos^4(\theta) - 4\sin^2(\theta) - 12\sin^2(\theta)\cos(\theta) - 8\cos^2(\theta)\sin^2(\theta))\, d\theta$$

All but three of these terms integrate to 0 over the interval $[0, 2\pi]$ leaving

$$\text{(area of } \mathcal{R}) = \int_0^{2\pi} (4\cos^4(\theta) + 2\sin^2(\theta) + \sin^2(2\theta))\, d\theta = 6\pi\,.$$

We have used the fact that $\int_0^{2\pi} \cos^4(\theta)\, d\theta = 3\pi/4$.

29. The region inside the curve $\mathcal{C}$ is $\mathcal{R} = \{(x, y) : 0 \le x \le 1, x^2 \le y \le \sqrt{x}\}$. Therefore, using Green's Theorem, the line integral I can be evaluated as follows.

$$I = \iint_{\mathcal{R}} \left(\frac{\partial}{\partial x}(2xy + \ln(1 + y^2)) - \frac{\partial}{\partial y}(y + \arcsin(x/2)) \right) dA$$

$$= \int_0^1 \int_{x^2}^{\sqrt{x}} (2y - 1)\, dy\, dx \;\; = \int_0^1 (x - \sqrt{x} - x^4 + x^2)\, dx = -1/30\,.$$

31. See the picture on the right. The integrand for the area integral, $y\, dx - x\, dy$, evaluates to

$$(\sin(2t) \cdot \cos(t) - \sin(t) \cdot 2\cos(2t))\, dt\,.$$

Since the curve is traced out in the *clockwise* direction for $0 \le t \le \pi$, the area integral sets up as follows.

$$A = \frac{1}{2} \int_0^\pi (\sin(2t)\cos(t) - 2\sin(t)\cos(2t))\, dt$$

Use the identity

$$\sin(a)\cos(b) = \tfrac{1}{2}(\sin(a + b) + \sin(a - b))$$

to convert this to

$$A = \frac{1}{2} \int_0^\pi (-(1/2)\sin(3t) + (3/2)\sin(t))\, dt$$

and evaluate: $A = 4/3$.

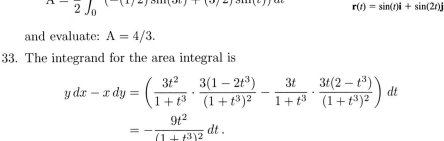

$\mathbf{r}(t) = \sin(t)\mathbf{i} + \sin(2t)\mathbf{j}$

33. The integrand for the area integral is

$$y\, dx - x\, dy = \left(\frac{3t^2}{1 + t^3} \cdot \frac{3(1 - 2t^3)}{(1 + t^3)^2} - \frac{3t}{1 + t^3} \cdot \frac{3t(2 - t^3)}{(1 + t^3)^2} \right) dt$$

$$= -\frac{9t^2}{(1 + t^3)^2}\, dt\,.$$

Therefore,

$$A = -\frac{1}{2} \int_0^\infty \frac{-9t^2}{(1 + t^3)^2}\, dt = -\frac{3}{2} \left(\frac{1}{1 + t^3} \right) \Big|_{t=0}^{t \to \infty} = \frac{3}{2}\,.$$

35. Let $v = 1$ in Green's Second Identity to obtain

$$\iint_{\mathcal{R}} (-\triangle u)\, dA = \int_{\mathcal{C}} (-u_x)\, dy + u_y\, dx\,.$$

If u is harmonic, then $\triangle u = 0$ and $\int_C u_y\, dx - u_x\, dy = 0$.

Calculator/Computer Exercises

37. Begin with the definition of the functions M and N for the vector field
    and the coordinate functions for the parametrizations of the two boundary
    curves.

    ```
    > M,N := unapply(sin(x+y^2),x,y),0:
      X1,Y1 := t,t^2:
      X2,Y2 := t,t:
    ```

    Calculate the integral around the boundary.

    ```
    > BoundaryIntegral = evalf(
        Int( M(X1,Y1)*diff(X1,t) + N(X1,Y1)*diff(Y1,t), t=0..1)
      + Int( M(X2,Y2)*diff(X2,t) + N(X2,Y2)*diff(Y2,t), t=1..0));
    ```

    $$BoundaryIntegral = -0.0703927038$$

    Calculate the integral over the enclosed region.

    ```
    > RegionIntegral = evalf( Int(Int(
      diff(N(x,y),x) - diff(M(x,y),y), y=x^2..x), x=0..1));
    ```

    $$RegionIntegral = -0.07039270459$$

39. As in Exercise 37.

    ```
    > M,N := unapply(x^2/(1+y^4),x,y),
            unapply(ln(1+x^4),x,y):
      X1,Y1 := t,0:
      X2,Y2 := t,sin(t):
    ```

    Calculate the integral around the boundary.

    ```
    > BoundaryIntegral = evalf(
        Int( M(X1,Y1)*diff(X1,t) + N(X1,Y1)*diff(Y1,t), t=0..Pi)
      + Int( M(X2,Y2)*diff(X2,t) + N(X2,Y2)*diff(Y2,t), t=Pi..0));
    ```

    $$BoundaryIntegral = 5.375521722$$

    Calculate the integral over the enclosed region.

    ```
    > RegionIntegral = evalf( Int(Int(
      diff(N(x,y),x) - diff(M(x,y),y), y=0..sin(x)), x=0..Pi));
    ```

    $$RegionIntegral = 5.375521722$$

13.6 Surface Integrals

Problems for Practice

1. Since $\sqrt{1 + f_x(x,y)^2 + f_y(x,y)^2} = \sqrt{1 + 9 + 16} = \sqrt{26}$, the surface area is

$$\iint_{\mathcal{R}} \sqrt{26}\,dA = \sqrt{26} \iint_{\mathcal{R}} dA = \sqrt{26}\,(\text{area of } \mathcal{R}) = 3\sqrt{26}\,.$$

3. Since $\sqrt{1 + f_x(x,y)^2 + f_y(x,y)^2} = \sqrt{1 + (x/\sqrt{x^2+y^2})^2 + (y/\sqrt{x^2+y^2})^2} = \sqrt{2}$ (see Example 2), the surface area is

$$\iint_{\mathcal{R}} \sqrt{2}\,dA = \sqrt{2} \iint_{\mathcal{R}} dA = \sqrt{2}\,(\text{area of } \mathcal{R}) = \sqrt{2}(9\pi - \pi) = 8\sqrt{2}\pi\,.$$

5. Since $\sqrt{1 + f_x(x,y)^2 + f_y(x,y)^2} = \sqrt{1 + (2x)^2 + (2y)^2} = \sqrt{1 + 4x^2 + 4y^2}$, the surface area evaluates as follows. Note the switch to polar coordinates.

$$\iint_{\mathcal{R}} \sqrt{1 + 4x^2 + 4y^2}\,dA = \int_0^{2\pi} \int_0^2 (\sqrt{1 + 4r^2}\,r\,dr\,d\theta$$

$$= \int_0^{2\pi} \left((1/12)(1 + 4r^2)^{3/2}\big|_0^2 \right) d\theta$$

$$= 1/12 \int_0^{2\pi} (17^{3/2} - 1)\,d\theta = (17^{3/2} - 1)\pi/6\,.$$

7. Since $\sqrt{1 + f_x(x,y)^2 + f_y(x,y)^2} = \sqrt{1 + x^2}$, the surface area is

$$\iint_{\mathcal{R}} \sqrt{1 + x^2}\,dA = \int_2^6 \int_0^x \sqrt{1 + x^2}\,dy\,dx$$

$$= \int_2^6 x\sqrt{1 + x^2}\,dx$$

$$= (1/3)(1 + x^2)^{3/2}\big|_2^6 = (37^{3/2} - 5^{3/2})/3\,.$$

9. Since $\sqrt{1 + f_x(x,y)^2 + f_y(x,y)^2} = \sqrt{1 + 1 + 36} = \sqrt{38}$, the surface area is

$$\iint_{\mathcal{R}} \sqrt{38}\,dA = \sqrt{38} \int_1^4 \int_2^{2x} dy\,dx$$

$$= \sqrt{38} \int_1^4 (2x - 2)\,dx = 9\sqrt{38}\,.$$

11. Since $\sqrt{1 + f_x(x,y)^2 + f_y(x,y)^2} = \sqrt{1 + (4x)^2 + (4y)^2} = \sqrt{1 + 16x^2 + 16y^2}$, the surface area evaluates as follows. Note the switch to polar coordinates.

$$\iint_{\mathcal{R}} \sqrt{1 + 16x^2 + 16y^2}\, dA = \int_0^{2\pi} \int_2^3 (\sqrt{1 + 16r^2}\, r\, dr\, d\theta$$

$$= \int_0^{2\pi} \left((1/48)(1 + 16r^2)^{3/2} \big|_2^3 \right) d\theta$$

$$= 1/48 \int_0^{2\pi} (145^{3/2} - 65^{3/2})\, d\theta$$

$$= (145^{3/2} - 65^{3/2})\pi/24\,.$$

13. The function defining the cone is $f(x,y) = 7 - (7/5)\sqrt{x^2 + y^2}$ over the region $\mathcal{R} = \{(x,y) : x^2 + y^2 < 25\}$. Thus $\sqrt{1 + f_x(x,y)^2 + f_y(x,y)^2} = \sqrt{1 + (49/25)(x^2/(x^2 + y^2) + y^2/(x^2 + y^2))} = \sqrt{74/25}$, and the surface area evaluates as follows.

$$\iint_{\mathcal{R}} \sqrt{74}/5\, dA = \sqrt{74}/5 \iint_{\mathcal{R}} dA = \sqrt{74}/5 \cdot (\text{area of } \mathcal{R}) = 5\pi\sqrt{74}\,.$$

15. The function defining the cone is $f(x,y) = \sqrt{x^2 + y^2}$ over the region $\mathcal{R} = \{(x,y) : 16 < x^2 + y^2 < 100\}$. Thus $\sqrt{1 + f_x(x,y)^2 + f_y(x,y)^2} = \sqrt{1 + x^2/(x^2 + y^2) + y^2/(x^2 + y^2)} = \sqrt{2}$, and the surface area evaluates as follows.

$$\iint_{\mathcal{R}} \sqrt{2}\, dA = \sqrt{2} \iint_{\mathcal{R}} dA = \sqrt{2} \cdot (\text{area of } \mathcal{R}) = 84\pi\sqrt{2}\,.$$

17. Since $\sqrt{1 + f_x(x,y)^2 + f_y(x,y)^2} = \sqrt{1 + x + y}$, the surface area is

$$\iint_{\mathcal{R}} \sqrt{1 + x + y}\, dA = \int_1^3 \int_2^5 \sqrt{1 + x + y}\, dy\, dx$$

$$= 2/3 \int_1^3 \left((1 + x + y)^{3/2} \big|_{y=2}^{y=5} \right) dx$$

$$= 2/3 \int_1^3 ((6 + x)^{3/2} - (3 + x)^{3/2})\, dx$$

$$= (4/15)((6 + x)^{5/2} - (3 + x)^{5/2}) \big|_1^3$$

$$= 4(275 - 6^{5/2} - 7^{5/2})/15\,.$$

19. The function defining the surface is $f(x,y) = 12 - 3x + 3y$. It lies over the region $\mathcal{R} = \{(x,y) : -1 \le y \le 1, -\sqrt{4 - 4y^2} \le x \le \sqrt{4 - 4y^2}\}$. Since $\sqrt{1 + f_x(x,y)^2 + f_y(x,y)^2} = \sqrt{1 + 9 + 9} = \sqrt{19}$, the surface area

calculation goes like this. Note the substitution $y = \sin(\theta)$, $dy = \cos(\theta)\, d\theta$ to go from line two to line three.

$$
\begin{aligned}
\iint_{\mathcal{R}} \sqrt{19}\, dA &= \sqrt{19} \int_{-1}^{1} \int_{-\sqrt{4-4y^2}}^{\sqrt{4-4y^2}} 1\, dx\, dy \\
&= \sqrt{19} \int_{-1}^{1} 2\sqrt{4 - 4y^2}\, dy \\
&= \sqrt{19} \int_{-\pi/2}^{\pi/2} 2\sqrt{4 - 4\sin^2(\theta)}\, \cos(\theta)\, d\theta \\
&= 4\sqrt{19} \int_{-\pi/2}^{\pi/2} \cos^2(\theta)\, d\theta = 4\sqrt{19} \cdot \pi/2 = 2\pi\sqrt{19}\,.
\end{aligned}
$$

21. The element of surface area is $dS = \sqrt{1 + f_x(x,y)^2 + f_y(x,y)^2}\, dA = \sqrt{1 + 4x^2 + 4y^2}\, dA$. Since $\phi(x, y, f(x,y)) = x - 2y + 3(x^2 + y^2 + 8)$, the surface integral sets up and can be split into two pieces as follows

$$
\begin{aligned}
\iint_{\mathcal{S}} \phi\, dS &= \iint_{\mathcal{R}} (x - 2y + 3x^2 + 3y^2 + 24)\sqrt{1 + 4x^2 + 4y^2}\, dA \\
&= \iint_{\mathcal{R}} (x - 2y)\sqrt{1 + 4x^2 + 4y^2}\, dA \\
&\quad + \iint_{\mathcal{R}} (3x^2 + 3y^2 + 24)\sqrt{1 + 4x^2 + 4y^2}\, dA\,.
\end{aligned}
$$

The first piece evaluates to 0 by symmetry, and polar coordinates help in the evaluation of what remains.

$$
\begin{aligned}
\iint_{\mathcal{S}} \phi\, dS &= \int_0^{2\pi} \int_0^{\sqrt{6}} (3r^2 + 24)\sqrt{1 + 4r^2}\, r\, dr\, d\theta \\
&= \int_0^{2\pi} \int_1^{25} \left(3 \cdot \frac{u - 1}{4} + 24 \right) \frac{\sqrt{u}}{8}\, du\, d\theta
\end{aligned}
$$

The substitution $u = 1 + 4r^2$, $du = 8r\, dr$ was used to obtain the second integral. It evaluates to $3574\pi/5$.

23. The element of surface area is $dS = \sqrt{1 + f_x(x,y)^2 + f_y(x,y)^2}\, dA = \sqrt{1 + 4x^2 + 4y^2}\, dA$. Since $\phi(x, y, f(x,y)) = (4x^2 + 4y^2 + 1)^{1/2}$, the surface integral calculation proceeds as follows. Note the switch to polar coordinates.

$$
\begin{aligned}
\iint_{\mathcal{S}} \phi\, dS &= \iint_{\mathcal{R}} \sqrt{4x^2 + 4y^2 + 1}\, \sqrt{1 + 4x^2 + 4y^2}\, dA \\
&= \int_0^{2\pi} \int_1^{2} (4r^2 + 1)\, r\, dr\, d\theta = 33\pi\,.
\end{aligned}
$$

25. The region is $\mathcal{R} = \{(x, y) : 0 \leq y \leq \sqrt{2}, 0 \leq x \leq y\}$. Since the element of surface area is $dS = \sqrt{1 + f_x(x, y)^2 + f_y(x, y)^2}\, dA = \sqrt{1 + 4y^2}\, dA$, and $\phi(x, y, f(x, y)) = 8$, the surface integral calculation goes like this.

$$\iint_{\mathcal{S}} \phi\, dS = \int_0^{\sqrt{2}} \int_0^y 8\sqrt{1 + 4y^2}\, dx\, dy$$

$$= 8 \int_0^{\sqrt{2}} y\sqrt{1 + 4y^2}\, dy$$

$$= 8 \cdot (1/12)(1 + 4y^2)^{3/2}\Big|_0^{\sqrt{2}} = 52/3.$$

27. The element of surface area is $dS = \sqrt{1 + f_x(x, y)^2 + f_y(x, y)^2}\, dA = \sqrt{1 + 9 + 49} = \sqrt{59}\, dA$. Since $\phi(x, y, f(x, y)) = x - xy + 3x - 7y$, the surface integral sets up as follows

$$\iint_{\mathcal{S}} \phi\, dS = \iint_{\mathcal{R}} (4x - xy - 7y)\sqrt{59}\, dA.$$

The region $\mathcal{R}$ is $\{(x, y) : 1 \leq x \leq 4, 1/2 \leq y \leq x/2\}$ so

$$\iint_{\mathcal{S}} \phi\, dS = \sqrt{59} \int_1^4 \int_{1/2}^{x/2} (4x - xy - 7y)\, dy\, dx$$

$$= \frac{1}{8}\sqrt{59} \int_1^4 (-x^3 + 9x^2 - 15x + 7)\, dx = \frac{135}{32}\sqrt{59}.$$

29. The parametrization is $\mathbf{r}(u, v) = \langle u^2 - v, u + v^2, v \rangle$. Therefore,

$$(\mathbf{r}_u \times \mathbf{r}_v)(u, v) = \det \left(\begin{bmatrix} \mathbf{i} & \mathbf{j} & \mathbf{k} \\ 2u & 1 & 0 \\ -1 & 2v & 1 \end{bmatrix} \right) = \langle 1, -2u, 4uv + 1 \rangle,$$

and the element of surface area is

$$\|\mathbf{r}_u \times \mathbf{r}_v\|\, du\, dv = \sqrt{1 + 4u^2 + (4uv + 1)^2}\, du\, dv.$$

Consequently, the surface area is

$$\iint_{\mathcal{R}} \|\mathbf{r}_u \times \mathbf{r}_v\|\, du\, dv = \int_0^1 \int_0^4 \sqrt{1 + 4u^2 + (4uv + 1)^2}\, dv\, du.$$

31. The parametrization is $\mathbf{r}(u, v) = \langle e^{u+v}, e^{u-v}, u \rangle$. Therefore,

$$(\mathbf{r}_u \times \mathbf{r}_v)(u, v) = \det \left(\begin{bmatrix} \mathbf{i} & \mathbf{j} & \mathbf{k} \\ e^{u+v} & e^{u-v} & 1 \\ e^{u+v} & -e^{u-v} & 0 \end{bmatrix} \right) = \langle e^{u-v}, e^{u+v}, -2e^{2u} \rangle,$$

and the element of surface area is

$$\|\mathbf{r}_u \times \mathbf{r}_v\| \, du \, dv = \sqrt{e^{2(u-v)} + e^{2(u+v)} + 4e^{4u}} \, du \, dv \,.$$

Consequently, the surface area is

$$\iint_{\mathcal{R}} \|\mathbf{r}_u \times \mathbf{r}_v\| \, du \, dv = \int_0^1 \int_{-1}^1 \sqrt{e^{2(u-v)} + e^{2(u+v)} + 4e^{4u}} \, dv \, du \,.$$

33. The parametrization is $\mathbf{r}(u,v) = \langle 2u - v, v + 2u, v - u \rangle$. Therefore,

$$(\mathbf{r}_u \times \mathbf{r}_v)(u,v) = \det\left(\begin{bmatrix} \mathbf{i} & \mathbf{j} & \mathbf{k} \\ 2 & 2 & -1 \\ -1 & 1 & 1 \end{bmatrix} \right) = \langle 3, -1, 4 \rangle \,,$$

and the element of surface area is

$$dS = \|\mathbf{r}_u \times \mathbf{r}_v\| \, du \, dv = \sqrt{9 + 1 + 16} \, du \, dv \,.$$

Since $\phi(\mathbf{r}(u,v)) = (2u - v) + (v + 2u) + (v - u) = 3u + v$, the surface integral evaluates as follows.

$$\iint_{\mathcal{S}} \phi \, dS = \iint_{\mathcal{R}} \phi(\mathbf{r}(u,v)) \, \|\mathbf{r}_u \times \mathbf{r}_v\| \, du \, dv$$
$$= \int_0^2 \int_0^3 (3u + v)\sqrt{26} \, dv \, du = 27\sqrt{26}$$

35. The parametrization is $\mathbf{r}(u,v) = \langle v, u, u + v \rangle$. Therefore,

$$(\mathbf{r}_u \times \mathbf{r}_v)(u,v) = \det\left(\begin{bmatrix} \mathbf{i} & \mathbf{j} & \mathbf{k} \\ 0 & 1 & 1 \\ 1 & 0 & 1 \end{bmatrix} \right) = \langle 1, 1, -1 \rangle \,,$$

and the element of surface area is

$$dS = \|\mathbf{r}_u \times \mathbf{r}_v\| \, du \, dv = \sqrt{1 + 1 + 1} \, du \, dv \,.$$

Since $\phi(\mathbf{r}(u,v)) = \sqrt{u^2} = u \ (u > 0)$, the surface integral evaluates as follows.

$$\iint_{\mathcal{S}} \phi \, dS = \iint_{\mathcal{R}} \phi(\mathbf{r}(u,v)) \, \|\mathbf{r}_u \times \mathbf{r}_v\| \, du \, dv$$
$$= \int_3^5 \int_2^3 u\sqrt{3} \, dv \, du = 8\sqrt{3}$$

Further Theory and Practice

37. The surface of revolution can be parametrized with

$$\mathbf{r}(x, \theta) = \langle x, f(x)\cos(\theta), f(x)\sin(\theta)\rangle$$

over the parameter domain $\mathcal{R} = \{(x, \theta) : a \le x \le b, 0 \le \theta \le 2\pi\}$. Therefore,

$$(\mathbf{r}_x \times \mathbf{r}_\theta)(x, \theta) = \det\left(\begin{bmatrix} \mathbf{i} & \mathbf{j} & \mathbf{k} \\ 1 & f'(x)\cos(\theta) & f'(x)\sin(\theta) \\ 0 & -f(x)\sin(\theta) & f(x)\cos(\theta) \end{bmatrix}\right)$$

$$= \langle f(x)f'(x)\cos^2(\theta) + f(x)f'(x)\sin^2(\theta), -f(x)\cos(\theta), -f(x)\sin(\theta)\rangle$$

$$= \langle f(x)f'(x), -f(x)\cos(\theta), -f(x)\sin(\theta)\rangle,$$

and the element of surface area is

$$\|\mathbf{r}_x \times \mathbf{r}_\theta\| \, dx \, d\theta = \sqrt{f(x)^2 f'(x)^2 + f(x)^2} \, dx \, d\theta.$$

Assuming that $f(x) \ge 0$, the surface area is

$$\iint_{\mathcal{R}} \|\mathbf{r}_x \times \mathbf{r}_\theta\| \, dx \, d\theta = \int_a^b \int_0^{2\pi} f(x)\sqrt{1 + f'(x)^2} \, d\theta \, dx$$

$$= 2\pi \int_a^b f(x)\sqrt{1 + f'(x)^2} \, dx.$$

39. Parametrize the surface with $\mathbf{r}(x, y) = \langle x, y, f(x, y)\rangle$. The vector $\mathbf{r}_x \times \mathbf{r}_y = \langle -f_x, -f_y, 1\rangle$ is normal to the surface and points upward. Therefore, the upward unit normal is $\mathbf{n} = (1 + (f_x)^2 + (f_y)^2)^{-1/2}\langle -f_x, -f_y, 1\rangle$ and

$$\cos(\gamma(x, y)) = \frac{1}{\sqrt{1 + f_x(x, y)^2 + f_y(x, y)^2}}.$$

Therefore, $\sec(\gamma(x, y)) = \sqrt{1 + f_x(x, y)^2 + f_y(x, y)^2}$ and

$$A(S) = \iint_{\mathcal{R}} \sqrt{1 + (f_x)^2 + (f_y)^2} \, dA = \iint_{\mathcal{R}} \sec(\gamma) \, dA.$$

Since the element of area is $dS = \sqrt{1 + (f_x)^2 + (f_y)^2} \, dA = \sec(\gamma) \, dA$ the integral of a function ϕ over the surface can be defined as

$$\iint_S \phi \, dS = \iint_{\mathcal{R}} \phi(\mathbf{r}(x, y))\sec(\gamma(x, y)) \, dA.$$

41. The surface should be $f(x, y) = (x^2 + y^2)/2$. The element of area on the surface is $dS = \sqrt{1 + x^2 + y^2} \, dx \, dy$. Since $\phi(x, y, f(x, y)) = 1/(x^2 + y^2 + 1)^3$, the integral of ϕ over the graph of f is

$$I = \lim_{R \to \infty}\left(\int_{-R}^R \int_{-R}^R \frac{1}{(x^2 + y^2 + 1)^3} \cdot \sqrt{1 + x^2 + y^2} \, dx \, dy\right).$$

Switch to polar coordinates to obtain

$$I = \lim_{R \to \infty} \left(\int_0^{2\pi} \int_0^R \frac{\sqrt{1+r^2}}{(r^2+1)^3} \cdot r \, dr \, d\theta \right) = 2\pi \lim_{R \to \infty} \int_0^R (r^2+1)^{-5/2} \cdot r \, dr$$

$$= 2\pi \lim_{R \to \infty} \left(-\frac{1}{3}(r^2+1)^{-3/2} \right) \Big|_0^R = \frac{2}{3}\pi \, .$$

43. By symmetry, $\bar{x} = \bar{y} = 0$. We may assume that the planar mass density of the cone is 1. Then $\bar{z} = M_{z=0}/M = \iint_{\mathcal{S}} z \, dS / \mathrm{A}(S)$.

The area computation

The element of surface area on the cone is

$$dS = \sqrt{1 + 4x^2/(x^2+y^2) + 4y^2/(x^2+y^2)} \, dx \, dy = \sqrt{5} \, dx \, dy \, .$$

Therefore,

$$\mathrm{A}(S) = \iint_{\mathcal{R}} \sqrt{5} \, dA = \sqrt{5} \cdot \pi \cdot 2^2 = 4\pi\sqrt{5} \, .$$

The computation of $M_{z=0}$

Since $z = 2\sqrt{x^2+y^2}$,

$$M_{z=0} = \int_{-2}^2 \int_{-\sqrt{4-x^2}}^{\sqrt{4-x^2}} 2\sqrt{x^2+y^2} \cdot \sqrt{5} \, dy \, dx$$

$$= 2\sqrt{5} \int_0^{2\pi} \int_0^2 r \cdot r \, dr \, d\theta = 32\pi\sqrt{5}/3 \, .$$

Therefore, $\bar{z} = \frac{32\pi\sqrt{5}/3}{4\pi\sqrt{5}} = 8/3$. The center of gravity of the cone is at the point $(0,0,8/3)$.

45. By symmetry, $\bar{x} = \bar{y} = 0$. We may assume that the planar mass density of the paraboloid is 1. Then $\bar{z} = M_{z=0}/M = \iint_{\mathcal{S}} z \, dS / \mathrm{A}(S)$.

The area computation

Note that the surface lies over the domain $\mathcal{R} = \{(x,y) : x^2 + y^2 < 4\}$. The element of surface area on the paraboloid is

$$dS = \sqrt{1 + 4x^2 + 4y^2} \, dx \, dy \, .$$

Therefore,

$$\mathrm{A}(S) = \int_{-2}^2 \int_{-\sqrt{4-x^2}}^{\sqrt{4-x^2}} \sqrt{1 + 4x^2 + 4y^2} \, dy \, dx$$

$$= \int_0^{2\pi} \int_0^2 \sqrt{1 + 4r^2} \cdot r \, dr \, d\theta$$

$$= 1/12 \int_0^{2\pi} \left((1+4r^2)^{3/2} \big|_0^2 \right) d\theta = \pi(17^{3/2} - 1)/6 \, .$$

The computation of $M_{z=0}$

Since $z = 4 - x^2 - y^2$,

$$M_{z=0} = \int_{-2}^{2} \int_{-\sqrt{4-x^2}}^{\sqrt{4-x^2}} (4 - x^2 - y^2) \cdot \sqrt{1 + 4x^2 + 4y^2}\, dy\, dx$$

$$= \int_{0}^{2\pi} \int_{0}^{2} (4 - r^2)\sqrt{1 + 4r^2} \cdot r\, dr\, d\theta$$

$$= 2\pi \int_{0}^{2} (4 - r^2)\sqrt{1 + 4r^2} \cdot r\, dr\,.$$

Continue with the substitution $u = 1 + 4r^2$, $du = 8r\, dr$. Noting that $4 - r^2 = (17 - u)/4$,

$$M_{z=0} = 2\pi \int_{1}^{17} \left(\frac{17 - u}{4}\right) \cdot \sqrt{u} \cdot \frac{1}{8}\, du$$

$$= \frac{\pi}{16} \int_{1}^{17} \left(17u^{1/2} - u^{3/2}\right) du$$

$$= \pi(289\sqrt{17} - 41)/60\,.$$

Therefore, $\bar{z} = \frac{\pi(289\sqrt{17}-41)/60}{\pi(17^{3/2}-1)/6} = \frac{289\sqrt{17}-41}{10(17\sqrt{17}-1)}$. The center of gravity of the cone is at the point $\left(0, 0, \frac{289\sqrt{17}-41}{10(17\sqrt{17}-1)}\right)$.

47. Use $F(x, y, z) = x^2 + y^2 + z^2 - a^2$. Then $\nabla F = \langle 2x, 2y, 2z \rangle$ so $\|\nabla F\| = 2\sqrt{x^2 + y^2 + z^2} = 2a$, $\nabla F \cdot \mathbf{k} = 2z = 2\sqrt{a^2 - x^2 - y^2}$, and the domain of integration is $\mathcal{R} = \{(x, y) : x^2 + y^2 < a^2\}$. Therefore,

$$\mathcal{A} = \iint_{\mathcal{R}} \frac{2a}{2\sqrt{a^2 - x^2 - y^2}}\, dA$$

$$= a \int_{-a}^{a} \int_{-\sqrt{a^2-x^2}}^{\sqrt{a^2-x^2}} \frac{1}{\sqrt{a^2 - x^2 - y^2}}\, dy\, dx$$

$$= a \int_{0}^{2\pi} \int_{0}^{a} \frac{1}{\sqrt{a^2 - r^2}} \cdot r\, dr\, d\theta$$

$$= a \int_{0}^{2\pi} \left(-(a^2 - r^2)^{1/2} \Big|_{r=0}^{r=a}\right) d\theta$$

$$= a \int_{0}^{2\pi} a\, d\theta = 2\pi a^2\,.$$

49. The formula in Exercise 46 can be used to calculate the area of the top half of the spheroid. Use $F(x, y, z) = x^2/a^2 + y^2/a^2 + z^2/c^2 - 1$. Then

$\nabla F = \langle 2x/a^2, 2y/a^2, 2z/c^2 \rangle$ so

$$\begin{aligned}
\|\nabla F\| &= 2\sqrt{x^2/a^4 + y^2/a^4 + z^2/c^4} \\
&= 2\sqrt{x^2/a^4 + y^2/a^4 + (1 - x^2/a^2 - y^2/a^2)/c^2} \\
&= \frac{2}{ac}\sqrt{a^2 + \left(\frac{c^2 - a^2}{a^2}\right)(x^2 + y^2)},
\end{aligned}$$

and

$$\nabla F \cdot \mathbf{k} = 2z/c^2 = \frac{2}{ac}\sqrt{a^2 - (x^2 + y^2)}.$$

Since the domain of integration is $\mathcal{R} = \{(x, y) : x^2 + y^2 < a^2\}$,

$$\begin{aligned}
\mathcal{A} &= 2\iint_{\mathcal{R}} \frac{\sqrt{a^2 + \left(\frac{c^2-a^2}{a^2}\right)(x^2 + y^2)}}{\sqrt{a^2 - (x^2 + y^2)}}\, dA \\
&= 2\int_0^{2\pi} \int_0^a \frac{\sqrt{a^2 + \alpha^2 r^2}}{\sqrt{a^2 - r^2}} \cdot r\, dr\, d\theta \\
&= 4\pi \int_0^a \frac{\sqrt{a^2 + \alpha^2 r^2}}{\sqrt{a^2 - r^2}} \cdot r\, dr,
\end{aligned}$$

where $\alpha = \sqrt{c^2 - a^2}/a$. Now substitute $u = a^2 + \alpha^2 r^2$, $du = 2\alpha^2 r\, dr$. Note that $a^2 - r^2 = (c^2 - u)/\alpha^2$ so

$$\begin{aligned}
\mathcal{A} &= 4\pi \int_{a^2}^{c^2} \frac{\sqrt{u}}{(\sqrt{c^2 - u})/\alpha} \cdot \frac{1}{2\alpha^2}\, du \\
&= \frac{2\pi}{\alpha} \int_{a^2}^{c^2} \frac{\sqrt{u}}{\sqrt{c^2 - u}}\, du \qquad \leftarrow \text{Substitute } u = c^2 \sin^2(v). \\
&= \frac{2\pi}{\alpha} \int_{\arcsin(a/c)}^{\pi/2} \frac{c\sin(v)}{c\cos(v)} \cdot 2c^2 \sin(v)\cos(v)\, dv \\
&= \frac{4\pi c^2}{\alpha} \int_{\arcsin(a/c)}^{\pi/2} \sin^2(v)\, dv \\
&= \frac{2\pi c^2}{\alpha} \int_{\arcsin(a/c)}^{\pi/2} (1 - \cos(2v))\, dv \\
&= \frac{2\pi c^2}{\alpha} \cdot (v - \sin(v)\cos(v))\Big|_{v=\arcsin(a/c)}^{v=\pi/2} \\
&= \frac{2\pi c^2}{\alpha} \cdot \left(\frac{\pi}{2} - \arcsin(a/c) + \frac{a}{c} \cdot \frac{\sqrt{c^2 - a^2}}{c}\right) \\
&= \frac{2\pi c^2}{\alpha} \cdot \left(\arcsin(\sqrt{c^2 - a^2}/c) + \frac{\alpha\, a^2}{c^2}\right) \\
&= 2\pi a^2 + 2\pi(ac/e)\arcsin(e),
\end{aligned}$$

where $e = \sqrt{c^2 - a^2}/c$.

51. The cylinder is parametrized with $\mathbf{r}(\theta, z) = \langle a\cos(\theta), b\sin(\theta), z \rangle$. Since

$$\mathbf{r}_\theta \times \mathbf{r}_z(\theta, z) = \det\left(\begin{bmatrix} \mathbf{i} & \mathbf{j} & \mathbf{k} \\ -a\sin(\theta) & a\cos(\theta) & 0 \\ 0 & 0 & 1 \end{bmatrix}\right) = \langle -a\cos(\theta), a\sin(\theta), 0 \rangle,$$

the surface area element is $dS = \|\mathbf{r}_\theta \times \mathbf{r}_z(\theta, z)\| \, dA = \sqrt{a^2\cos^2(\theta) + a^2\sin^2(\theta)} \, dA = a \, d\theta \, dz$.

53. Using the parametrization $x = 4\cos(\theta)$, $y = y$, $z = 4\sin(\theta)$, the surface is $\mathcal{S} = \{(4\cos(\theta), y, 4\sin(\theta)) : 0 \leq \theta \leq \pi/2, 2 \leq y \leq 4\}$. We wish to integrate $\phi(x, y, z) = xyz = 4\cos(\theta) \cdot y \cdot 4\sin(\theta)$ and, according to Exercise 51, $dS = 4 \, dy \, d\theta$. Therefore, the surface integral is

$$I = \int_0^{\pi/2} \int_2^4 16y\cos(\theta)\sin(\theta) \cdot 4 \, dy \, d\theta = 32 \int_0^{\pi/2} \left(y^2\cos(\theta)\sin(\theta)\right)\Big|_{y=2}^{y=4} d\theta$$

$$= 32 \cdot 12 \int_0^{\pi/2} \cos(\theta)\sin(\theta) \, d\theta = 32 \cdot 12 \cdot \frac{1}{2}\sin^2(\theta)\Big|_0^{\pi/2} = 192.$$

Calculator/Computer Exercises

55. Define the function f, the element of surface area, and then calculate the area.

```
> f := exp(x^2):
  dS := sqrt(1+diff(f,x)^2 + diff(f,y)^2):
  Area = evalf( Int(Int( dS, x=-1..1), y=-1..1));
```

$$Area = 8.510465659$$

57. Define the function f, the element of surface area, and then calculate the area.

```
> f := 2 - x^2*y^4:
  dS := sqrt(1+diff(f,x)^2 + diff(f,y)^2):
  Area = evalf( Int(Int( dS, x=-1..1), y=-1..1));
```

$$Area = 4.741207007$$

13.7   Stokes's Theorem

Problems for Practice

1. <u>Surface integral calculation.</u> The curl of $\mathbf{F}$ is

$$\mathbf{curl}(\mathbf{F}) = \det\left(\begin{bmatrix} \mathbf{i} & \mathbf{j} & \mathbf{k} \\ \frac{\partial}{\partial x} & \frac{\partial}{\partial y} & \frac{\partial}{\partial z} \\ z^3 & -xy & xz \end{bmatrix}\right) = \langle 0, 3z^2 - z, -y \rangle\,.$$

An upward normal for the surface is given by $\mathbf{N} = \langle -2, 1, 1 \rangle$. Therefore, on the surface, where $z = 2x - y$,

$$\mathbf{curl}(\mathbf{F}) \cdot \mathbf{n}\, dS = \langle 0, 3z^2 - z, -y \rangle \cdot \left(\frac{\mathbf{N}}{\|\mathbf{N}\|}\right) \|\mathbf{N}\|\, dA$$
$$= (3(2x - y)^2 - 2x)\, dA$$

and

$$\iint_{\mathcal{S}} \mathbf{curl}(\mathbf{F}) \cdot \mathbf{n}\, dS = \int_{-2}^{2} \int_{-1}^{1} (3(2x - y)^2 - 2x)\, dy\, dx$$
$$= \int_{-2}^{2} (24x^2 - 4x + 2)\, dx = 136\,.$$

<u>Line integral calculation.</u> The boundary of the surface $\mathcal{S}$ consists of four line segments. In the counter-clockwise direction (viewed from the positive z-axis) they go from $P = (-2, -1, -3)$ to $Q = (2, -1, 5)$ to $R = (2, 1, 3)$ to $S = (-2, 1, -5)$, and then back to P.

The computation of the line integral from P to Q is shown below. The other three computations are carried out in a similar fashion. $\overline{PQ}$ can be parametrized with $x = -2 + 4t$, $y = -1$, and $z = -3 + 8t$, $0 \le t \le 1$. Therefore,

$$\int_{\overline{PQ}} \mathbf{F} \cdot d\mathbf{r} = \int_0^1 \left((-3 + 8t)^3 \cdot 4 - (-2 + 4t)(-1) \cdot 0 + (-2 + 4t)(-3 + 8t) \cdot 8\right) dt$$
$$= \int_0^1 (2048t^3 - 2048t^2 + 640t - 60)\, dt = 268/3\,.$$

You can verify that

$$\int_{\overline{QR}} \mathbf{F} \cdot d\mathbf{r} = -16\,, \quad \int_{\overline{RS}} \mathbf{F} \cdot d\mathbf{r} = 140/3\,, \quad \int_{\overline{SP}} \mathbf{F} \cdot d\mathbf{r} = 16\,.$$

3. <u>Surface integral calculation.</u> The curl of $\mathbf{F}$ is

$$\mathbf{curl}(\mathbf{F}) = \det\left(\begin{bmatrix} \mathbf{i} & \mathbf{j} & \mathbf{k} \\ \frac{\partial}{\partial x} & \frac{\partial}{\partial y} & \frac{\partial}{\partial z} \\ e^z & 2 & -y \end{bmatrix}\right) = \langle -1, e^z, 0 \rangle\,.$$

An upward normal for the surface is given by $\mathbf{N} = \langle -1, 2, 1 \rangle$. Therefore, on the surface, where $z = x - 2y$,

$$\mathbf{curl}(\mathbf{F}) \cdot \mathbf{n} \, dS = \langle -1, e^{x-2y}, 0 \rangle \cdot \left(\frac{\mathbf{N}}{\|\mathbf{N}\|} \right) \|\mathbf{N}\| \, dA$$

$$= (1 + 2e^{x-2y}) \, dA$$

and

$$\iint_{\mathcal{S}} \mathbf{curl}(\mathbf{F}) \cdot \mathbf{n} \, dS = \int_0^1 \int_{\ln(2)}^{\ln(4)} (1 + 2e^{x-2y}) \, dx \, dy$$

$$= \int_0^1 (\ln(2) + 4e^{-2y}) \, dx = \ln(2) - 2e^{-2} + 2 .$$

Line integral calculation. The boundary of the surface $\mathcal{S}$ consists of four line segments. In the counter-clockwise direction (viewed from the positive z-axis) they go from $P = (\ln(2), 0, \ln(2))$ to $Q = (\ln(4), 0, \ln(4))$ to $R = (\ln(4), 1, \ln(4) - 2)$ to $S = (\ln(2), 1, \ln(2) - 2)$, and then back to P.

The computation of the line integral from P to Q is shown below. The other three computations are carried out in a similar fashion. $\overline{PQ}$ can be parametrized with $x = \ln(2) + t \ln(2)$, $y = 0$, and $z = \ln(2) + t \ln(2)$, $0 \leq t \leq 1$. Therefore,

$$\int_{\overline{PQ}} \mathbf{F} \cdot d\mathbf{r} = \int_0^1 \left(e^{\ln(2) + t \ln(2)} \cdot \ln(2) + 2 \cdot 0 - 0 \cdot \ln(2) \right) dt$$

$$= \ln(2) \int_0^1 2^{t+1} \, dt = 2 .$$

You can verify that

$$\int_{\overline{QR}} \mathbf{F} \cdot d\mathbf{r} = 3 , \quad \int_{\overline{RS}} \mathbf{F} \cdot d\mathbf{r} = \ln(2) - 2e^{-2} , \quad \int_{\overline{SP}} \mathbf{F} \cdot d\mathbf{r} = -3 .$$

5. Surface integral calculation. The curl of $\mathbf{F}$ is

$$\mathbf{curl}(\mathbf{F}) = \det \left(\begin{bmatrix} \mathbf{i} & \mathbf{j} & \mathbf{k} \\ \frac{\partial}{\partial x} & \frac{\partial}{\partial y} & \frac{\partial}{\partial z} \\ xz & -x & yz \end{bmatrix} \right) = \langle z, x, -1 \rangle .$$

An upward normal for the surface is given by

$$\mathbf{N} = \langle x(1 - x^2 - y^2)^{-1/2}, y(1 - x^2 - y^2)^{-1/2}, 1 \rangle .$$

Therefore, on the surface, where $z = (1 - x^2 - y^2)^{1/2}$,

$$\mathbf{curl}(\mathbf{F}) \cdot \mathbf{n} \, dS = \langle (1 - x^2 - y^2)^{1/2}, x, -1 \rangle \cdot \left(\frac{\mathbf{N}}{\|\mathbf{N}\|} \right) \|\mathbf{N}\| \, dA$$

$$= (x + xy(1 - x^2 - y^2)^{-1/2} - 1) \, dA$$

and

$$\iint_{\mathcal{S}} \mathbf{curl}(\mathbf{F}) \cdot \mathbf{n} \, dS = \iint_{\mathcal{R}} (x + xy(1 - x^2 - y^2)^{-1/2} - 1) \, dx \, dy$$
$$= \iint_{\mathcal{R}} (-1) \, dx \, dy = -\pi \, .$$

Note that the first two terms in the integrand evaluate to 0 by symmetry.

Line integral calculation. The boundary curve $\mathcal{C}$ for the surface $\mathcal{S}$ can be parametrized with $\mathbf{r}(t) = \langle \cos(t), \sin(t), 0 \rangle$, $0 \leq t \leq 2\pi$. Since $\mathbf{F}(\mathbf{r}(t)) = \langle 0, -\cos(t), 0 \rangle$ and $\mathbf{r}'(t) = \langle -\sin(t), \cos(t), 0 \rangle$, $\mathbf{F}(\mathbf{r}(t)) \cdot \mathbf{r}'(t) = -\cos^2(t)$ and

$$\oint_{\mathcal{C}} \mathbf{F} \cdot d\mathbf{r} = \int_0^{2\pi} (-\cos^2(t)) \, dt = -\pi \, .$$

7. <u>Surface integral calculation.</u> The curl of $\mathbf{F}$ is

$$\mathbf{curl}(\mathbf{F}) = \det \left(\begin{bmatrix} \mathbf{i} & \mathbf{j} & \mathbf{k} \\ \frac{\partial}{\partial x} & \frac{\partial}{\partial y} & \frac{\partial}{\partial z} \\ 7z & -5x & 3y \end{bmatrix} \right) = \langle 3, 7, -5 \rangle \, .$$

An upward normal for the surface is given by $\mathbf{N} = \langle -2x, 0, 1 \rangle$. Therefore, on the surface, where $z = x^2$,

$$\mathbf{curl}(\mathbf{F}) \cdot \mathbf{n} \, dS = \langle 3, 7, -5 \rangle \cdot \left(\frac{\mathbf{N}}{\|\mathbf{N}\|} \right) \|\mathbf{N}\| \, dA$$
$$= (-6x - 5) \, dA$$

and

$$\iint_{\mathcal{S}} \mathbf{curl}(\mathbf{F}) \cdot \mathbf{n} \, dS = \int_0^3 \int_1^2 (-6x - 5) \, dy \, dx$$
$$= \int_0^3 (-6x - 5) \, dx = -42 \, .$$

Line integral calculation. The boundary of the surface $\mathcal{S}$ consists of four curves. Two are straight line segments and two are parabolas. In the counter-clockwise direction (viewed from the positive z-axis) the parabola $\mathcal{C}_1$ is from $P = (0, 1, 0)$ to $Q = (3, 1, 9)$ parametrized by $\mathbf{r}_1(t) = \langle 3t, 1, 9t^2 \rangle$, $0 \leq t \leq 1$.

$$\int_{\mathcal{C}_1} \mathbf{F} \cdot d\mathbf{r} = \int_0^1 (63t^2 \cdot 3 - 15t \cdot 0 + 3 \cdot 18t) \, dt = 90 \, .$$

This is followed by the line segment from Q to $R = (3, 2, 9)$ parametrized by $\mathbf{r}_2(t) = \langle 3, 1 + t, 9 \rangle$, $0 \leq t \leq 1$.

$$\int_{\overline{QR}} \mathbf{F} \cdot d\mathbf{r} = \int_0^1 \left(63 \cdot 0 - 15 \cdot 1 + 3(1 + t) \cdot 0 \right) dt = -15 \,.$$

You can verify that the other two line integrals evaluate as follows.

The parabola $\mathcal{C}_2$ from R to $S = (0, 2, 0)$, $\mathbf{r}_3(t) = \langle 3 - 3t, 2, (3 - 3t)^2 \rangle$, $0 \leq t \leq 1$:

$$\int_{\mathcal{C}_2} \mathbf{F} \cdot d\mathbf{r} = -117 \,.$$

The line segment from S to back to P:

$$\int_{\overline{SP}} \mathbf{F} \cdot d\mathbf{r} = 0 \,.$$

9. <u>Surface integral calculation.</u> The curl of $\mathbf{F}$ is

$$\mathbf{curl}(\mathbf{F}) = \det \left(\begin{bmatrix} \mathbf{i} & \mathbf{j} & \mathbf{k} \\ \frac{\partial}{\partial x} & \frac{\partial}{\partial y} & \frac{\partial}{\partial z} \\ x & y - z^2 & 1 \end{bmatrix} \right) = \langle -2z, 0, 0 \rangle \,.$$

An upward normal for the surface is given by $\mathbf{N} = \langle -1, -1, 1 \rangle$. Therefore, on the surface, where $z = x + y + 3$,

$$\mathbf{curl}(\mathbf{F}) \cdot \mathbf{n} \, dS = \langle -2(x + y + 3), 0, 0 \rangle \cdot \left(\frac{\mathbf{N}}{\|\mathbf{N}\|} \right) \|\mathbf{N}\| \, dA$$

$$= (-2x - 2y - 6) \, dA$$

and

$$\iint_{\mathcal{S}} \mathbf{curl}(\mathbf{F}) \cdot \mathbf{n} \, dS = \int_0^1 \int_{-y}^y (-2x - 2y - 6) \, dx \, dy$$

$$= \int_0^1 (-4y^2 - 12y) \, dx = -22/3 \,.$$

<u>Line integral calculation.</u> The boundary of the surface $\mathcal{S}$ consists of three line segments. In the counter-clockwise direction (viewed from the positive z-axis) they go from $P = (0, 0, 3)$ to $Q = (1, 1, 5)$ to $R = (-1, 1, 3)$, and then back to P.

The computation of the line integral from P to Q is shown below. The other two computations are carried out in a similar fashion. $\overline{PQ}$ can be parametrized with $x = t$, $y = t$, and $z = 3 + 2t$, $0 \leq t \leq 1$. Therefore,

$$\int_{\overline{PQ}} \mathbf{F} \cdot d\mathbf{r} = \int_0^1 \left(t \cdot 1 + (t - (3 + 2t)^2) \cdot 1 + 1 \cdot 2 \right) dt = -40/3 \,.$$

You can verify that

$$\int_{\overline{QR}} \mathbf{F} \cdot d\mathbf{r} = -2 , \quad \int_{\overline{RP}} \mathbf{F} \cdot d\mathbf{r} = 8 .$$

11. <u>Surface integral calculation</u>. The curl of $\mathbf{F}$ is

$$\mathbf{curl}(\mathbf{F}) = \det\left(\begin{bmatrix} \mathbf{i} & \mathbf{j} & \mathbf{k} \\ \frac{\partial}{\partial x} & \frac{\partial}{\partial y} & \frac{\partial}{\partial z} \\ z & 0 & -x \end{bmatrix}\right) = \langle 0, -2, 0 \rangle .$$

An upward normal for the surface is given by $\mathbf{N} = \langle -1, -3, 1 \rangle$. Therefore, on the surface, where $z = x + 3y$,

$$\mathbf{curl}(\mathbf{F}) \cdot \mathbf{n} \, dS = \langle 0, -2, 0 \rangle \cdot \left(\frac{\mathbf{N}}{\|\mathbf{N}\|}\right) \|\mathbf{N}\| \, dA$$

$$= 6 \, dA$$

and

$$\iint_{\mathcal{S}} \mathbf{curl}(\mathbf{F}) \cdot \mathbf{n} \, dS = \int_{-1}^{1} \int_{-1}^{1} (-6) \, dy \, dx = -24 .$$

<u>Line integral calculation</u>. The boundary of the surface $\mathcal{S}$ consists of four line segments. In the counter-clockwise direction (viewed from the positive z-axis) they go from $P = (-1, -1, -4)$ to $Q = (1, -1, -2)$ to $R = (1, 1, 4)$ to $S = (-1, 1, 2)$, and then back to P.

The computation of the line integral from P to Q is shown below. The other three computations are carried out in a similar fashion. $\overline{PQ}$ can be parametrized with $x = -1 + 2t$, $y = -1$, and $z = -4 + 2t$, $0 \le t \le 1$. Therefore,

$$\int_{\overline{PQ}} \mathbf{F} \cdot d\mathbf{r} = \int_{0}^{1} \left((-4 + 2t) \cdot 2 + 0 \cdot 0 - (-1 + 2t) \cdot 2\right) dt = -6 .$$

You can verify that

$$\int_{\overline{QR}} \mathbf{F} \cdot d\mathbf{r} = -6 , \quad \int_{\overline{RS}} \mathbf{F} \cdot d\mathbf{r} = -6 , \quad \int_{\overline{SP}} \mathbf{F} \cdot d\mathbf{r} = -6 .$$

13. The surface is the part of the sphere that lies above the xy-plane. Since the normal is upward, its boundary: $\mathcal{C} = \{(x, y, 0) : x^2 + y^2 = 9\}$, should be traversed in the counter-clockwise direction as viewed from the positive z-axis; parametrize it with $\mathbf{r}(t) = \langle 3\cos(t), 3\sin(t), 0 \rangle$. Since $\mathbf{F}(\mathbf{r}(t)) =$

$\langle 3\sin(t), 0, 1\rangle$ and $\mathbf{r}'(t) = \langle -3\sin(t), 3\cos(t), 0\rangle$, $\mathbf{F}(\mathbf{r}(t)) \cdot \mathbf{r}'(t) = -9\sin^2(t)$. The integral of the vector field $\mathbf{F}$ around the boundary is

$$\int_{\mathcal{C}} \mathbf{F} \cdot d\mathbf{r} = \int_0^{2\pi} (-9\sin^2(t))\, dt = -9\pi\,.$$

Therefore, $\iint_{\mathcal{S}} \mathbf{curl}(\mathbf{F}) \cdot \mathbf{n}\, dS = \int_{\mathcal{C}} \mathbf{F} \cdot d\mathbf{r} = -9\pi\,.$

15. The surface is a hollow cylinder oriented with an inward normal. The top boundary curve $\mathcal{C}_1$ ($z = 5$) is a circle which must be traversed in the counter-clockwise direction when viewed from above. One parametrization is $x = 3\cos(t)$, $y = 3\sin(t)$, $z = 5$, $0 \le t \le 2\pi$. The integral of the vector field $\mathbf{F}$ around this circle is

$$\int_{\mathcal{C}_1} \mathbf{F} \cdot d\mathbf{r} = \int_0^{2\pi} 15\cos(t) \cdot (3\cos(t))\, dt = 45\pi\,.$$

The bottom boundary curve $\mathcal{C}_2$ ($z = -2$) is another circle, and it must be traversed in the clockwise direction when viewed from above. One parametrization is $x = 3\cos(t)$, $y = -3\sin(t)$, $z = -2$, $0 \le t \le 2\pi$. The integral of the vector field $\mathbf{F}$ around the bottom circle is

$$\int_{\mathcal{C}_2} \mathbf{F} \cdot d\mathbf{r} = \int_0^{2\pi} (-6\cos(t)) \cdot (-3\cos(t))\, dt = 18\pi\,.$$

Therefore, $\iint_{\mathcal{S}} \mathbf{curl}(\mathbf{F}) \cdot \mathbf{n}\, dS = \int_{\mathcal{C}_1} \mathbf{F} \cdot d\mathbf{r} + \int_{\mathcal{C}_2} \mathbf{F} \cdot d\mathbf{r} = 63\pi\,.$

17. Let $\mathcal{C}$ denote the boundary of the surface $\mathcal{S}$. It consists of the line segments from $P = (0,0,0)$ to $Q = (2,0,0)$ to $R = (2,2,0)$ to $S = (0,2,0)$, and then back to P. Therefore, the line integrals are in the (x,y)-plane ($z = 0$). Each parametrization in the table below is for $0 \le t \le 2$.

$\mathcal{C}$	Parametrization	$\int_{\mathcal{C}} y^2\, dx + x\, dy$
$\overline{PQ}$	$x = t,\ y = 0$	$\int_0^2 (0 \cdot 1 + t \cdot 0)\, dt = 0$
$\overline{QR}$	$x = 2,\ y = t$	$\int_0^2 (t^2 \cdot 0 + 2 \cdot 1)\, dt = 4$
$\overline{RS}$	$x = 2 - t,\ y = 2$	$\int_0^2 (4 \cdot (-1) + (2 - t) \cdot 0)\, dt = -8$
$\overline{SP}$	$x = 0,\ y = 2 - t$	$\int_0^2 ((2 - t)^2 \cdot 0 + 0 \cdot (-1))\, dt = 0$

Therefore, $\iint_{\mathcal{S}} \mathbf{curl}(\mathbf{F}) \cdot \mathbf{n}\, dS = \int_{\mathcal{C}} \mathbf{F} \cdot d\mathbf{r} = -4\,.$

Further Theory and Practice

19. Surface integral calculation. The curl of $\mathbf{F}$ is

$$\mathbf{curl}(\mathbf{F}) = \det\left(\begin{bmatrix} \mathbf{i} & \mathbf{j} & \mathbf{k} \\ \frac{\partial}{\partial x} & \frac{\partial}{\partial y} & \frac{\partial}{\partial z} \\ y^3 & z^3 & x^3 \end{bmatrix}\right) = \langle -3z^2, -3x^2, -3y^2\rangle\,.$$

The surface can be parametrized by $\mathbf{r}(z, x) = \langle x, g(x, z), z \rangle$ where $g(x, z) = (1 - x^2 - z^2)^{1/2}$. The normal is $\mathbf{N} = \langle -g_x, 1, -g_z \rangle = \langle x(1 - x^2 - z^2)^{-1/2}, 1, z(1 - x^2 - z^2)^{-1/2} \rangle$. Therefore, on the surface, where $y = (1 - x^2 - z^2)^{1/2}$,

$$\mathbf{curl}(\mathbf{F}) \cdot \mathbf{n}\, dS = \langle -3z^2, -3x^2, -3(1 - x^2 - z^2) \rangle \cdot \left(\frac{\mathbf{N}}{\|\mathbf{N}\|} \right) \|\mathbf{N}\|\, dA$$

$$= (-3xz^2(1 - x^2 - z^2)^{-1/2} - 3x^2 - 3z(1 - x^2 - z^2)^{1/2})\, dA$$

and, with $\mathcal{R} = \{(x, z) : -1 \le x \le 1, -\sqrt{1 - x^2} \le z \le \sqrt{1 - x^2}\}$,

$$\iint_{\mathcal{S}} \mathbf{curl}(\mathbf{F}) \cdot \mathbf{n}\, dS = \iint_{\mathcal{R}} (-3xz^2(1 - x^2 - z^2)^{-1/2} - 3x^2 - 3z(1 - x^2 - z^2)^{1/2})\, dA$$

$$= \int_{-1}^{1} \int_{-\sqrt{1-x^2}}^{\sqrt{1-x^2}} (-3x^2)\, dz\, dx$$

$$= -6 \int_{-1}^{1} x^2 \sqrt{1 - x^2}\, dx \qquad \leftarrow \text{Substitute } x = \sin(u).$$

$$= -3\pi/4.$$

Note that the first and third terms in the first integral yeild 0 by symmetry.

Line integral calculation. The boundary $\mathcal{C}$ of the hemisphere is a circle in the xz plane, $\mathbf{r}(t) = \langle \sin(t), 0, \cos(t) \rangle$, $0 \le t \le 2\pi$, gives the correct orientation. Since $\mathbf{F}(\mathbf{r}(t)) = \langle 0, \cos^3(t), \sin^3(t) \rangle$ and $\mathbf{r}'(t) = \langle \cos(t), 0, -\sin(t) \rangle$, $\mathbf{F}(\mathbf{r}(t)) \cdot \mathbf{r}'(t) = -\sin^4(t)$ and

$$\oint_{\mathcal{C}} \mathbf{F} \cdot d\mathbf{r} = \int_{0}^{2\pi} (-\sin^4(t))\, dt = -3\pi/4.$$

The identities $\sin^4(t) = (1 - \cos^2(t)) \sin^2(t) = \sin^2(t) - \frac{1}{4} \sin^2(2t)$ can be used to evaluate the integral.

21. Surface integral calculation. The curl of $\mathbf{F}$ is

$$\mathbf{curl}(\mathbf{F}) = \det \left(\begin{bmatrix} \mathbf{i} & \mathbf{j} & \mathbf{k} \\ \frac{\partial}{\partial x} & \frac{\partial}{\partial y} & \frac{\partial}{\partial z} \\ y & -x & 3 \end{bmatrix} \right) = \langle 0, 0, -2 \rangle.$$

The surface can be parametrized by $\mathbf{r}(u, \theta) = \langle u \cos(\theta), u \sin(\theta), u \rangle$. The induced normal is $\mathbf{r}_u \times \mathbf{r}_\theta = \langle -u \cos(\theta), -u \sin(\theta), u \rangle$ (verify). The correctly oriented normal is: $\mathbf{N} = \langle u \cos(\theta), u \sin(\theta), -u \rangle$. Therefore,

$$\mathbf{curl}(\mathbf{F}) \cdot \mathbf{n}\, dS = \langle 0, 0, -2 \rangle \cdot \left(\frac{\mathbf{N}}{\|\mathbf{N}\|} \right) \|\mathbf{N}\|\, dA$$

$$= 2u\, dA$$

and, with $\mathcal{R} = \{(u, \theta) : 0 \le u \le 2, 0 \le \theta \le 2\pi\}$,

$$\iint_{\mathcal{S}} \mathbf{curl}(\mathbf{F}) \cdot \mathbf{n}\, dS = \iint_{\mathcal{R}} 2u\, dA$$

$$= \int_0^{2\pi} \int_0^2 2u\, du\, d\theta = 8\pi\,.$$

Line integral calculation. The boundary $\mathcal{C}$ of the cone is a circle in the $z = 2$ plane; $\mathbf{r}(t) = \langle 2\sin(t), 2\cos(t), 2 \rangle$, $0 \le t \le 2\pi$, gives the correct orientation. Since $\mathbf{F}(\mathbf{r}(t)) = \langle 2\cos(t), -2\sin(t), 3 \rangle$ and $\mathbf{r}'(t) = \langle 2\cos(t), -2\sin(t), 0 \rangle$, $\mathbf{F}(\mathbf{r}(t)) \cdot \mathbf{r}'(t) = 4\cos^2(t) + 4\sin^2(t) = 4$ and

$$\oint_{\mathcal{C}} \mathbf{F} \cdot d\mathbf{r} = \int_0^{2\pi} 4\, dt = 8\pi\,.$$

23. Surface integral calculation. The curl of $\mathbf{F}$ is

$$\mathbf{curl}(\mathbf{F}) = \det\left(\begin{bmatrix} \mathbf{i} & \mathbf{j} & \mathbf{k} \\ \frac{\partial}{\partial x} & \frac{\partial}{\partial y} & \frac{\partial}{\partial z} \\ z & y & z \end{bmatrix} \right) = \langle 0, 1, 0 \rangle\,.$$

A correctly oriented normal to the surface is $\mathbf{N} = \langle 2x, 2y, -1 \rangle$. Therefore, since $z = x^2 + y^2$, on the surface,

$$\mathbf{curl}(\mathbf{F}) \cdot \mathbf{n}\, dS = \langle 0, 1, 0 \rangle \cdot \left(\frac{\mathbf{N}}{\|\mathbf{N}\|} \right) \|\mathbf{N}\|\, dA$$

$$= 2y\, dA\,.$$

The surface is the portion of the paraboloid above the circluar disk $x^2 + y^2 \le x + y$ in the xy-plane. In polar coordinates, this region is $\mathcal{R} = \{(r, \theta) : -\pi/4 \le 3\pi/4, 0 \le r \le \sin(\theta) + \cos(\theta)\}$ (draw a picture). Therefore,

$$\iint_{\mathcal{S}} \mathbf{curl}(\mathbf{F}) \cdot \mathbf{n}\, dS = \iint_{\mathcal{R}} 2y\, dA$$

$$= \int_{-\pi/4}^{3\pi/4} \int_0^{\sin(\theta)+\cos(\theta)} 2r\sin(\theta) \cdot r\, dr\, d\theta$$

$$= 2/3 \int_{-\pi/4}^{3\pi/4} \sin(\theta)(\sin(\theta) + \cos(\theta))^3\, d\theta$$

$$= \pi/2$$

The integral calculation is straight-forward, but rather long. Note that the integrand expands to the four pieces shown below and each one of them integrates easily with the application of the appropriate identities.

$$\sin^4(\theta) + 3\sin^3(\theta)\cos(\theta) + 3\sin^2(\theta)\cos^2(\theta) + \sin(\theta)\cos^3(\theta)\,.$$

Line integral calculation. The boundary curve $\mathcal{C}$ can be parametrized using a circle parametrization in the xy-plane, and lifting the z-coordinate to the plane $z = x + y$ that contains $\mathcal{C}$:

$$\mathbf{r}(t) = \langle 1/2 + (\sqrt{2}/2)\sin(t), 1/2 + (\sqrt{2}/2)\cos(t), 1 + (\sqrt{2}/2)(\sin(t) + \cos(t)) \rangle,$$

$0 \le t \le 2\pi$, gives the correct orientation.

Another straightforward, but long, computation will show that

$$\mathbf{F}(\mathbf{r}(t)) \cdot \mathbf{r}'(t) = \sqrt{2}\cos(t) - (3\sqrt{2}/4)\sin(t) + (3/2)\cos^2(t) - 1/2.$$

Therefore,

$$\oint_{\mathcal{C}} \mathbf{F} \cdot d\mathbf{r} = \int_0^{2\pi} (\sqrt{2}\cos(t) - (3\sqrt{2}/4)\sin(t) + (3/2)\cos^2(t) - 1/2)\,dt$$
$$= (3/2) \cdot \pi - (1/2) \cdot 2\pi = \pi/2.$$

25. <u>Surface integral calculation.</u> The curl of $\mathbf{F}$ is

$$\mathbf{curl}(\mathbf{F}) = \det \left(\begin{bmatrix} \mathbf{i} & \mathbf{j} & \mathbf{k} \\ \frac{\partial}{\partial x} & \frac{\partial}{\partial y} & \frac{\partial}{\partial z} \\ 0 & xy^2 & -xz \end{bmatrix} \right) = \langle 0, z, y^2 \rangle.$$

Parametrizing the sphere as the graph of $f(x, y) = (25 - x^2 - y^2)^{1/2}$ the normal is $\mathbf{N} = \langle x(25 - x^2 - y^2)^{-1/2}, y(25 - x^2 - y^2)^{-1/2}, 1 \rangle$. Therefore, on the surface, where $z = (25 - x^2 - y^2)^{1/2}$,

$$\mathbf{curl}(\mathbf{F}) \cdot \mathbf{n}\,dS = \langle 0, (25 - x^2 - y^2)^{1/2}, y^2 \rangle \cdot \left(\frac{\mathbf{N}}{\|\mathbf{N}\|} \right) \|\mathbf{N}\|\,dA$$
$$= (y + y^2)\,dA.$$

The domain of integration, in rectangular coordinates, is $\mathcal{R} = \{(x, y) : 9 \le x^2 + y^2 \le 16\}$ and the surface integral sets up as shown below. The integral is evaluated in polar coordinates after the y term is dropped because it integrates to 0 by symmetry.

$$\iint_{\mathcal{S}} \mathbf{curl}(\mathbf{F}) \cdot \mathbf{n}\,dS = \iint_{\mathcal{R}} (y + y^2)\,dA$$
$$= \int_0^{2\pi} \int_3^4 r^2 \sin^2(\theta) \cdot r\,dr\,d\theta$$
$$= 175/4 \int_0^{2\pi} \sin^2(\theta)\,d\theta = 175\pi/4.$$

Line integral calculation. The boundary consists of two circles. One of them, $\mathcal{C}_1$, has radius 4, is in the plane $z = 3$, and is oriented positively. The other, $\mathcal{C}_2$, is in the plane $z = 4$, has radius 3, and is oriented negatively.

The computation for the positively oriented circle goes like this.

$\mathbf{r}(t) = \langle 4\cos(t), 4\sin(t), 3 \rangle$, $0 \le t \le 2\pi$, gives the correct orientation. Since $\mathbf{F}(\mathbf{r}(t)) = \langle 0, 64\cos(t)\sin^2(t), -3\cos(t) \rangle$ and $\mathbf{r}'(t) = \langle -4\sin(t), 4\cos(t), 0 \rangle$, $\mathbf{F}(\mathbf{r}(t)) \cdot \mathbf{r}'(t) = 256\cos^2(t)\sin^2(t)$ and

$$\oint_{\mathcal{C}_1} \mathbf{F} \cdot d\mathbf{r} = \int_0^{2\pi} 256\cos^2(t)\sin^2(t)\, dt$$

$$= 64 \int_0^{2\pi} \sin^2(2t)\, dt = 64\pi\,.$$

You can verify that the computation for the negatively oriented circle yields $\oint_{\mathcal{C}_2} \mathbf{F} \cdot d\mathbf{r} = -81\pi/4$.

27. <u>Surface integral calculation.</u> The curl of $\mathbf{F}$ is

$$\mathbf{curl}(\mathbf{F}) = \det\left(\begin{bmatrix} \mathbf{i} & \mathbf{j} & \mathbf{k} \\ \frac{\partial}{\partial x} & \frac{\partial}{\partial y} & \frac{\partial}{\partial z} \\ x^3 & -z & y \end{bmatrix} \right) = \langle 2, 0, 0 \rangle\,.$$

Parametrizing the cone using $\mathbf{r}(z, y) = \langle (y^2 + z^2)^{1/2}, y, z \rangle$, the normal is $\mathbf{N} = \mathbf{r}_z \times \mathbf{r}_y = \langle -1, y(y^2 + z^2)^{-1/2}, z(y^2 + z^2)^{-1/2} \rangle$ (verify). Therefore, on the surface, where $x = (y^2 + z^2)^{1/2}$,

$$\mathbf{curl}(\mathbf{F}) \cdot \mathbf{n}\, dS = \langle 2, 0, 0 \rangle \cdot \left(\frac{\mathbf{N}}{\|\mathbf{N}\|} \right) \|\mathbf{N}\|\, dA$$

$$= -2\, dA\,.$$

The domain of integration, in rectangular coordinates, is $\mathcal{R} = \{(y, z) : 1 \le y^2 + z^2 \le 4\}$ and the surface integral sets up as shown below.

$$\iint_{\mathcal{S}} \mathbf{curl}(\mathbf{F}) \cdot \mathbf{n}\, dS = \iint_{\mathcal{R}} (-2)\, dA$$

$$= -2 \cdot (\text{Area of } \mathcal{R}) = -6\pi\,.$$

<u>Line integral calculation.</u> The boundary consists of two circles. One of them, $\mathcal{C}_1$, has radius 2, is in the plane $x = 2$, and is oriented negatively. The other, $\mathcal{C}_2$, is in the plane $x = 1$, has radius 1, and is oriented positively.

The computation for the negatively oriented circle goes like this.

$\mathbf{r}(t) = \langle 2, 2\sin(t), 2\cos(t) \rangle$, $0 \le t \le 2\pi$, gives the correct orientation. Since $\mathbf{F}(\mathbf{r}(t)) = \langle 8, -2\cos(t), 2\sin(t) \rangle$ and $\mathbf{r}'(t) = \langle 0, 2\cos(t), -2\sin(t) \rangle$, $\mathbf{F}(\mathbf{r}(t)) \cdot \mathbf{r}'(t) = -4\cos^2(t) - 4\sin^2(t) = -4$ and

$$\oint_{\mathcal{C}_1} \mathbf{F} \cdot d\mathbf{r} = \int_0^{2\pi} (-4)\, dt = -8\pi\,.$$

You can verify that the computation for the positively oriented circle yields $\oint_{\mathcal{C}_2} \mathbf{F} \cdot d\mathbf{r} = 2\pi$.

29. <u>Surface integral calculation.</u> The curl of $\mathbf{F}$ is

$$\mathbf{curl}(\mathbf{F}) = \det\left(\begin{bmatrix} \mathbf{i} & \mathbf{j} & \mathbf{k} \\ \frac{\partial}{\partial x} & \frac{\partial}{\partial y} & \frac{\partial}{\partial z} \\ y^2 & -xz & x \end{bmatrix} \right) = \langle x, -1, -z - 2y \rangle .$$

Parametrizing the cylinder using $\mathbf{r}(\theta, z) = \langle 2\cos(\theta), 2\sin(\theta), z \rangle$, the normal is $\mathbf{N} = \mathbf{r}_\theta \times \mathbf{r}_z = \langle 2\cos(\theta), 2\sin(\theta), 0 \rangle$ (verify). Therefore, on the surface, where $x = 2\cos(\theta)$ and $y = 2\sin(\theta)$,

$$\mathbf{curl}(\mathbf{F}) \cdot \mathbf{n}\, dS = \langle 2\cos(\theta), -1, -z - 4\sin(\theta) \rangle \cdot \left(\frac{\mathbf{N}}{\|\mathbf{N}\|} \right) \|\mathbf{N}\|\, dA$$

$$= (4\cos^2(\theta) - 2\sin(\theta))\, dA .$$

The domain of integration, is $\mathcal{R} = \{(\theta, z) : 0 \le \theta \le 2\pi, -2 \le z \le 3\}$ and the surface integral sets up as shown below.

$$\iint_{\mathcal{S}} \mathbf{curl}(\mathbf{F}) \cdot \mathbf{n}\, dS = \iint_{\mathcal{R}} (4\cos^2(\theta) - 2\sin(\theta))\, dA$$

$$= \int_0^{2\pi} \int_{-2}^{3} (4\cos^2(\theta) - 2\sin(\theta))\, dz\, d\theta$$

$$= 5\int_0^{2\pi} (4\cos^2(\theta) - 2\sin(\theta))\, d\theta = 20\pi .$$

<u>Line integral calculation.</u> The boundary consists of two circles. Both have radius 2. One of them, $\mathcal{C}_1$, is in the plane $z = -2$, and is oriented positively. The other, $\mathcal{C}_2$, is in the plane $z = 3$, and is oriented negatively.

The computation for the positively oriented circle goes like this.

$\mathbf{r}(t) = \langle 2\cos(t), 2\sin(t), -2 \rangle$, $0 \le t \le 2\pi$, gives the correct orientation. Since $\mathbf{F}(\mathbf{r}(t)) = \langle 4\sin^2(t), 4\cos(t), 2\cos(t) \rangle$ and $\mathbf{r}'(t) = \langle -2\sin(t), 2\cos(t), 0 \rangle$, $\mathbf{F}(\mathbf{r}(t)) \cdot \mathbf{r}'(t) = -8\sin^3(t) + 8\cos^2(t)$ and

$$\oint_{\mathcal{C}_1} \mathbf{F} \cdot d\mathbf{r} = \int_0^{2\pi} (-8\sin^3(t) + 8\cos^2(t))\, dt = 8\pi .$$

You can verify that the computation for the negatively oriented circle yields $\oint_{\mathcal{C}_2} \mathbf{F} \cdot d\mathbf{r} = 12\pi$.

31. The curve $\mathcal{C} = \{(x, y, 0) : x^2 + y^2 = 1\}$ that bounds the spherical surface $\mathcal{S}$ also bounds the unit disk: $\mathcal{D} = \{(x, y, 0) : x^2 + y^2 < 1\}$. Orienting the circle positively,

$$\iint_{\mathcal{S}} \mathbf{curl}(\mathbf{F}) \cdot \mathbf{n}\, dS = \oint_{\mathcal{C}} \mathbf{F} \cdot d\mathbf{r} = \iint_{\mathcal{D}} \mathbf{curl}(\mathbf{F}) \cdot \mathbf{k}\, dS .$$

Since

$$\mathbf{curl}(\mathbf{F}) = \det\left(\begin{bmatrix} \mathbf{i} & \mathbf{j} & \mathbf{k} \\ \frac{\partial}{\partial x} & \frac{\partial}{\partial y} & \frac{\partial}{\partial z} \\ y\cos(xz) & -xy\sin(z^2) & e^{x^2} \end{bmatrix}\right),$$

$\mathbf{curl}(\mathbf{F}) \cdot \mathbf{k} = -y\sin(z^2) - \cos(xz) = -1$ on $\mathcal{D}$. Therefore,

$$\iint_{\mathcal{D}} \mathbf{curl}(\mathbf{F}) \cdot \mathbf{k}\, dS = -\pi.$$

33. The equatorial circle $\{(x, y, 0) : x^2 + y^2 = 1\}$ bounds both the top hemisphere, $\mathcal{S}_1$, and the bottom hemisphere, $\mathcal{S}_2$. Let $\mathcal{C}$ denote the circle with a positive orientation, and orient $\mathcal{S}_1$ with the upward normal and $\mathcal{S}_2$ with the downward normal.

According to Stokes's Theorem,

$$\iint_{\mathcal{S}_1} \mathbf{curl}(\mathbf{F}) \cdot \mathbf{n}\, dS = \oint_{\mathcal{C}} \mathbf{F} \cdot d\mathbf{r} \quad \text{and} \quad \iint_{\mathcal{S}_2} \mathbf{curl}(\mathbf{F}) \cdot \mathbf{n}\, dS = \oint_{-\mathcal{C}} \mathbf{F} \cdot d\mathbf{r}.$$

Consequently,

$$\iint_{\mathcal{S}} \mathbf{curl}(\mathbf{F}) \cdot \mathbf{n}\, dS = \iint_{\mathcal{S}_1} \mathbf{curl}(\mathbf{F}) \cdot \mathbf{n}\, dS + \iint_{\mathcal{S}_2} \mathbf{curl}(\mathbf{F}) \cdot \mathbf{n}\, dS$$

$$= \oint_{\mathcal{C}} \mathbf{F} \cdot d\mathbf{r} + \oint_{-\mathcal{C}} \mathbf{F} \cdot d\mathbf{r}$$

$$= 0.$$

35. $\mathbf{curl}(\mathbf{F}) = \det\left(\begin{bmatrix} \mathbf{i} & \mathbf{j} & \mathbf{k} \\ \frac{\partial}{\partial x} & \frac{\partial}{\partial y} & \frac{\partial}{\partial z} \\ 2yz & xz & 0 \end{bmatrix}\right) = \langle -x, 2y, -z \rangle$. Parametrize the cylinder using $\mathbf{r}(\theta, z) = \langle 3\cos(\theta), 3\sin(\theta), z \rangle$, $0 \le \theta \le 2\pi$, $0 \le z \le 2$. The normal $\mathbf{N} = \mathbf{r}_\theta \times \mathbf{r}_z = \langle 3\cos(\theta), 3\sin(\theta), 0 \rangle$ is outward (verify). Therefore, using (13.7.6),

$$\iint_{\mathcal{S}} \mathbf{curl}(\mathbf{F}) \cdot \mathbf{n}\, dS = \iint_{\mathcal{S}} \mathbf{curl}(\mathbf{F}) \cdot (\mathbf{r}_\theta \times \mathbf{r}_z)\, d\theta\, dz$$

$$= \int_0^{2\pi} \int_0^2 \langle -3\cos(\theta), 6\sin(\theta), -z \rangle \cdot \langle 3\cos(\theta), 3\sin(\theta), 0 \rangle\, dz\, d\theta$$

$$= \int_0^{2\pi} \int_0^2 (-9\cos^2(\theta) + 18\sin^2(\theta))\, dz\, d\theta$$

$$= 2 \int_0^{2\pi} (-9\cos^2(\theta) + 18\sin^2(\theta))\, d\theta$$

$$= 2 \cdot (-9\pi + 18\pi) = 18\pi.$$

<u>Line integral calculations.</u> The boundary of the cylinder consists of two circles.

$$\mathcal{C}_1 : \mathbf{r}_1(\theta) = \langle 3\cos(\theta), 3\sin(\theta), 0 \rangle \text{ and } \mathcal{C}_2 : \mathbf{r}_2(\theta) = \langle -3\cos(\theta), 3\sin(\theta), 2 \rangle.$$

The bottom circle, $\mathcal{C}_1$, is in the plane $z = 0$, and is oriented counterclockwise. The top circle, $\mathcal{C}_2$, is in the plane $z = 2$, has radius 3, and is oriented clockwise.

Since $\mathbf{F}(\mathbf{r}_1(\theta)) \cdot \mathbf{r}_1'(\theta) = 0$, $\int_{\mathcal{C}_1} \mathbf{F} \cdot d\mathbf{r}_1 = 0$.

Since $\mathbf{F}(\mathbf{r}_2(\theta)) \cdot \mathbf{r}_2'(\theta) = 36\sin^2(\theta) - 18\cos^2(\theta)$,

$$\int_{\mathcal{C}_2} \mathbf{F} \cdot d\mathbf{r}_2 = \int_0^{2\pi} (36\sin^2(\theta) - 18\cos^2(\theta))\, d\theta = 36\pi - 18\pi = 18\pi.$$

Therefore, $\int_{\mathcal{C}_1} \mathbf{F} \cdot d\mathbf{r}_1 + \int_{\mathcal{C}_2} \mathbf{F} \cdot d\mathbf{r}_2 = 18\pi$, which verifies the surface integral calculation above.

37. $\mathbf{curl(F)} = \det\left(\begin{bmatrix} \mathbf{i} & \mathbf{j} & \mathbf{k} \\ \frac{\partial}{\partial x} & \frac{\partial}{\partial y} & \frac{\partial}{\partial z} \\ x & \frac{2xz}{x^2+y^2} & y \end{bmatrix} \right) = \langle 1 - \frac{2x}{x^2+y^2}, 0, -\frac{2z(x^2-y^2)}{(x^2+y^2)^2} \rangle$. As

in Exercise 35, parametrize the cylinder using $\mathbf{r}(\theta, z) = \langle 3\cos(\theta), 3\sin(\theta), z \rangle$, $0 \le \theta \le 2\pi$, $0 \le z \le 2$. The normal $\mathbf{N} = \mathbf{r}_\theta \times \mathbf{r}_z = \langle 3\cos(\theta), 3\sin(\theta), 0 \rangle$ is outward. A straightforward substitution will show that

$$\mathbf{curl(F)}(\mathbf{r}(\theta, z)) = \langle 1 - \tfrac{2}{3}\cos(\theta), 0, -\tfrac{2}{9}z(2\cos^2(\theta) - 1) \rangle.$$

Therefore, $\mathbf{curl(F)}(\mathbf{r}(\theta, z)) \cdot (\mathbf{r}_\theta \times \mathbf{r}_z) = 3\cos(\theta) - 2\cos^2(\theta)$, and using (13.7.6),

$$\iint_{\mathcal{S}} \mathbf{curl(F)} \cdot \mathbf{n}\, dS = \iint_{\mathcal{S}} \mathbf{curl(F)} \cdot (\mathbf{r}_\theta \times \mathbf{r}_z)\, d\theta\, dz$$
$$= \int_0^{2\pi} \int_0^2 (3\cos(\theta) - 2\cos^2(\theta))\, dz\, d\theta$$
$$= 2\int_0^{2\pi} (3\cos(\theta) - 2\cos^2(\theta))\, d\theta = -4\pi.$$

<u>Line integral calculations.</u> The boundary of the cylinder consists of two circles.

$$\mathcal{C}_1 : \mathbf{r}_1(\theta) = \langle 3\cos(\theta), 3\sin(\theta), 0 \rangle \text{ and } \mathcal{C}_2 : \mathbf{r}_2(\theta) = \langle -3\cos(\theta), 3\sin(\theta), 2 \rangle.$$

The bottom circle, $\mathcal{C}_1$, is in the plane $z = 0$, and is oriented counterclockwise. The top circle, $\mathcal{C}_2$, is in the plane $z = 2$, has radius 3, and is oriented clockwise.

Since $\mathbf{F}(\mathbf{r}_1(\theta)) \cdot \mathbf{r}_1'(\theta) = -9\cos(\theta)\sin(\theta)$,

$$\int_{C_1} \mathbf{F} \cdot d\mathbf{r}_1 = \int_0^{2\pi} (-9\cos(\theta)\sin(\theta))\, d\theta = 0\,.$$

Since $\mathbf{F}(\mathbf{r}_2(\theta)) \cdot \mathbf{r}_2'(\theta) = -9\cos(\theta)\sin(\theta) - 4\cos^2(\theta)$,

$$\int_{C_2} \mathbf{F} \cdot d\mathbf{r}_2 = \int_0^{2\pi} (-9\cos(\theta)\sin(\theta) - 4\cos^2(\theta))\, d\theta = -4\pi\,.$$

Therefore, $\int_{C_1} \mathbf{F}\cdot d\mathbf{r}_1 + \int_{C_2} \mathbf{F}\cdot d\mathbf{r}_2 = -4\pi$, which verifies the surface integral calculation above.

Calculator/Computer Exercises

39. Begin by loading the VectorCalculus package to have access to the Curl procedure. The next entries define the vector field $\mathbf{F}$, the function f that determines the surface, then calculates the appropriate normal vector $\mathbf{N}$.

```
> with(VectorCalculus):
  F := (x,y,z) -> <0,x*exp(-y),0>:
  f := (x,y) -> exp(x^2+y^2):
  N := <D[1](f)(x,y),D[2](f)(x,y),-1>:
```

Now calculate the integrand in the surface integral: $\mathbf{curl}(\mathbf{F}) \cdot \mathbf{N}$, and then the integral.

```
> CurlNdA := convert(Curl(F)(x,y,f(x,y)),Vector).N:
  Int(Int(CurlNdA,y=-sqrt(1-x^2)..sqrt(1-x^2)),x=-1..1) =
  evalf(
  Int(Int(CurlNdA,y=-sqrt(1-x^2)..sqrt(1-x^2)),x=-1..1)
      );
```

$$\int_{-1}^{1} \int_{-\sqrt{1-x^2}}^{\sqrt{1-x^2}} \left(-e^{-y}\right)\, dy\, dx = -3.550999378$$

Now evaluate the line integral. Begin with a parametrization of the boundary curve.

```
> r := <sin(t),cos(t),0>:
  Int(F(r[1],r[2],r[3]).diff(r,t),t=0..2*Pi) =
  evalf(
  Int(F(r[1],r[2],r[3]).diff(r,t),t=0..2*Pi)
      );
```

$$\int_0^{2\pi} \left(-\sin(t)^2 e^{-\cos(t)}\right)\, dt = -3.550999378$$

41. As in Exercise 39, except the normal points upward.

```
> F := (x,y,z) -> <0,exp(z^2),0>:
  f := (x,y) -> 1-x-y:
  N := <-D[1](f)(x,y),-D[2](f)(x,y),1>:
```

Now calculate the integrand in the surface integral: $\mathbf{curl}(\mathbf{F}) \cdot \mathbf{N}$, and then the integral.

```
> CurlNdA := convert(Curl(F)(x,y,f(x,y)),Vector).N:
  Int(Int(CurlNdA,y=0..1-x),x=0..1) =
  evalf(
  Int(Int(CurlNdA,y=0..1-x),x=0..1)
       );
```

$$\int_0^1 \int_1^{1-x} \left(-2(1-x-y)e^{(1-x-y)^2}\right) \, dy \, dx = -.4626517459$$

Evaluate the line integral. Begin the definition of a procedure called LI having the property that $LI(F, P, Q) = \int_{\overline{PQ}} \mathbf{F} \cdot d\mathbf{r}$.

```
> LI := proc(F,P,Q)
    local r;
    r := P + t*(Q-P);
    int( F(r[1],r[2],r[3]).diff(r,t),t=0..1);
  end proc:
```

Using this, the line integral calculation can proceed as follows.

```
> P,Q,R := <1,0,0>,<0,1,0>,<0,0,1>:
  T := evalf([LI(F,P,Q),LI(F,Q,R),LI(F,R,P)]);
  LinInt = add(k,k=T);
```

$$T := [\, 1., \; -1.462651746, \; 0. \,]$$

$$LinInt = -.462651746$$

13.8 Flux and the Divergence Theorem

Problems for Practice

1. Surface integral calculation

   The solid $\mathcal{U}$ is a cylinder. Its boundary $\mathcal{S}$ consists of three pieces: top, $\mathcal{S}_t = \{(x,y,z) : x^2 + y^2 < 9, z = 2\}$, bottom, $\mathcal{S}_b = \{(x,y,z) : x^2 + y^2 < 9, z = -1\}$, side, $\mathcal{S}_s = \{(x,y,z) : -1 < z < 2, x^2 + y^2 = 9\}$.

   The outward unit normal on the top of the cylinder is $\mathbf{n} = \mathbf{k}$. Therefore $\mathbf{F} \cdot \mathbf{n} = (5\mathbf{k}) \cdot \mathbf{k} = 5$, and $\int_{\mathcal{S}_t} \mathbf{F} \cdot \mathbf{n} \, dS = \iint_{\mathcal{S}_t} 5 \, dS = 5 \cdot (\text{area of top}) = 5 \cdot (\pi \cdot 3^2) = 45\pi$.

The outward unit normal on the bottom of the cylinder is $\mathbf{n} = -\mathbf{k}$ and, proceeding as above, $\iint_{\mathcal{S}_b} \mathbf{F} \cdot \mathbf{n} \, dS = -45\pi$.

The outward unit normal at a point (x, y, z) on the side of the cylinder is $\mathbf{n}(x, y, z) = (1/3)(x\mathbf{i} + y\mathbf{j})$ (verify with a picture!). Consequently, $\mathbf{F} \cdot \mathbf{n} = (1/3)(x\mathbf{i} + y\mathbf{j}) \cdot \mathbf{k} = 0$, and $\iint_{\mathcal{S}_s} \mathbf{F} \cdot \mathbf{n} \, dS = \iint_{\mathcal{S}_s} 0 \, dS = 0$.

Adding we have $\iint_{\mathcal{S}} \mathbf{F} \cdot \mathbf{n} \, dS = 0$.

Using the Divergence Theorem

Since $\operatorname{div}(\mathbf{F}) = 0$, $\iint_{\mathcal{S}} \mathbf{F} \cdot \mathbf{n} \, dS = \iiint_{\mathcal{U}} \operatorname{div}(\mathbf{F}) \, dV = 0$.

3. Surface integral calculation

The solid $\mathcal{U}$ is a cylinder. Its boundary $\mathcal{S}$ consists of three pieces: top, $\mathcal{S}_t = \{(x, y, z) : x^2 + y^2 < 1, z = 1\}$, bottom, $\mathcal{S}_b = \{(x, y, z) : x^2 + y^2 < 1, z = 0\}$, side, $\mathcal{S}_s = \{(x, y, z) : 0 < z < 1, x^2 + y^2 = 1\}$.

The outward unit normal on the top of the cylinder is $\mathbf{n} = \mathbf{k}$. Therefore $\mathbf{F} \cdot \mathbf{n} = (z\mathbf{k}) \cdot \mathbf{k} = z$, and $\int_{\mathcal{S}_t} \mathbf{F} \cdot \mathbf{n} \, dS = \iint_{\mathcal{S}_t} z \, dS = \iint_{\mathcal{S}_t} 1 \, dS = \text{area of top} = \pi$.

The outward unit normal on the bottom of the cylinder is $\mathbf{n} = -\mathbf{k}$ and, proceeding as above, $\iint_{\mathcal{S}_b} \mathbf{F} \cdot \mathbf{n} \, dS = -\iint_{\mathcal{S}_b} z \, dS = -\iint_{\mathcal{S}_t} 0 \, dS = 0$.

The outward unit normal at a point (x, y, z) on the side of the cylinder is $\mathbf{n}(x, y, z) = x\mathbf{i} + y\mathbf{j}$ (verify with a picture!). Consequently, $\mathbf{F} \cdot \mathbf{n} = z\mathbf{k} \cdot (x\mathbf{i} + y\mathbf{j}) = 0$, and $\iint_{\mathcal{S}_s} \mathbf{F} \cdot \mathbf{n} \, dS = \iint_{\mathcal{S}_s} 0 \, dS = 0$.

Adding we have $\iint_{\mathcal{S}} \mathbf{F} \cdot \mathbf{n} \, dS = \pi$.

Using the Divergence Theorem

Since $\operatorname{div}(\mathbf{F}) = 1$, $\iint_{\mathcal{S}} \mathbf{F} \cdot \mathbf{n} \, dS = \iiint_{\mathcal{U}} \operatorname{div}(\mathbf{F}) \, dV = \iiint_{\mathcal{U}} 1 \, dV = (\text{volume of } V) = \pi$.

5. Surface integral calculation

The solid $\mathcal{U}$ is a cylinder. Its boundary $\mathcal{S}$ consists of three pieces: top, $\mathcal{S}_t = \{(x, y, z) : x^2 + y^2 < 1, z = 1\}$, bottom, $\mathcal{S}_b = \{(x, y, z) : x^2 + y^2 < 1, z = 0\}$, side, $\mathcal{S}_s = \{(x, y, z) : 0 < z < 1, x^2 + y^2 = 1\}$.

The outward unit normal on the top of the cylinder is $\mathbf{n} = \mathbf{k}$. Therefore $\mathbf{F} \cdot \mathbf{n} = (x\mathbf{i}) \cdot \mathbf{k} = 0$, and $\int_{\mathcal{S}_t} \mathbf{F} \cdot \mathbf{n} \, dS = \iint_{\mathcal{S}_t} 0 \, dS = 0$.

The outward unit normal on the bottom of the cylinder is $\mathbf{n} = -\mathbf{k}$ and, proceeding as above, $\iint_{\mathcal{S}_b} \mathbf{F} \cdot \mathbf{n} \, dS = 0$.

The outward unit normal at a point (x, y, z) on the side of the cylinder is $\mathbf{n}(x, y, z) = x\mathbf{i} + y\mathbf{j}$ (verify with a picture!). Consequently, $\mathbf{F} \cdot \mathbf{n} = x\mathbf{i} \cdot (x\mathbf{i} + y\mathbf{j}) = x^2$, and $\iint_{\mathcal{S}_s} \mathbf{F} \cdot \mathbf{n} \, dS = \iint_{\mathcal{S}_s} x^2 \, dS$. The side surface can be parametrized with $\mathbf{r}(\theta, z) = \langle \cos(\theta), \sin(\theta), z \rangle$. Since $\mathbf{r}_\theta \times \mathbf{r}_z = \langle \cos(\theta), \sin(\theta), 0 \rangle$ (verify), $dS = \|\mathbf{r}_\theta \times \mathbf{r}_z\| \, d\theta \, dz = d\theta \, dz$ and

$$\iint_{\mathcal{S}_s} x^2 \, dS = \int_0^{2\pi} \int_0^1 \cos^2(\theta) \, dz \, d\theta = \pi \, .$$

Adding we have $\iint_{\mathcal{S}} \mathbf{F} \cdot \mathbf{n} \, dS = \pi$.

Using the Divergence Theorem

Since $\operatorname{div}(\mathbf{F}) = 1$, $\iint_{\mathcal{S}} \mathbf{F} \cdot \mathbf{n} \, dS = \iiint_{\mathcal{U}} \operatorname{div}(\mathbf{F}) \, dV = \iiint_{\mathcal{U}} 1 \, dV = $ volume of $V = \pi$.

7. Surface integral calculation

The solid $\mathcal{U}$ is a triangular pyramid. Its boundary $\mathcal{S}$ consists of four triangles: top, $\mathcal{S}_t = \{(x, y, z) : z = 1 - x - y, 0 < x < 1, 0 < y < 1 - x\}$, bottom, $\mathcal{S}_b = \{(x, y, 0) : 0 < x < 1, 0 < y < 1 - x\}$, side 1, $\mathcal{S}_1 = \{(0, y, z) : 0 < y < 1, 0 < z < 1 - y\}$, and side 2, $\mathcal{S}_2 = \{(x, 0, z) : 0 < x < 1, 0 < z < 1 - x\}$.

Since the top is the graph of $f(x, y) = 1 - x - y$, an appropriately oriented normal is $\mathbf{N} = \langle 1, 1, 1 \rangle$. In addition, on the top surface $\mathbf{F}(x, y, z) = \langle 7x, y, -2(1 - x - y) \rangle$, so $\mathbf{F} \cdot \mathbf{N} = 7z + y - 2(1 - x - y)$ and

$$\mathbf{F} \cdot \mathbf{n} \, dS = \mathbf{F} \cdot \mathbf{N} \, dx \, dy = (9x + 3y - 2) \, dx \, dy \, .$$

Therefore,

$$\int_{\mathcal{S}_t} \mathbf{F} \cdot \mathbf{n} \, dS = \int_0^1 \int_0^{1-x} (9x + 3y - 2) \, dy \, dx$$
$$= \int_0^1 (8x - (15/2)x^2 - 1/2)) \, dx = 1 \, .$$

The outward unit normal on the bottom is $\mathbf{n} = -\mathbf{k}$ and $\mathbf{F} \cdot \mathbf{n} = 2z$. Since $z = 0$, $\iint_{\mathcal{S}_b} \mathbf{F} \cdot \mathbf{n} \, dS = 0$.

The outward unit normal on the side 1 is $\mathbf{n} = -\mathbf{i}$ and $\mathbf{F} \cdot \mathbf{n} = -7x$. Since $x = 0$, $\iint_{\mathcal{S}_1} \mathbf{F} \cdot \mathbf{n} \, dS = 0$.

The outward unit normal on the side 2 is $\mathbf{n} = \mathbf{j}$ and $\mathbf{F} \cdot \mathbf{n} = y$. Since $y = 0$, $\iint_{\mathcal{S}_2} \mathbf{F} \cdot \mathbf{n} \, dS = 0$.

Adding we have $\iint_{\mathcal{S}} \mathbf{F} \cdot \mathbf{n} \, dS = 1$.

Using the Divergence Theorem

Since $\operatorname{div}(\mathbf{F}) = 6$, $\iint_{\mathcal{S}} \mathbf{F} \cdot \mathbf{n} \, dS = \iiint_{\mathcal{U}} \operatorname{div}(\mathbf{F}) \, dV = \iiint_{\mathcal{U}} 6 \, dV = 6 \cdot$ (volume of V) $= 6 \cdot (1/3) \cdot (1/2) \cdot 1 = 1$ (recall that the volume of a pyramid is $1/3 \cdot$ area of base $\cdot$ height).

9. Surface integral calculation

The solid $\mathcal{U}$ is a rectangular box. Its boundary $\mathcal{S}$ consists of the six sides of the box: top, $\mathcal{S}_t = \{(x, y, z) : |x| < 2, |y| < 2, z = 1\}$, bottom, $\mathcal{S}_b = \{(x, y, z) : |x| < 2, |y| < 2, z = 0\}$, and four sides:

$$\text{side 1, } \mathcal{S}_1 = \{(x, y, z) : x = 2, |y| < 2, 0 < z < 1\},$$

$$\text{side 2, } \mathcal{S}_2 = \{(x, y, z) : x = -2, |y| < 2, 0 < z < 1\},$$

$$\text{side 3, } \mathcal{S}_3 = \{(x, y, z) : |x| < 2, y = 2, 0 < z < 1\},$$

and

$$\text{side 4, } \mathcal{S}_4 = \{(x, y, z) : |x| < 2, y = -2, 0 < z < 1\}.$$

The surface integrals can be calculated in pairs.

Top and bottom. The element of area for the top and bottom is $dS = dx\,dy$ and the outward normals are $\mathbf{k}$ and $-\mathbf{k}$ respectively. Since $z = 1$ on the top, $\iint_{\mathcal{S}_t} \mathbf{F} \cdot \mathbf{n}\,dS$ evaluates as follows.

$$\int_{-2}^{2} \int_{-2}^{2} (y^2 \mathbf{i} - y\mathbf{j} + x\mathbf{k}) \cdot \mathbf{k}\,dx\,dy = \int_{-2}^{2} \int_{-2}^{2} x\,dx\,dy = 0$$

The integral over the bottom is also 0, but for a different reason. Since $z = 0$ on the bottom,

$$\int_{-2}^{2} \int_{-2}^{2} (y^2 \mathbf{i} - 0 \cdot \mathbf{j} + 0 \cdot \mathbf{k}) \cdot (-\mathbf{k})\,dx\,dy = \int_{-2}^{2} \int_{-2}^{2} 0\,dx\,dy = 0.$$

Sides 1 and 2. The element of area for these two sides is $dS = dy\,dz$ and the outward normals are $\mathbf{i}$ and $-\mathbf{i}$ respectively. Since $x = 2$ on $\mathcal{S}_1$, $\iint_{\mathcal{S}_1} \mathbf{F} \cdot \mathbf{n}\,dS$ evaluates as follows.

$$\int_{0}^{1} \int_{-2}^{2} y^2 \mathbf{i} - yz\mathbf{j} + 2z\mathbf{k}) \cdot \mathbf{i}\,dy\,dz = \int_{0}^{1} \int_{-2}^{2} y^2\,dy\,dz = 16/3$$

On side 2, $x = -2$ and $\iint_{\mathcal{S}_2} \mathbf{F} \cdot \mathbf{n}\,dS$ evaluates as follows.

$$\int_{0}^{1} \int_{-2}^{2} (y^2 \mathbf{i} - yz\mathbf{j} - 2z\mathbf{k}) \cdot (-\mathbf{i})\,dy\,dz = \int_{0}^{1} \int_{-2}^{2} (-y^2)\,dy\,dz = -16/3$$

Sides 3 and 4. The element of area for these two sides is $dS = dz\,dx$ and the outward normals are $\mathbf{j}$ and $-\mathbf{j}$ respectively. Since $y = 2$ on $\mathcal{S}_3$, $\iint_{\mathcal{S}_3} \mathbf{F} \cdot \mathbf{n}\,dS$ evaluates as follows.

$$\int_{-2}^{2} \int_{0}^{1} (4\mathbf{i} - 2z\mathbf{j} + xz\mathbf{k}) \cdot \mathbf{j}\,dz\,dx = \int_{-2}^{2} \int_{0}^{1} (-2z)\,dz\,dx = -4$$

On side 4, $y = -2$ and $\iint_{\mathcal{S}_4} \mathbf{F} \cdot \mathbf{n} \, dS$ evaluates as follows.

$$\int_{-2}^{2} \int_0^1 (4\mathbf{i} + 2z\mathbf{j} + xz\mathbf{k}) \cdot (-\mathbf{j}) \, dz \, dx = \int_{-2}^{2} \int_0^1 (-2z) \, dz \, dx = -4$$

Adding we get $\iint_{\mathcal{S}} \mathbf{F} \cdot \mathbf{n} \, dS = -8$.

Using the Divergence Theorem

Since $\text{div}(\mathbf{F}) = -z + x$,

$$\iint_{\mathcal{S}} \mathbf{F} \cdot \mathbf{n} \, dS = \iiint_{\mathcal{U}} \text{div}(\mathbf{F}) \, dV$$
$$= \int_0^1 \int_{-2}^{2} \int_{-2}^{2} (x - z) \, dx \, dy \, dz$$
$$= \int_0^1 \int_{-2}^{2} (-4z) \, dy \, dz$$
$$= \int_0^1 (-16z) \, dz = -8.$$

11. Surface integral calculation

   The solid $\mathcal{U}$ is a region with three boundary surfaces. The top, $\mathcal{S}_t = \{(x, y, z) : z = x^2 - y^2, 0 < x < 1, -x < y < x\}$, the bottom, $\mathcal{S}_b = \{(x, y, 0) : 0 < x < 1, -x < y < x\}$, and one side, $\mathcal{S}_s = \{(1, y, z) : -1 < y < 1, 0 < z < 1 - y^2\}$.

   The top is the graph of $f(x, y) = x^2 - y^2$, an appropriately oriented normal is $\mathbf{N} = \langle -2x, 2y, 1 \rangle$. In addition, on the top surface, $z = x^2 - y^2$, so $\mathbf{F}(x, y, z) = \langle x(x^2 - y^2), -y(x^2 - y^2), x(x^2 - y^2) \rangle$, and $\mathbf{F} \cdot \mathbf{N} = -2x^2(x^2 - y^2) - 2y^2(x^2 - y^2) + x(x^2 - y^2)$. Therefore,

   $$\mathbf{F} \cdot \mathbf{n} \, dS = \mathbf{F} \cdot \mathbf{N} \, dx \, dy = (-2x^4 + 2y^4 + x^3 - xy^2) \, dx \, dy.$$

   Consequently,

   $$\int_{\mathcal{S}_t} \mathbf{F} \cdot \mathbf{n} \, dS = \int_0^1 \int_{-x}^{x} (-2x^4 + 2y^4 + x^3 - xy^2) \, dy \, dx$$
   $$= \int_0^1 ((4/3)x^4 - (16/5)x^5) \, dx = -4/15.$$

   The outward unit normal on the bottom is $\mathbf{n} = -\mathbf{k}$ and $\mathbf{F} \cdot \mathbf{n} = -xz$. Since $z = 0$, $\iint_{\mathcal{S}_b} \mathbf{F} \cdot \mathbf{n} \, dS = 0$.

The outward unit normal on the side is $\mathbf{n} = \mathbf{i}$ and $\mathbf{F} \cdot \mathbf{n} = xz$. Since $x = 1$ and the element of area is $dS = dy\,dz$,

$$\iint_{\mathcal{S}_s} \mathbf{F} \cdot \mathbf{n}\,dS = \int_{-1}^{1} \int_{0}^{1-y^2} z\,dz\,dy$$
$$= \int_{-1}^{1} (1/2)(1-y^2)^2\,dy = 8/15\,.$$

Adding we have $\iint_{\mathcal{S}} \mathbf{F} \cdot \mathbf{n}\,dS = 4/15$.

Using the Divergence Theorem

Since $\operatorname{div}(\mathbf{F}) = z - z + x$,

$$\iint_{\mathcal{S}} \mathbf{F} \cdot \mathbf{n}\,dS = \iiint_{\mathcal{U}} \operatorname{div}(\mathbf{F})\,dV = \iiint_{\mathcal{U}} x\,dV$$
$$= \int_{0}^{1} \int_{-x}^{x} \int_{0}^{x^2-y^2} x\,dz\,dy\,dx$$
$$= \int_{0}^{1} \int_{-x}^{x} x(x^2-y^2)\,dy\,dx$$
$$= \int_{0}^{1} (4/3)x^4\,dx = 4/15\,.$$

13. Surface integral calculation

The solid $\mathcal{U}$ is a quarter of a cylinder. Its boundary $\mathcal{S}$ is in five pieces: the top, $\mathcal{S}_t = \{(x,y,z) : z = \sqrt{9-x^2}, 0 < x < 3, 1 < y < 2\}$, the bottom, $\mathcal{S}_b = \{(x,y,0) : 0 < x < 3, 1 < y < 2\}$, side 1, $\mathcal{S}_1 = \{(x,2,z) : 0 < x < 3, 0 < z < \sqrt{9-x^2}\}$, side 2, $\mathcal{S}_2 = \{(x,1,z) : 0 < x < 3, 0 < z < \sqrt{9-x^2}\}$, and side 3, $\mathcal{S}_3 = \{(0,y,z) : 1 < y < 1, 0 < z < 3\}$.

The top is the graph of $f(x,y) = \sqrt{9-x^2}$, an appropriately oriented normal is $\mathbf{N} = \langle x(9-x^2)^{-1/2}, 0, 1 \rangle$. In addition, on the top surface, $z = \sqrt{9-x^2}$, so $\mathbf{F}(x,y,z) = \langle x - \sqrt{9-x^2}, y^2, x + \sqrt{9-x^2} \rangle$, and $\mathbf{F} \cdot \mathbf{N} = x^2(9-x^2)^{-1/2} + \sqrt{9-x^2}$. Therefore,

$$\mathbf{F} \cdot \mathbf{n}\,dS = \mathbf{F} \cdot \mathbf{N}\,dx\,dy = (x^2(9-x^2)^{-1/2} + \sqrt{9-x^2}\,)\,dx\,dy\,.$$

Consequently,

$$\int_{\mathcal{S}_t} \mathbf{F} \cdot \mathbf{n} \, dS = \int_0^3 \int_1^2 \left(x^2 (9 - x^2)^{-1/2} + \sqrt{9 - x^2} \right) dy \, dx$$

$$= \int_0^3 \left(x^2 (9 - x^2)^{-1/2} + \sqrt{9 - x^2} \right) dx$$

$$= \int_0^{\pi/2} \left(3\sin^2(\theta)/\cos(\theta) + 3\cos(\theta) \right) \cdot 3\cos(\theta) \, d\theta$$

$$= \int_0^{\pi/2} \left(9\sin^2(\theta) + 9\cos^2(\theta) \right) d\theta = 9\pi/2 \,.$$

Note the substitution $x = 3\sin(\theta)$.

The outward unit normal on the bottom is $\mathbf{n} = -\mathbf{k}$ and $\mathbf{F} \cdot \mathbf{n} = -(x + z)$. Since $z = 0$ and $dS = dx \, dy$,

$$\iint_{\mathcal{S}_b} \mathbf{F} \cdot \mathbf{n} \, dS = \iint_{\mathcal{S}_b} (-x) \, dS$$

$$= \int_0^3 \int_1^2 (-x) \, dy \, dx = -9/2 \,.$$

The outward unit normal on side 1 is $\mathbf{n} = \mathbf{j}$ and $\mathbf{F} \cdot \mathbf{n} = y^2$. Since $y = 2$, $\iint_{\mathcal{S}_1} \mathbf{F} \cdot \mathbf{n} \, dS = \iint_{\mathcal{S}_1} 4 \, dS = 4 \cdot (\text{area of side } 1) = 9\pi$.

The outward unit normal on side 2 is $\mathbf{n} = -\mathbf{j}$ and $\mathbf{F} \cdot \mathbf{n} = -y^2$. Since $y = 1$, $\iint_{\mathcal{S}_1} \mathbf{F} \cdot \mathbf{n} \, dS = \iint_{\mathcal{S}_1} (-1) \, dS = -(\text{area of side } 2) = -9\pi/4$.

The outward unit normal on side 3 is $\mathbf{n} = -\mathbf{i}$ and $\mathbf{F} \cdot \mathbf{n} = -(x - z)$. Since $x = 0$, and $dS = dy \, dz$,

$$\iint_{\mathcal{S}_3} \mathbf{F} \cdot \mathbf{n} \, dS = \iint_{\mathcal{S}_3} z \, dS$$

$$= \int_0^3 \int_1^2 z \, dy \, dz = 9/2 \,.$$

Adding we have $\iint_{\mathcal{S}} \mathbf{F} \cdot \mathbf{n} \, dS = 45\pi/4$.

Using the Divergence Theorem

Since $\operatorname{div}(\mathbf{F}) = 2 + 2y$,

$$\iint_S \mathbf{F} \cdot \mathbf{n}\, dS = \iiint_{\mathcal{U}} \operatorname{div}(\mathbf{F})\, dV = \iiint_{\mathcal{U}} (2 + 2y)\, dV$$

$$= \int_0^3 \int_1^2 \int_0^{\sqrt{9-x^2}} (2 + 2y)\, dz\, dy\, dx$$

$$= \int_0^3 \int_1^2 (2 + 2y)\sqrt{9 - x^2}\, dy\, dx$$

$$= \int_0^3 5\sqrt{9 - x^2}\, dx \qquad \leftarrow \text{Substitute } x = 3\sin(\theta).$$

$$= 5 \int_0^{\pi/2} 3\cos(\theta) \cdot 3\cos(\theta)\, d\theta = 45\pi/4 \,.$$

15. Surface integral calculation

The solid $\mathcal{U}$ is six-sided box, each side is a parallelogram. Viewing it from the positive x-axis, the top, bottom, front, and rear parallelograms are all parallel to the coordinate planes. They are

$$S_t = \{(x, y, 3): \ 4 < x < 5, x + 6 < y < 14 + 2x\},$$
$$S_b = \{(x, y, -2): 4 < x < 5, x - 4 < y < 2 + 2x\},$$
$$S_f = \{(5, y, z): -2 < z < 3, 5 + 2z < y < 18 + 3z\},$$
$$S_r = \{(4, y, z): -2 < z < 3, 4 + 2z < y < 16 + 3z\}.$$

The left and right sides are also parallelograms, described as follows.

$$S_{\text{left}} = \{(x, y, z): y = x + 2z, 4 < x < 5, -2 < z < 3\},$$
$$S_{\text{rght}} = \{(x, y, z): y = 8 + 2x + 3z, 4 < x < 5, -2 < z < 3\}$$

The flux through the top and the left side are calculated below.

Top

The outward normal on the top surface is $\mathbf{n} = \mathbf{k}$ so $\mathbf{F} \cdot \mathbf{n} = -9y$. The element of area is $dS = dx\, dy$ and

$$\int_{S_t} \mathbf{F} \cdot \mathbf{n}\, dS = \int_4^5 \int_{x+6}^{14+2x} (-9y)\, dy\, dx$$

$$= -9/2 \int_4^5 ((14 + 2x)^2 - (x + 6)^2)\, dx = -3771/2 \,.$$

Left side

An outward normal for the left side must have a negative y-componant. The surface is $y = x + 2z$, so $\mathbf{N} = \langle 1, -1, 2 \rangle$. On this surface $\mathbf{F} =$

$\langle x + 2z - 2z, x + 2z - 5x, -9(x + 2z) \rangle = \langle x, 2z - 4x, -9x - 18z \rangle$ which yields

$$\mathbf{F} \cdot \mathbf{n}\, dS = \mathbf{F} \cdot \mathbf{N}\, dx\, dz = (-13x - 38z)\, dx\, dz\,.$$

Therefore,

$$\int_{\mathcal{S}_{\text{left}}} \mathbf{F} \cdot \mathbf{n}\, dS = \int_4^5 \int_{-2}^3 (-13x - 38z)\, dz\, dx$$

$$= \int_4^5 (-65x - 95)\, dx = -775/2\,.$$

The other four flux integrals are calculated in a similar fashion.

Using the Divergence Theorem

Since $\text{div}(\mathbf{F}) = 1$,

$$\iint_{\mathcal{S}} \mathbf{F} \cdot \mathbf{n}\, dS = \iiint_{\mathcal{U}} \text{div}(\mathbf{F})\, dV = \iiint_{\mathcal{U}} 1\, dV$$

$$= \int_4^5 \int_{-2}^3 \int_{x+2z}^{8+2x+3z} 1\, dy\, dz\, dx$$

$$= \int_4^5 \int_{-2}^3 (8 + x + z)\, dz\, dx$$

$$= \int_4^5 (5x + 85/2)\, dx = 65\,.$$

17. Surface integral calculation

The solid $\mathcal{U}$ is a cylinder with a spherical top and bottom. The top is $\mathcal{S}_t = \{(x, y, z) : z = \sqrt{4 - x^2 - y^2}, x^2 + y^2 < 1\}$, the bottom is $\mathcal{S}_b = \{(x, y, z) : z = -\sqrt{4 - x^2 - y^2}, x^2 + y^2 < 1\}$, and the side surface is $\mathcal{S}_s = \{(x, y, z) : x^2 + y^2 = 1, -\sqrt{3} < z < \sqrt{3}\}$.

The top is the graph of $f(x, y) = \sqrt{4 - x^2 - y^2}$, an appropriately oriented normal is $\mathbf{N} = \langle x(4 - x^2 - y^2)^{-1/2}, y(4 - x^2 - y^2)^{-1/2}, 1 \rangle$. In addition, on the top surface, $z = (4 - x^2 - y^2)^{1/2}$, so

$$\mathbf{F}(x, y, z) = \langle xy, y^2, (4 - x^2 - y^2)^{1/2} \rangle\,.$$

Therefore,

$\mathbf{F} \cdot \mathbf{n}\, dS = \mathbf{F} \cdot \mathbf{N}\, dx\, dy$
$$= (x^2 y (4 - x^2 - y^2)^{-1/2} + y^3 (4 - x^2 - y^2)^{-1/2} + (4 - x^2 - y^2)^{1/2})\, dx\, dy\,.$$

Symmetry in the domain: $\mathcal{R} = \{(x, y) : x^2 + y^2 < 1\}$, and in the integrands, imply that the first and second terms will integrate to 0. The remaining term can be handled using polar coordinates.

$$\iint_{\mathcal{S}_t} \mathbf{F} \cdot \mathbf{n} \, dS = \iint_{\mathcal{R}} (4 - x^2 - y^2)^{1/2} \, dy \, dx$$

$$= \int_0^{2\pi} \int_0^1 (4 - r^2)^{1/2} \cdot r \, dr \, d\theta$$

$$= -1/3 \int_0^{2\pi} \left((4 - r^2)^{3/2} \big|_0^1 \right) d\theta = (16 - 6\sqrt{3})\pi/3 \,.$$

A similar calculation will show that $\iint_{\mathcal{S}_b} \mathbf{F} \cdot \mathbf{n} \, dS = (16 - 6\sqrt{3})\pi/3$ also.

The side surface can be parametrized with $\mathbf{r}(\theta, z) = \langle \cos(\theta), \sin(\theta), z \rangle$. The outward normal is $\mathbf{N} = \mathbf{r}_\theta \times \mathbf{r}_z = \langle \cos(\theta), \sin(\theta), 0 \rangle$. Since $\mathbf{F}(r, \theta) = \langle \sin(\theta) \cos(\theta), \sin^2(\theta), z \rangle$,

$$\mathbf{F} \cdot \mathbf{n} \, dS = \mathbf{F} \cdot \mathbf{N} \, d\theta \, dz = (\sin(\theta) \cos^2(\theta) + \sin^3(\theta)) \, d\theta \, dz \,,$$

and

$$\iint_{\mathcal{S}_s} \mathbf{F} \cdot \mathbf{n} \, dS = \int_0^{2\pi} \int_{-\sqrt{3}}^{\sqrt{3}} (\sin(\theta) \cos^2(\theta) + \sin^3(\theta)) \, dz \, d\theta$$

$$= 2\sqrt{3} \int_0^{2\pi} (\sin(\theta) \cos^2(\theta) + \sin^3(\theta)) \, d\theta$$

$$= 2\sqrt{3} \cdot ((1/3) \cos^3(\theta)) \Big|_0^{2\pi} = 0 \,.$$

Adding we have $\iint_{\mathcal{S}} \mathbf{F} \cdot \mathbf{n} \, dS = (32 - 12\sqrt{3})\pi/3$.

Using the Divergence Theorem

Since $\text{div}(\mathbf{F}) = 3y + 1$, the calculation goes like this. Note the switch to cylindrical coordinates.

$$\iint_{\mathcal{S}} \mathbf{F} \cdot \mathbf{n} \, dS = \iiint_{\mathcal{U}} \text{div}(\mathbf{F}) \, dV = \iiint_{\mathcal{U}} (3y + 1) \, dV$$

$$= \int_0^1 \int_{-\sqrt{1-x^2}}^{\sqrt{1-x^2}} \int_{-\sqrt{4-x^2-y^2}}^{\sqrt{4-x^2-y^2}} (3y + 1) \, dz \, dy \, dx$$

$$= \int_0^1 \int_0^{2\pi} \int_{-\sqrt{4-r^2}}^{\sqrt{4-r^2}} (3r \sin(\theta) + 1) \cdot r \, dz \, d\theta \, dr$$

$$= 2 \int_0^1 \int_0^{2\pi} (3r \sin(\theta) + 1) \sqrt{4 - r^2} \cdot r \, d\theta \, dr$$

$$= 4\pi \int_0^1 \sqrt{4 - r^2} \cdot r \, dr$$

$$= 4\pi (-(1/3)(4 - r^2)^{3/2}) \Big|_0^1 = (32 - 12\sqrt{3})\pi/3 \,.$$

19. Surface integral calculation

The solid $\mathcal{U}$ is one half of a circular paraboloid. The boundary surfaces are the top, $\mathcal{S}_t = \{(x, y, 4) : -2 < x < 2, 0 < y < \sqrt{4 - x^2}\}$, the bottom, $\mathcal{S}_b = \{(x, y, z) : z = x^2 + y^2, -2 < x < 2, 0 < y < \sqrt{4 - x^2}\}$, and a side surface, $\mathcal{S}_s = \{(x, 0, z) : -2 < x < 2, x^2 < z < 4\}$.

The outward normal on the top is $\mathbf{n} = \mathbf{k}$. Since $z = 4$, $\mathbf{F}(x, y, z) = \langle x^2 - 8x, y^2 - 3, 16 \rangle$, and $\mathbf{F} \cdot \mathbf{n}\, dS = 16\, dx\, dy$. Therefore,

$$\iint_{\mathcal{S}_t} \mathbf{F} \cdot \mathbf{n}\, dS = \int_{-2}^{2} \int_{0}^{\sqrt{4-x^2}} 16\, dy\, dx$$

$$= 16 \int_{-2}^{2} \sqrt{4 - x^2}\, dx = 32\pi\,.$$

The outward normal on the side surface is $\mathbf{n} = -\mathbf{j}$. Since $y = 0$ on this surface, $\mathbf{F} \cdot \mathbf{n}\, dS = 3\, dx\, dz$ and

$$\iint_{\mathcal{S}_s} \mathbf{F} \cdot \mathbf{n}\, dS = \int_{-2}^{2} \int_{x^2}^{4} 3\, dz\, dx$$

$$= 3 \int_{-2}^{2} (4 - x^2)\, dx = 32\,.$$

The bottom surface is the graph of $z = x^2 + y^2$. An outward normal is $\mathbf{N} = \langle 2x, 2y, -1 \rangle$. Since $\mathbf{F} \cdot \mathbf{N} = 2x(x^2 - 2xz) - 2y(3 - y^2) - z^2$ and $z = x^2 + y^2$,

$$\mathbf{F} \cdot \mathbf{n}\, dS = \mathbf{F} \cdot \mathbf{N}\, dx\, dy$$
$$= (2x^3 - 4x^2(x^2 + y^2) - 6y + 2y^3 - (x^2 + y^2)^2)\, dx\, dy\,,$$

and

$$\iint_{\mathcal{S}_s} \mathbf{F} \cdot \mathbf{n}\, dS = \int_{-2}^{2} \int_{0}^{\sqrt{4-x^2}} (2x^3 - 4x^2(x^2 + y^2) - 6y + 2y^3 - (x^2 + y^2)^2)\, dy\, dx$$

$$= \int_{0}^{\pi} \int_{0}^{2} (2r^3 \cos^3(\theta) - 4r^4 \cos^2(\theta) - 6r\sin(\theta) + 2r^3 \sin^3(\theta) - r^4) \cdot r\, dr\, d\theta$$

$$= \int_{0}^{\pi} ((64/5)(\cos^3(\theta) + \sin^3(\theta)) - (128/3)\cos^2(\theta) - 16\sin(\theta) - 32/3)\, d\theta$$

$$= -32\pi - 224/15\,.$$

You can fill in the details in the last step. (Note that $\cos^3(\theta)$ integrates to 0 over the interval $[0, \pi]$.)

Adding the three flux integrals, we have $\iint_{\mathcal{S}} \mathbf{F} \cdot \mathbf{n}\, dS = 256/15$.

Using the Divergence Theorem

Since $\operatorname{div}(\mathbf{F}) = 2x + 2y$,

$$\iint_{\mathcal{S}} \mathbf{F} \cdot \mathbf{n} \, dS = \iiint_{\mathcal{U}} \operatorname{div}(\mathbf{F}) \, dV = \iiint_{\mathcal{U}} (2x + 2y) \, dV$$

$$= \int_0^\pi \int_0^2 \int_{r^2}^4 (2r\cos(\theta) + 2r\sin(\theta)) \cdot r \, dz \, dr \, d\theta$$

$$= \int_0^\pi \int_0^2 (2r\cos(\theta) + 2r\sin(\theta))(4 - r^2) \cdot r \, dr \, d\theta$$

$$= 128/15 \int_0^\pi (\cos(\theta) + \sin(\theta)) \, d\theta = 256/15 \,.$$

Further Theory and Practice

21. The flux is 0 because $\operatorname{div}(\mathbf{F}) = 0$.

23. Let Σ_a be a sphere of radius a centered at the origin with an outward-pointing unit normal. Choose a small enough so that Σ_a is inside of $\mathcal{U}$. Let $\mathcal{S}$ be the boundary surface of $\mathcal{U}$, also oriented with an outward normal. Because the divergence of $\mathbf{F}$ is 0, $\iint_{\mathcal{S}} \mathbf{F} \cdot \mathbf{n} \, dS + \iint_{-\Sigma_a} \mathbf{F} \cdot \mathbf{n} \, dS = 0$, so $\iint_{\mathcal{S}} \mathbf{F} \cdot \mathbf{n} \, dS = \iint_{\Sigma_a} \mathbf{F} \cdot \mathbf{n} \, dS$. Apply Exercise 22.

25. The divergence of $\mathbf{G}$ is zero (Theorem 1, Section 13.4).

27. The integration will be carried out in spherical coordinates. Since $\operatorname{div}(\mathbf{F}) = 1 - 2y + x$,

$$\iint_{\mathcal{S}} \mathbf{F} \cdot \mathbf{n} \, dS = \iiint_{\mathcal{U}} \operatorname{div}(\mathbf{F}) \, dV = \iiint_{\mathcal{U}} (1 - 2y + x) \, dV$$

$$= \int_0^\pi \int_0^{2\pi} \int_1^2 (1 - 2\rho\sin(\theta)\sin(\phi) + \rho\cos(\theta)\sin(\phi)) \cdot \rho^2 \sin(\phi) \, d\rho \, d\theta \, d\phi$$

$$= \int_0^\pi \int_0^{2\pi} ((7/3)\sin(\phi) - (15/2)\sin(\theta)\sin^2(\phi) + (15/4)\cos(\theta)\sin^2(\phi)) \, d\theta \, d\phi$$

$$= \int_0^\pi (14\pi/3)\sin(\phi) \, d\phi = 28\pi/3 \,.$$

29. Since $\operatorname{div}(\mathbf{F}) = 1$,

$$\iint_{\mathcal{S}} \mathbf{F} \cdot \mathbf{n} \, dS = \iiint_{\mathcal{U}} \operatorname{div}(\mathbf{F}) \, dV = \iiint_{\mathcal{U}} 1 \, dV = (\text{Volume of } \mathcal{U})$$

31. According to Theorem 1 of Section 15.4,

$$\operatorname{div}(u\nabla u) = u \operatorname{div}(\nabla u) + \nabla u \cdot \nabla u \,.$$

But $\nabla u \cdot \nabla u = \|\nabla u\|^2$ and, because u is harmonic, $\mathrm{div}(\nabla u) = 0$. Consequently $\mathrm{div}(u\nabla u) = \||\nabla u\|^2$ and, applying the Divergence Theorem,

$$\iint_{\mathcal{S}} (\nabla u \cdot \mathbf{n}) u \, dS = \iint_{\mathcal{S}} (u\nabla u) \cdot \mathbf{n} \, dS$$

$$= \iiint_{\mathcal{U}} \mathrm{div}(u\nabla u) \, dV$$

$$= \iiint_{\mathcal{U}} \|\nabla u\|^2 \, dV \,.$$

33. Stokes's Theorem does not apply in this scenario because the surface $\mathcal{S}$, as the boundary of a solid region, has no boundary curves.

35. Since $\nabla f(x,y,z) = z\mathbf{i} + x\mathbf{k}$, $\iiint_{\mathcal{U}} \nabla f \, dV = \iiint_{\mathcal{U}} z \, dV \mathbf{i} + \iiint_{\mathcal{U}} x \, dV \mathbf{k}$. The second volume integral evaluates to 0 by symmetry, half the time x is positive and half the time it is negative. Using cylindrical coordinates, $\mathcal{U} = \{(r,\theta,z) : 0 \le \theta \le 2\pi, 0 \le r \le 1, 0 \le z \le 1 - r^2\}$, and

$$\iiint_{\mathcal{U}} z \, dV = \int_0^{2\pi} \int_0^1 \int_0^{1-r^2} z \cdot r \, dz \, dr \, d\theta = \frac{1}{2} \int_0^{2\pi} \int_0^1 r(1-r^2)^2 \, dr \, d\theta$$

$$= -\frac{1}{4} \int_0^{2\pi} \left(\frac{1}{3}(1-r^2)^3 \right) \Big|_0^1 \, d\theta = \frac{1}{12} \int_0^{2\pi} d\theta = \frac{\pi}{6} \,.$$

Therefore, $\iiint_{\mathcal{U}} \nabla f \, dV = \frac{\pi}{6} \mathbf{i}$.

For the left side there are two surface integrals to evaluate. One over the top surface and the other over the bottom which is the unit disk in the xy-plane. However, the integral over the bottom is 0 because $z = 0$. The outward unit normal at a point (x,y,z) on the top surface $\mathcal{S}_t$ is $\mathbf{n} = \frac{1}{\sqrt{1+4x^2+4y^2}} \langle 2x, 2y, 1 \rangle$ so $f(x,y,z)\mathbf{n} = \frac{1}{\sqrt{1+4x^2+4y^2}} \langle 2x^2 z, 2xyz, xz \rangle$. Since $dS = \sqrt{1 + 4x^2 + 4y^2} \, dx \, dy$,

$$\iint_{\mathcal{S}_t} f(x,y,z)\mathbf{n} \, dS = \iint_{\mathcal{R}} 2x^2(1 - x^2 - y^2) \, dx \, dy \, \mathbf{i}$$

$$+ \iint_{\mathcal{R}} 2xy(1 - x^2 - y^2) \, dx \, dy \, \mathbf{j} + \iint_{\mathcal{R}} x(1 - x^2 - y^2) \, dx \, dy \, \mathbf{k} \,,$$

where $\mathcal{R}$ is the unit disk in the xy-plane. The second and third integrals evaluate to 0 by symmetry. Half the time the integrands are positive and half the time they are negative. The first integral evaluates easily in polar coordinates.

$$\iint_{\mathcal{R}} 2x^2(1 - x^2 - y^2)\,dx\,dy = \int_0^{2\pi} \int_0^1 2r^2\cos^2(\theta)(1 - r^2) \cdot r\,dr\,d\theta$$

$$= 2\int_0^{2\pi} \cos^2(\theta)\,d\theta \int_0^1 (r^3 - r^5)\,dr$$

$$= 2\pi \cdot \left(\frac{1}{4}r^4 - \frac{1}{6}r^6\right)\Big|_0^1 = \frac{\pi}{6}\,.$$

Therefore, $\iint_{\mathcal{S}} f\mathbf{n}\,dS = \dfrac{\pi}{6}\,\mathbf{i}$.

37. Since $\nabla f(x,y,z) = 3\mathbf{k}$, $\iiint_{\mathcal{U}} \nabla f\,dV = \iiint_{\mathcal{U}} 3\,dV\mathbf{k} = 3 \cdot \frac{4}{3}\pi\mathbf{k} = 4\pi\mathbf{k}$.

For the left side, the outward unit normal at a point (x, y, z) on the unit sphere $\mathcal{S}$ is $\mathbf{n} = \langle x, y, z\rangle$ so $f(x, y, z)\mathbf{n} = 3\langle xz, yz, z^2\rangle$ and

$$\iint_{\mathcal{S}} f(x,y,z)\mathbf{n}\,dS = 3\iint_{\mathcal{S}} xz\,dS\,\mathbf{i} + 3\iint_{\mathcal{S}} yz\,dS\,\mathbf{j} + 3\iint_{\mathcal{S}} z^2\,dS\,\mathbf{k}.$$

The first two surface integrals evaluate to 0 by symmetry. Each integrand is positive half the time and negative half the time. The third integral can be evaluated using the spherical coordinate element of area on a sphere centered at the origin of radius ρ: $dS = \rho^2\sin(\phi)\,d\theta\,d\phi$. [2]

$$3\iint_{\mathcal{S}} z^2\,dS = 3\int_0^\pi \int_0^{2\pi} \cos^2(\phi) \cdot \sin(\phi)\,d\theta\,d\phi = -2\pi \cdot \cos^3(\phi)\Big|_0^\pi = 4\pi\,.$$

Therefore, $\iint_{\mathcal{S}} f(x,y,z)\mathbf{n}\,dS = 4\pi\mathbf{k}$.

Calculator/Computer Exercises

39. Define the field $\mathbf{F}$, calculate its divergence, and integrate.

```
> F := <ln(y^2+z^2),ln(x^2+z^2),z^2*ln(2+y)>:
  divF := diff(F[1],x) + diff(F[2],y) + diff(F[3],z);
  Int(Int(Int( divF, z=1..2-x^2-y^2),
              y=-sqrt(1-x^2)..sqrt(1-x^2)), x=-1..1) =
  evalf[3](
  Int(Int(Int( divF, z=1..2-x^2-y^2),
              y=-sqrt(1-x^2)..sqrt(1-x^2)), x=-1..1)
          );
```

$$divF := 2z\ln(x + y)$$

$$\int_{-1}^1 \int_{-\sqrt{1-x^2}}^{\sqrt{1-x^2}} \int_1^{2-x^2-y^2} 2z\ln(2+y)\,dz\,dy\,dx = 2.82$$

---

[2]This can be motivated using a carefully drawn picture or, better yet, use the parametrization $\mathbf{R}(\theta, \phi) = \langle \rho\sin(\phi)\cos(\theta), \rho\sin(\phi)\sin(\theta), \rho\cos(\phi)\rangle$, $dS = \|\mathbf{R}_\theta \times \mathbf{R}_\phi\|\,d\theta\,d\phi$.

Maple's numerical integration routine is only able to find three digits. Let's try it in polar coordinates.

```
> Int(Int(Int( 2*z*ln(2+r*sin(theta))*r, z=1..2-r^2),
              r=0..1), theta=0..2*Pi) =
  evalf(
  Int(Int(Int( 2*z*ln(2+r*sin(theta))*r, z=1..2-r^2),
              r=0..1), theta=0..2*Pi)
      );
```

$$\int_0^{2\pi} \int_0^1 \int_1^{2-r^2} 2z \ln(2 + r\sin(\theta))\, r\, dz\, dr\, d\theta = 2.817619304$$

41. As in Exercise 39. Note that Maple can find 6 digits in this approximation (rectangular coordinates).

```
> F := <(1+z)^y,(1+x)^z,exp(z)*arctan(1+x)>:
  divF := diff(F[1],x) + diff(F[2],y) + diff(F[3],z);
  Int(Int(Int( divF, z=0..1-x^2-y^2),
              y=-sqrt(1-x^2)..sqrt(1-x^2)), x=-1..1) =
  evalf[6](
  Int(Int(Int( divF, z=0..1-x^2-y^2),
              y=-sqrt(1-x^2)..sqrt(1-x^2)), x=-1..1)
          );
```

$$divF := e^z \arctan(1 + x)$$

$$\int_{-1}^1 \int_{-\sqrt{1-x^2}}^{\sqrt{1-x^2}} \int_0^{1-x^2-y^2} e^z \arctan(1 + x)\, dz\, dy\, dx = 1.68773$$